PROGRESS IN

Nucleic Acid Research
and Molecular Biology

Volume 57

PROGRESS IN
Nucleic Acid Research and Molecular Biology

edited by

KIVIE MOLDAVE
Department of Molecular Biology and Biochemistry
University of California, Irvine
Irvine, California

Volume 57

ACADEMIC PRESS
San Diego London Boston New York
Sydney Tokyo Toronto

This book is printed on acid-free paper.

Copyright © 1997 by ACADEMIC PRESS

All Rights Reserved.
No part of this publication may be reproduced or transmitted in any form or by any means, electronic or mechanical, including photocopy, recording, or any information storage and retrieval system, without permission in writing from the Publisher.
The appearance of the code at the bottom of the first page of a chapter in this book indicates the Publisher's consent that copies of the chapter may be made for personal or internal use of specific clients. This consent is given on the condition, however, that the copier pay the stated per copy fee through the Copyright Clearance Center, Inc. (222 Rosewood Drive, Danvers, Massachusetts 01923), for copying beyond that permitted by Sections 107 or 108 of the U.S. Copyright Law. This consent does not extend to other kinds of copying, such as copying for general distribution, for advertising or promotional purposes, for creating new collective works, or for resale. Copy fees for pre-1997 chapters are as shown on the title pages, if no fee code appears on the title page, the copy fee is the same as for current chapters.
0079-6603/97 $25.00

Academic Press
a division of Harcourt Brace & Company
525 B Street, Suite 1900, San Diego, California 92101-4495, USA
http://www.apnet.com

Academic Press Limited
24-28 Oval Road, London NW1 7DX, UK
http://www.hbuk.co.uk/ap/

International Standard Book Number: 0-12-540057-8

PRINTED IN THE UNITED STATES OF AMERICA
97 98 99 00 01 02 BC 9 8 7 6 5 4 3 2 1

Contents

PREFACE ix
SOME ARTICLES PLANNED FOR FUTURE VOLUMES xi

Mismatch Base Pairs in RNA 1
Stefan Limmer

I. G·U Base Pairs 2
II. G·A Base Pairs 15
III. U·U and U·C Base Pairs 23
IV. Other Mismatches 30
V. Possible Biological Roles of Mismatch Pairs 33
References 35

The Mechanism of 3' Cleavage and Polyadenylation of Eukaryotic Pre-mRNA 41
Elmar Wahle and Uwe Kühn

I. Pre-mRNA Sequences Directing 3' End Formation 42
II. Cleavage and Polyadenylation in Animal Cells 48
III. Cleavage and Polyadenylation in Yeast 56
IV. Regulation of Polyadenylation 57
V. Coupling of 3' End Formation to Transcription and Splicing 64
VI. Conclusion 65
References 66

Stimulation of Kinase Cascades by Growth Hormone: A Paradigm for Cytokine Signaling 73
Timothy J. J. Wood, Lars-Arne Haldosén, Daniel Sliva, Michael Sundström, and Gunnar Norstedt

I. The Cytokines 74
II. The Hematopoietic Cytokine Receptor Superfamily 76

III. Growth Hormone Receptor Signal Transduction via Phosphorylation
　　　　 Cascades 80
　　IV. The SPI 2.1 Promoter as a Model System for Cytokine Signaling 87
　　V. Conclusion 89
　　　　 Note Added in Proof 91
　　　　 References 91

Oligonucleotides and Polynucleotides as Biologically Active Compounds 95

V. V. Vlassov, I. E. Vlassova, and L. V. Pautova

　　I. Oligonucleotide Derivatives for Targeting Biopolymers 96
　　II. Pharmacokinetics of Oligonucleotides 105
　　III. Biological Activities of Oligonucleotides and Polynucleotides 116
　　IV. Concluding Remarks 132
　　　　 References 132

Replication Control of Plasmid P1 and Its Host Chromosome: The Common Ground 145

Dhruba K. Chattoraj and Thomas D. Schneider

　　I. Replication Control: Definition 146
　　II. P1 Basic Replicon: Isolation and Structure 147
　　III. Iterons and Their Evolution 150
　　IV. Multiple Functions of Initiator Proteins 155
　　V. Mode of Replication: Similarity of P1 and *oriC* 158
　　VI. Negative Control of Replication 159
　　VII. Positive Control and Coordination with Cell Cycle 168
　　VIII. Summary Remarks and the Future 179
　　　　 References 181

Changes in Gene Structure and Regulation of E-Cadherin during Epithelial Development, Differentiation, and Disease 187

Janusz A. Jankowski, Fiona K. Bedford, and Young S. Kim

　　I. Cadherin Structure 188
　　II. Diversity of Action and Interaction 194

 III. Cellular Properties of E-Cadherins Conserved during Development,
   Differentiation, and Disease 198
 IV. Decreased or Defective E-Cadherin Expression in
   Colorectal Disease 204
 V. Conclusions 208
   References 209

The Formation of DNA Methylation Patterns and the Silencing of Genes 217

Jean-Pierre Jost and Alain Bruhat

 I. The Formation of DNA Methylation Patterns 217
 II. DNA Methylation and the Silencing of Genes 234
 III. Conclusions 243
   References 244

The Role of mRNA Stability in the Control of Globin Gene Expression 249

J. Eric Russell, Julia Morales, and Stephen A. Liebhaber

 I. General Aspects of Eukaryotic mRNA Stability 250
 II. Specific Aspects of Globin mRNA Stability 253
 III. Structural Determinants of α-Globin mRNA Stability 263
 IV. Stability of Other Globin mRNAs 276
 V. Potential Therapeutic Applications 283
   References 285

Self-Glucosylating Initiator Proteins and Their Role in Glycogen Biosynthesis 289

Peter J. Roach and Alexander V. Skurat

 I. Glycogenin and the Pathway of Glycogen Biogenesis 290
 II. Biochemistry of Glycogenin and the Glg Proteins 293
 III. Molecular Biology of Glycogenin and Glycogenin-like Proteins 301
 IV. Physiological Function of Glycogenin and the Glg Proteins 309
 V. Conclusion 314
   References 315

Molecular Genetics of Yeast TCA Cycle Isozymes 317
Lee McAlister-Henn and W. Curtis Small

I. Malate Dehydrogenases	319
II. Isocitrate Dehydrogenases	326
III. Citrate Synthases	331
IV. Aconitase and Fumarase Isozymes	333
V. Conclusions and Perspective	335
References	337

INDEX 341

Preface

After many years of devoted service to this series, Dr. Waldo Cohn has decided to retire from the editorship. The series, initially titled *Progress in Nucleic Acid Research*, was founded and co-edited by J. N. Davidson and Waldo Cohn; the first volume appeared in 1962. On Dr. Davidson's death in 1972, Waldo Cohn assumed sole editorship. Sometime after that, the series was renamed *Progress in Nucleic Acid Research and Molecular Biology*, and in 1983, after 29 volumes, Kivie Moldave became a co-editor.

Waldo Cohn was a rigorous, demanding editor who insisted on the highest standards for the series. Although he was, on occasion, tough with authors, most if not all came to appreciate and thank him for his improvements to their contributions. Under Cohn's editorship, *Progress* established a well-deserved reputation and a broad audience in the scientific community. After 56 volumes, he has certainly earned the right to "rest on his laurels" and we wish him the best.

As initially set out by the founding editors, the major aim of this series will remain the same: to "encourage the writing of essays in circumscribed areas in which recent developments in particular aspects . . . are discussed by workers provided with an opportunity for more personal interpretation than is normally provided in review articles. More discussion and speculation than is customary are encouraged, as is the expression of points of view that are perhaps controversial and individualistic."

Kivie Moldave

Some Articles Planned for Future Volumes

Structure and Transcription Regulation of Nuclear Genes for the Mouse Mitochondrial Cytochome c Oxidase
  Narayan G. Avadhani, A. Basu, C. Sucharov, and N. Lenka

Impaired Folding and Subunit Assembly as Disease Mechanisms
  Peter Bross et al.

Regulation of Translational Initiation during Cellular Responses to Stress
  Margaret A. Brostrom

Tissue Transglutaminase—Retinoid Regulation and Gene Expression
  Peter J. A. Davies and Shakid Mian

Genetic Approaches to Structural Analysis of Membrane Transport Systems
  Wolfgang Epstein

Intron-Encoded snRNAs
  Maurille J. Fournier and E. Stuart Maxwell

Molecular Analyses of Metallothionine Gene Regulation
  Lashitew Gedamu et al.

Mechanisms for the Selectivity of the Cell's Proteolytic Machinery
  Alfred Goldberg, Michael Sherman, and Oliver Coux

Mechanisms of RNA Editing
  Stephen L. Hajduk and Susan Madison-Antenucci

The Hairpin Ribozyme: Discovery, Development, and Applications for Regulation of Gene Expression
  Arnold Hampel

Structure/Function Relationships of Phosphoribulokinase and Ribulosebisphosphate Carboxylase/Oxygenase
  Fred C. Hartman and Hillel K. Brandes

Molecular Biology of Trehalose and the Trehalases in the Yeast *S. cerevisiae*
  Helmut Holzer and Solomon Nwaka

The Nature of DNA Replication Origins in Higher Eukaryotic Organisms
  Joel A. Huberman and William C. Burhans

Synthesis of DNA Precursors in *Lactobacillus acidophilus* R-26
  David H. Ives

SOME ARTICLES PLANNED FOR FUTURE VOLUMES

A Kaleidoscopic View of the Transcriptional Machinery in the Nucleolus
SAMSON T. JACOB

Sphingomyelinases in Cytokine Signaling
MARTIN KRONKE

Mammalian DNA Polymerase Delta: Structure and Function
MARIETTA Y. W. T. LEE

Reverse Transcriptase and Implications for Retroviral Replication
STUART F. J. LE GRICE AND ERIC ARTS

DNA Helicases: Roles in DNA Metabolism
STEVEN W. MATSON AND DANIEL W. BEAM

Lactose Repressor Protein: Perspectives on Structure and Function
KATHLEEN SHIVE MATTHEWS AND JEFFRY NICHOLS

Haparan Sulfate-Fibroblast Growth Factor Family
WALLACE L. MCKEEHAN AND MIKIO KAN

Molecular Biology of Snake Toxins: Is the Functional Diversity of Snake Toxins Associated with a Mechanism of Accelerated Evolution?
ANDRE MENEZ et al.

Inosine Monophosphate Dehydrogenase: Role in Cell Division and Differentiation
BEVERLY S. MITCHELL

Specificity of Eukaryotic Type II Topoisomerase: Influence of Drugs, DNA Structure, and Local Sequence
MARK T. MULLER AND JEFFREY SPITZNER

Localization and Movement of tRNAs on the Ribosome during Protein Synthesis
KNUD H. NIERHAUS

Immunoanalysis of DNA Damage and Repair Using Monoclonal Antibodies
MANFRED F. RAJEWSKY

Mechanism of Transcriptional Regulation by the Retinoblastoma Tumor Suppressor Gene Product
PAUL D. ROBBINS AND JON HOROWITZ

Transcriptional Regulation of Small Nuclear RNA Genes
WILLIAM E. STUMPH

Organization and Expression of the Chicken α-Globin Genes
KLAUS SCHERRER AND FELIX R. TARGA

Physico-chemical Studies on DNA Triplexes and Quadruplexes
RICHARD H. SHAFER

Bacillus subtilis as I Know It
  NOBORU SUEOKA

Molecular Genetic Approaches to Understanding Drug Resistance in Protozoan Parasites
  DYANN WIRTH *et al.*

Mismatch Base Pairs in RNA

STEFAN LIMMER

Laboratorium für Biochemie der
Universität Bayreuth
D-95440 Bayreuth, Germany

I. G·U Base Pairs		2
A. General Features and X-Ray Structure Analyses of tRNAs and Short Duplexes		2
B. NMR Spectroscopy and Thermodynamic Stability		6
II. G·A Base Pairs		15
A. General Features and X-Ray Structure Analyses		15
B. NMR Studies and Thermodynamic Stability		18
III. U·U and U·C Base Pairs		23
A. General Features and Crystal Structures of Mismatch Duplexes		23
B. NMR Investigations and Thermodynamic Stability		27
IV. Other Mismatches		30
V. Possible Biological Roles of Mismatch Pairs		33
References		35

During the past few years, there have been several reviews dealing with general features of RNA structure and folding (1–6) as well as with the prediction and determination of secondary and tertiary structures (6–11). Whereas the first years of research on nucleic acids were marked by interest mainly in the elucidation of general structural and conformational features of regular or canonical DNA and RNA helices, improved methods of structure analysis have subsequently permitted detection of the subtle details associated with sequence-dependent structural and conformational variations of the helical parameters in regular Watson–Crick helices (e.g., cf. Ref. 12).

The characterization of structural modifications introduced by nonregular or non-Watson–Crick pairs, also termed "mismatches" or "mispairs," now appears more attractive and important, in particular in view of the relevance of these mispairs to the intermolecular recognition of DNA and RNA. This is also documented by reviews dealing with crystal structure (13, 13a) and NMR solution structure (14) analyses of DNA duplexes containing mismatch base pairs. However, no comparable representation of mismatches in RNA has yet been published.

The regular, canonical A-RNA double-helical geometry involving standard Watson–Crick G·C and A·U base pairs is well documented (1, 12). However, the existence of deviating base-pairing schemes was proposed in 1966 to account for the possibility of the formation of different ("wobble") base pairs between the 3′-terminal nucleotide of an mRNA codon ("wobble base") and the corresponding 5′ nucleotide of the tRNA anticodon (15). Later, certain non-Watson–Crick pairs (G·U pairs) were found in many tRNA secondary structures (16). Interest in non-Watson–Crick pairs, in particular G·U pairs, was subsequently stimulated considerably by the observation that a specific G·U pair in the acceptor arm of the tRNA[Ala] from *Escherichia coli* represents the major recognition element ("identity element") for the interaction of this tRNA with its cognate aminoacyl-tRNA synthetase (17–19).

In the present review, we focus on the effect of non-Watson–Crick base pairs, or mismatches, being the simplest nonregular structural units in RNA duplexes, on the local geometry and stability of the corresponding oligoribonucleotides.

Since about 1990, knowledge about mismatches and their influence on local helix structure and duplex stability has distinctly increased. In addition to analyses of the crystal structures of several tRNAs (20–26), which were solved and refined in the 1970s and 1980s, systematic studies of different mismatch base pairs in varying sequence contexts have contributed many new insights into the peculiar structures brought about by certain mismatches, as well as their corresponding stability increments. In many cases, these investigations have benefited by the establishment and the increased use of synthetic chemical oligonucleotides, which enabled production of the relatively large quantities (milligrams) of pure RNA necessary for both NMR and X-ray crystallographic investigations.

On the basis of the available data, I attempt in the following work to analyze the general features of the most frequent and sufficiently well-characterized mismatch base pairs, with respect both to their geometry and to the effect on the stability of RNA duplexes. Moreover, the biological relevance of certain mismatches, i.e., their functional implications for recognition by and/or interaction with ligands (proteins, nucleic acids, metal ions), is discussed briefly.

I. G·U Base Pairs

A. General Features and X-Ray Structure Analyses of tRNAs and Short Duplexes

G·U mismatches represent the most frequently occurring kind of non-canonical base pairs in both tRNA (16) and rRNA secondary structures (27–31). It was originally pointed out (32) and later corroborated (31, 33) that these G·U pairs are not statistically distributed over all the possible locations

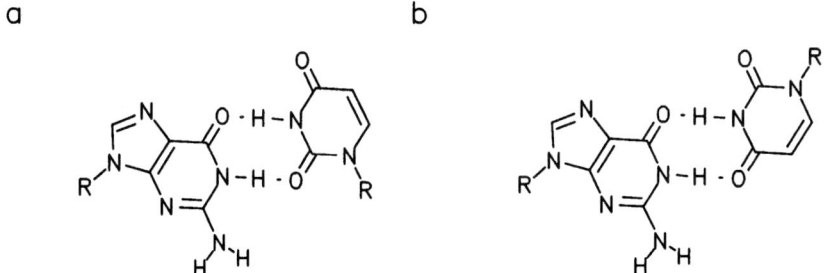

FIG. 1. Base pair geometry of the G·U mismatch pair. (a) Wobble pair; (b) reverse wobble pair.

within the RNA sequences, but that there are rather distinctly preferred locations (sequence contexts) for the appearance of this mismatch type. In particular, if a G·U pair forms the terminal pair of a helical stem, it is preferably oriented in such a way that the loop is 5′ of the mismatch G (or, respectively, 3′ of U) (31, 32). A plausible explanation for this observation is based on the special base pair geometry of this mismatch, which deviates from that of canonical Watson–Crick pairs. As compared to regular base pairs, the two bases are staggered (Fig. 1), with the guanine being displaced toward the minor groove. This gives rise to a change of the glycosyl bond angles, which are about 52° for both nucleotides in Watson–Crick pairs, but adopt values of 38° and 59° for the G and U, respectively, in the G·U mismatch [G4·U69 in the acceptor stem of yeast tRNA[Phe] (21)].

Though these modifications of the base pair geometry do not significantly perturb the backbone conformation of the helix or the glycosyl torsion angles and sugar pucker (32), it was reasoned that they should have effects on the stacking between bases of the G·U pair and its neighboring base pairs in both 5′ and 3′ directions (31–33). In particular, it was derived from the analysis of G·U base pairs in yeast tRNA[Phe] (20–24) that the stacking, i.e., the degree of base plane overlap, between the G of the mismatch and its 3′ neighbor base is much better than that with its 5′ neighbor (32, 33). There is likewise a loss of stacking between the mismatch U and its 3′ neighbor. Accordingly, a G·U pair at the end of a helical stem occurs preferentially in an orientation with the G of the mismatch appearing 3′ of the loop, so that the G·U pair is well stacked above the preceding base pair of the stem (31–33). Otherwise, there is no obvious bias for the occurrence of G·U pairs at stem–loop junctions (31).

These stacking features for three consecutive base pairs with a G·U wobble pair sandwiched between two regular Watson–Crick pairs were also found for different G·U pairs in the anticodon and acceptor stems of tRNA[Asp]

(G3·U40, U5·G68) (25, 26, 34), i.e., extensive base overlap between G of the wobble pair and its 3′ neighbor base, and of the U with its 5′ neighbor, whereas there is only poor stacking with the other adjacent base pair. A quite similar base stacking pattern was also found for a G·U pair in the P1 helix from the so-called group I self-splicing intron, which is strictly conserved in all sequences known so far. A detailed NMR structural analysis (see Section I,B) (35) at high precision showed that base stacking in the vicinity of the G·U pair complies perfectly with the previously described stacking pattern.

However, a thorough comparison of the structural details of several G·U pairs in varying sequence contexts led Moral et al. (34) to conclude that the sequence context is more important for stacking properties than are the geometric peculiarities of the G·U mismatch. It is pointed out that all the G·U pairs that display the stacking pattern just discussed occur in sequence stacks where the G of the mismatch is flanked by pyrimidine nucleotides, and, correspondingly, the U is base-paired to the mismatch G by purines. If, however, the G·U pair is accommodated in a different sequence context, such as R-G-R (and Y-U-Y, respectively), the base stacking is significantly altered. In particular, this applies to the G4·U69 base pair in the acceptor stem of yeast tRNA[Phe], which is located between the G3·U70 and A5·U68 base pairs. Moras et al. (34) convincingly demonstrate that the stacking of these bases does not significantly differ from that found for the base pairs A29·U41, G30·m^{5}C40, and A31·Ψ39, where the G of the "inner" G·C pair is flanked by two purines (A29·G30·A31). However, if sequence tracks R-Y-R (base-paired to a Y-R-Y track in a regular Watson–Crick mode) from aminoacyl and TΨC stems of yeast tRNA[Phe] are considered, it turns out that, even for these regular double-helical regions, the base stacking between the inner base pair and the two adjacent base pairs is different in the two directions: the base overlap between the purine of the central base pair and its 3′ neighbor, as well as that of the pyrimidine of the middle base pair and its 3′ neighbor, is distinctly greater than stacking on the other side (34).

Consequently, the sequence context affects the base stacking—even for the G·U pair—to a greater extent than do the modifications of the base pair geometry introduced by the altered arrangement of the two bases.

Though it is generally difficult to obtain sufficiently large and well-diffracting crystals of small oligoribonucleotides suited for a high-resolution X-ray structure determination, several X-ray structures of small RNA duplexes containing mismatch base pairs have been published since 1991 (36–41). Two of the analyzed oligoribonucleotides, 5′-GGAUUCGGUCC-3′ (UUCG dodecamer) (36) and 5′-GGACUUUGGUCC-3′ (UUUG dodecamer) (40), which differ only in one position (U and C, respectively, in position 6), form hairpins in solution and have been structurally characterized by NMR (42, 43). In the crystal, however, they form double-helical stems, each of which includes a track of four consecutive mismatch base pairs [U·G, U·C, C·U, and G·U in the UUCG dodecamer (36), and U·G, U·U, U·U, and G·U in the UUUG dodecamer (40)]. The monomer duplex 5'-GCUUCGGG-brU-3' (brU is 5'-bromo-2'-deoxyuridine) likewise crystallizes as a duplex, again containing four central mismatch base pairs in the interior of the stem (U·G, U·C, C·U, G·U) (37), whereas the non-self-complementary domain A sequence from a 5S rRNA, with 24 nucleotides, accommodates a 5'-GU-3'/5'-GU-3' tandem mismatch near one end of the double-helical stem (38).

In all cases, as in the crystal structures of yeast tRNA^{Phe} (21, 23, 24) and tRNA^{Asp} (25, 26), the G·U pairs adopt the wobble conformation with the expected hydrogen bond pattern (Fig. 1a). Generally, the mismatch base pairs—and in particular the G·U pairs—are accommodated in the helices without strong distortions of the A-RNA sugar-phosphate backbone, though the major groove in the UUCG dodecamer is significantly widened as compared to both a canonical A-RNA helix and the UUUG dodecamer (36, 40). In addition, the helical twist between the two inner U·U pairs in the UUUG dodecamer is unusually large (55°) in comparison to the average twist angle of 33–37° found for this duplex (40).

Superposition of the two dodecamer structures onto a canonical A-RNA helix gives root-mean-square deviations of only 1.0 Å for the UUUG dodecamer (40), and about 1.5 Å for the UUCG dodecamer (36). As expected, the glycosyl bond angles are generally around 68° for uridine, and about 43° for guanosine (37). The G·U pairs appear to be well stacked on the preceding standard Watson–Crick base pairs, whereas the stacking with the adjacent U·U or U·C pairs, respectively, is at least in part reduced (37, 40). In particular, U18 of the U7·U18 mismatch stacks on the G8·U17 base pair in the UUUG duplex (40), i.e., on the 3' side of the U in a G·U base pair, demonstrating that the postulated stacking rule (32, 33) (see above) demanding poor stacking 3' of U is not generally applicable. This observation again stresses the importance of the consideration of the sequence context for the assessment of stacking interactions (34).

In all studies describing G·U mismatches (36–38, 40), the critical role of highly ordered water molecules for the stabilization of the base pair is emphasized. In particular, there is a tightly bound water molecule bridging the O-2' of the uridine ribose and the N-2 amino group of the guanine (36, 37). This is also seen in the G·U pair in the acceptor stem of yeast tRNA^{Phe} (21), however, not in the G·U pairs of the UUUG dodecamer (40) or in the domain A helix of the 5S rRNA (38). In the latter, the three water molecules detected are bound to the G·U/U·G tandem mismatch in a variant way.

Finally, it should be noted that the average atomic positions of nonhydrogen atoms can be determined by X-ray crystallography with an accuracy that is defined by the resolution. Consequently, hydrogen bonds (and hence the base pairing) are not directly visible but rather are inferred from appropriate distances and orientations between the "heavy" (oxygen, nitrogen) atoms forming the hydrogen donors and acceptors, respectively.

By contrast, NMR spectroscopy permits immediate detection of intact, stable hydrogen bonds, and this provides direct information about base-pairing patterns and secondary structures (see later). However, structural information is gained in a more indirect way than in X-ray crystallography, mainly by using a set of (not precisely defined) distances between protons from different parts of the molecule as well as dihedral and torsion angles from which the conformation of the molecule can be derived (10, 44).

B. NMR Spectroscopy and Thermodynamic Stability

As already mentioned, proton NMR spectroscopy is a feasible method for the detection of hydrogen atoms involved in stable hydrogen bonds. Though both amino protons and imino protons of the bases that form base pairs can be observed, the latter are of greater use in the analysis of secondary structures. Unlike the amino resonances (typical chemical shifts around 8.5 to 9.0 ppm for the proton taking part in the hydrogen bond, and ~6.5 ppm for the second amino proton not directly involved in hydrogen bonding) that can at least partly overlap with the aromatic spectral region, the imino resonances (N-1H of guanine, N-3H of uracil) are shifted considerably downfield and thus are well separated from the normally fairly crowded (at least for oligoribonucleotides with more than about 15 residues) aromatic and ribose resonance parts of the spectrum. (It should be noted that these spectra have to be recorded with samples dissolved in H_2O buffers instead of the more frequently used D_2O buffers to prevent the exchange of these so-called labile protons for deuterium nuclei).

Favorably, each regular Watson–Crick base pair contributes one imino resonance to the proton spectrum: an A·U (N-3H of U) pair typically gives rise to an imino signal between about 14.8 and 13.5 ppm, whereas the G·C pair imino proton resonates at about 13.5 to 12.0 ppm.

Due to the particular base-pairing scheme of a G·U pair (compare Fig. 1), with the imino protons of both guanine and uracil being involved in hydrogen bonds, usually two imino signals are observed. The U N-3H resonance generally appears at lower field (~12.5–11.5 ppm), whereas the G N-1H imino peak is shifted distinctly upfield—as compared to the resonance positions usually found for imino protons—with typical chemical shift values of ~11.0–10.0 ppm.

Usually, the two imino resonances of a stable G·U pair in the "normal" wobble form (Fig. 1a) can be easily identified by their extremely strong (negative) nuclear Overhauser effect (NOE) (20–25%); i.e., irradiation of one imino resonance of the G·U pair gives rise to an extraordinarily strong attenuation of the peak from the other imino proton of the base pair. NOEs to the imino protons of adjacent base pairs are much weaker. Such a strong NOE is only possible if the proton–proton distance is less than 3 Å, a requirement met by the G·U wobble conformation, but not, however, by regular Watson–Crick pairs (even if they are directly adjacent in sequence).

In contrast to the case of the resonance lines of nonexchangeable protons (as, e.g., aromatic and ribose protons), the linewidth of the imino signals directly reflects the exchange rate of the imino protons with the protons of the surrounding water and consequently represents a measure of the base pair stability (45–47). From the temperature dependence of the imino linewidth, an activation enthalpy for this proton-exchange process can be calculated. However, this does not directly and solely reflect the stability of the hydrogen bond, from which the observed imino resonance originates, but rather is affected by the environment of the given base pair, i.e., by the sequence context (45–48).

The first G·U pairs that were detected by NMR measurements in the late 1970s and early 1980s were the G4·U69 pair in the acceptor arm of yeast tRNA^{Phe} (49–51), the G64·U50 pair in the TΨC stem of *E. coli* tRNA^{Val} (52), the G22·Ψ13 and G10·U25 pairs in the dihydrouridine stem of yeast tRNA^{Asp} (53), and the G68·U5 (acceptor stem) and G49·Ψ65 (hU stem) pairs of *E. coli* tRNA$_1^{Ile}$ (54). All of them could be uniquely identified by their extraordinarily strong intrapair NOEs between the imino protons of the base-paired G and U, respectively, as previously explained.

Later, fragments of prokaryotic 5S rRNA (55, 56), 16S rRNA (57), and eukaryotic 5.8S rRNA (58) were studied by NMR, and the imino resonances of these comparatively large RNAs have been partially assigned. In particular, several G·U pairs identified could be located afterward in certain secondary structure models of the 5S rRNA.

For the 5S and 16S rRNA fragments (55–57) "one-sided" NOEs between the G·U imino proton resonances and the imino resonances of the base pair 3' of G and 5' of U, respectively, were observed. However, all of the G·U pairs whose imino resonances could be safely sequentially assigned are accommodated in sequence tracks 5'-Y-G-Y-3' (and, correspondingly, 5'-R-U-R-3' on the opposite strand). This sequence pattern indeed gives rise to a largely destacked arrangement of the G·U pair and the base pair on the 3' side of the mismatch U, as well as that on the 5' side of the G, as was pointed out in Section I,A.

Interestingly, NOEs to both sides of the G·U pair were found if the mismatch pairs were in a different sequence context, as for the G4·U69 and U5·G68 base pairs in the acceptors stems of yeast tRNA^{Phe} (51) and *E. coli* tRNA$_1^{Ile}$ (54), respectively, or in the 5.8S rRNA (58). In these cases, the G of the G·U pair is embedded between two purines (always both A and G), which obviously warrants a more or less regular stacking of the involved bases. Two-sided NOE contacts are also found in a two-dimensional NOE spectroscopy (2D-NOESY) study of *E. coli* tRNA^{Val} for the G50·U64 pair of the TΨC stem, where G50 occurs in a sequence 5'-G49·G50·C51-3' (59).

Otherwise, only one-sided NOEs for a G·U pair in a quite similar sequential context were reported in an NMR investigation of helix I from *E. coli* 5S rRNA (60). However, here the G of the mismatch pair is followed on its 3' side by another G, whereas its 5' preceding neighbor is a C (5'-C-G-G-3'). Here, the lack of NOEs between either of the imino protons of the G·U pair and the imino proton of the G on the 3' side of the mismatch U and the structural model of the duplex both clearly indicate a destacking between the G·U pair and the C·G pair 3' of the mismatch U (5' of G) (60). Interestingly, the helical twist between these two base pairs was unexpectedly found to be lower than the standard value for a regular A helix. According to the detailed analysis of the helical parameters of the structural model, the distinct destacking is mainly brought about by a fairly large displacement (translation) of the whole base pair parallel to the base pair axis away from the helix axis (60). Thereby, the C of the C·G pair adjacent to the G·U pair is moved slightly into the major groove, and at the same time the base-paired G is shifted toward the exterior of the helix, whereas the U of the mismatch pair comes closer to the helix center.

Another sequence that contains a functionally relevant G·U pair is that of the acceptor stem from *E. coli* tRNA^{Ala} (17–19). The imino proton spectra of this 18-nucleotide duplex with seven base pairs and four single-strand nucleotides (including the 3'-CCA terminus, which all tRNAs have in common), as well as those of several sequence variants, display distinct and narrow imino signals from both the G and U of the mismatch (48), indicating stable base pairing. The stability of the corresponding hydrogen bonds, as assessed by the activation enthalpies for the exchange of the concerned proton with the solvent, is comparable to that of Watson–Crick pairs in the interior of helical stems for the G imino proton of the mismatch pair, whereas it is generally distinctly less for the U imino proton in all the G·U pair-containing sequences studied so far (S. Limmer, G. Ott, and H.-P. Hofmann, unpublished results). This hints either at an asymmetric opening mechanism of the G·U base pair and/or different accessibility of the two imino protons for proton exchange-catalyzing ions (46, 47).

If the G in the mismatch pair is replaced by inosine (I), which lacks the 2-amino group, the thermodynamic stability (melting temperature, negative free energy of duplex formation) of the corresponding duplexes is markedly decreased (48). As has been discussed, in an X-ray structure analysis of G·U pair-containing duplexes (36, 37), a well-ordered water molecule linking the 2'-OH group of the uridine ribose and the 2-amino group of the base-paired guanidine was detected. This tightly bound hydrogen-bonded water bridge could be responsible—at least in part—for the increased stability of the G·U duplexes as compared to their I·U variants.

Extending these investigations, a 2D-NMR analysis of the tRNA^{Ala} acceptor stem and several variants has been performed (61). Here, in particular the differences in chemical shifts of the original wild-type sequence with the G·U pair and of the sequence variants with G·C and I·U pairs were compared instead.

Significant shift differences are found only at the position 3·70 of the wobble base pair and in its immediate neighborhood. In particular, there are marked shift differences for the H-5 protons of U70 and C70, as well as of C71, comparing the wild-type G·U duplex and its variant with a regular G3·C70 pair. This difference is still increased (as to U/C 70) with respect to the intrinsic chemical shift differences between uridine and cytidine (61). The chemical shift changes of the protons in nucleic acids on transition from the totally disordered coil state to the well-ordered helical duplex are mainly caused by the ring current effects of the stacked bases (62, 63). Thus, the chemical shift data can be interpreted in terms of altered stacking between the second G2·C71 and the third G3·U/C70 base pairs such that the C71 displays poorer stacking above U70 in the wild-type G·U duplex as compared to the duplex with a regular G3·U70 base pair (61). This assumption is corroborated by the NOESY data as well as a comparison of the chemical shift differences between ordered duplexes and coiled single-stranded sequences at high temperature (S. Limmer, B. Reif, and G. Ott, unpublished).

Thus in this special sequence context (four consecutive Gs in one strand, with the mismatch G at the third position beginning from the 5' end), the G·U pair evidently introduces a deviation from the helical geometry and the stacking pattern that is observed for a similar duplex having a regular G·C pair in place of the G·U wobble pair. This is possibly of relevance in intermolecular recognition (see later).

In an NMR study of the equine infectious anemia virus (EIAV) transactivation response (TAR) RNA element, which has 25 nucleotides and forms a stem–loop structure, the arrangement of two consecutive G·U pairs is discussed (64, 65). One of these G·U pairs forms the terminal base pair of the stem and borders on the four-nucleotide loop. From the NMR data and the derived structural model (65) it can be concluded that the whole stem—including the two terminal G·U pairs—does not deviate from the canonical A-helical geometry; however, certain alterations in the base-stacking pattern of the G·U pairs as compared to a regular A-helix are apparent. In particular, the bases of the "inner" G·U pair, i.e., the pair that does not border on the loop, are stacked more directly above the bases 3' of the mismatch G (U) and 5' of the mismatch U (A) than is found in a regular A-helical geometry

(65). On the other hand, all the NOE contacts to be expected for a more or less regular A-RNA geometry are detected throughout the whole stem, including the two consecutive G·U base pairs (64). This corroborates the proposed almost regular A-helix structure of the stem, with certain structural modifications introduced by the G·U pairs, which possibly could play a functional role in the interaction with the tat protein (64, 65).

The structure of a G·U pair that terminates the stem in a stem–loop (hairpin) sequence with a three-nucleotide loop has been characterized by NMR (66). Here, too, both imino resonances to be expected for a wobble G·U pair are indeed observed, indicating stable base pair formation. The G·U pair is obviously well stacked on the preceding A·U pair (i.e., to the 3′ side of the G and the 5′ side of the U); in addition, the loop uridine residue preceding the G on the 5′ side of the mismatch G·U pair appears to be at least partly stacked on the mismatch G.

Another hairpin sequence with a 5′-CUUG-3′ tetraloop and a 4-bp stem has been structurally characterized by NMR. It contains a G·U pair at the second position in the stem (67). Again the NOE pattern to be expected for a usual A-RNA geometry was found, including all sequential imino–imino NOE contacts. However, some minor deviations from regular A-helix geometry are detected, which may also be affected by incorporation into the stem of a C·G base pair that is formed by the first and the fourth nucleotides of the loop sequence (67).

Employing homogeneously ^{13}C- and ^{15}N-labeled RNA, the structure of the 20-nucleotide P1 helix from a group I intron ribozyme was determined by means of very high-precision homo- and heteronuclear multidimensional NMR analysis (35). This sequence forms a hairpin in solution with a 5′-UUCG-3′ tetraloop and a G·U base pair in the stem. The mismatch G occurs in the sequence 5′-U-G-U-3′. As extensively discussed above, in such a context a stacking "asymmetry" on both sides of the G·U pair can be expected (32–34) and is indeed observed. The bases of the A·U pair, with the A on the 5′ side of the mismatch U and the U 3′ of the mismatch G, display extensive overlap with the bases of the mismatch pair, whereas the bases of the A·U pair on the other side of the G·U pair are almost totally destacked above the wobble pair. Moreover, the G·U pair opens toward the minor groove (35).

Moreover, the extraordinarily high precision of this NMR structure permitted a reexamination of the conformation of one of the most frequent tetraloop sequences (5′-UUCG-3′). In earlier NMR investigations (42, 43, 68), the formation of a stable G·U base pair between the first and the last (fourth) nucleotides of the loop sequence has been determined. The most remarkable finding in these studies, the unusual *syn* conformation of the mismatch guanosine, was confirmed in a more recent analysis (35). Both bases of the loop mismatch G·U pair are stacked on the bases of their adjacent stem nucleotides. However, in the earlier structural models (42, 43), a reverse wobble base pair (Fig. 1b) was proposed, whereas the more precise recent analysis clearly indicates a peculiar arrangement of the two nucleotides, including nonstandard interactions between a G amino proton and O-2 of the uracil, and the 2'-OH of the uridine and O-6 of guanine. These interactions should contribute to the extraordinary stability.

The importance of the sequence context of a (tandem) G·U mismatch base pair, not only for the stability of the wobble pair but also for the whole duplex, can be simply demonstrated by the analysis of the imino proton spectra of two self-complementary oligoribonucleotides with identical nucleotide sequences, but of opposite polarity (i.e., one sequence is 5'-ACGUCUG-GACGU-3', the other one is 3'-ACGUCUGGACGU-5'). These oligoribonucleotides can form duplexes with two central GU/UG base pairs and the same "topological" arrangement of base pairs, yet different sequences (5' → 3' and 3' → 5', respectively). Owing to the inversion symmetry of the duplexes, in principle, only six distinct base pairs are detectable in the NMR spectra, one of which (the first A·U pair) is not visible even at low temperature due to extensive fraying (base pair opening). The corresponding imino resonance regions of the two sequences are shown in Fig. 2.

The differences in the chemical shifts of the imino resonances of corresponding base pairs are obvious. Moreover, for the 3' → 5' sequence the resonances are significantly broader than for the 5' → 3' sequence. The greatest shift change (~1 ppm) is observed for the G imino resonance of the mismatch pair, indicating the different stacking geometries in the two duplexes at and in the neighborhood of the tandem mismatch. Part of the shift changes may be explained by different shielding effects of 5'- and 3'-neighbored bases on the protons of a given residue (62). However, this cannot account for the total, very large shift difference of the G imino resonances of the G·U mismatch pair in the two different duplexes. Hence structural differences, especially in the vicinity of the tandem G·U mismatch, must be assumed.

This assumption is corroborated by analysis of the activation enthalpies for the imino proton exchange as determined from the temperature dependence of the imino linewidths (S. Limmer and G. Ott, unpublished results). This activation enthalpy is ~390 kJ mol^{-1} for the imino proton of the fifth G·C base pair in the 5' → 3' sequence (Fig. 2a), whereas it is distinctly less (~240 kJ mol^{-1}) for the same base pair in the 3' → 5' sequence. Likewise, the activation enthalpy for the U imino proton exchange of the mismatch pair is greater by a factor of more than 1.6 for the 5' → 3' sequence (~135 kJ mol^{-1}) than that for the 3' → 5' duplex (~82 kJ mol^{-1}). Surprisingly, the activation enthalpies for the G imino protons of the G·U pair are almost equal (~230 kJ mol^{-1}).

From these findings it can be concluded that the proton exchange kinet-

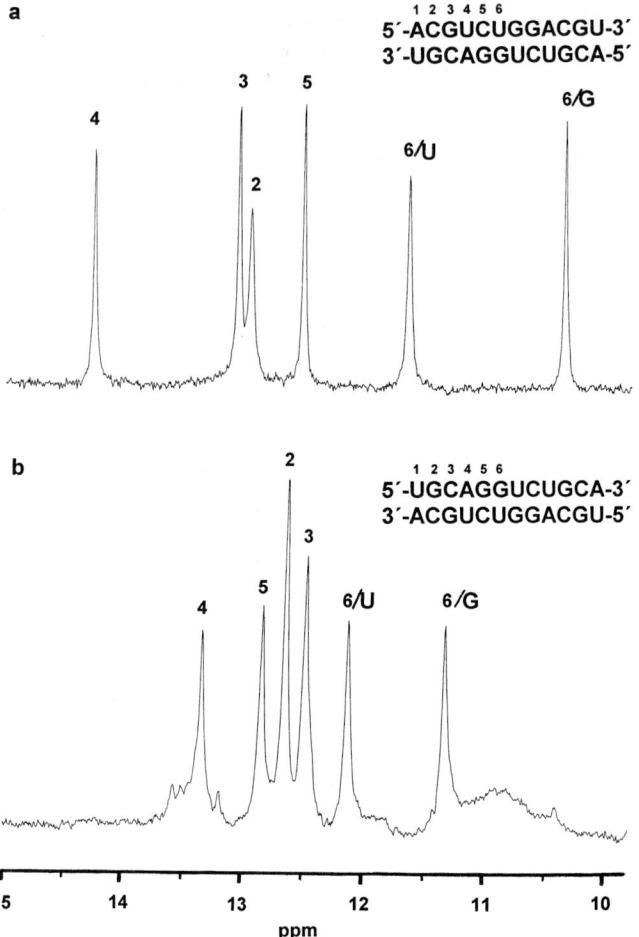

FIG. 2. Imino proton regions of the proton spectra of two self-complementary strands; (a) 5'-ACGUCUGGACGU-3' sequence; (b) 3'-ACGUCUGGACGU-5' sequence. Both spectra were recorded at 20°C, RNA single-strand concentration ~1 mM (10 mM sodium phosphate buffer, pH 6.5, 100 mM NaCl). The peak numbers indicate the assignment of imino resonances to individual base pairs according to the numbering in the duplex sequences.

ics (and hence hydrogen bond stability) is different for 5'-GU-3' and 5'-UG-3' tandem mismatches, in full agreement with extensive and thorough thermodynamic analyses of the influence on duplex stability of G·U mismatches in varying sequence contexts (69, 70). In addition, the type and orientation of the flanking base pair affect tandem mismatch stability (70). According to these analyses the tandem G·U mismatch in the 5' → 3' sequence should be energetically favored as compared to the mismatch in the 3' → 5' sequence by a Gibbs free energy gain of about -2.5 kJ mol^{-1} (70).

Systematic stability analyses of duplexes with isolated G·U pairs and tandem G·U mismatches (69, 70) have derived free-energy increments for G·U mismatch formation, and in particular for the formation of tandem G·U mispairs [as well as further tandem mismatches (70)]. It is shown that in general the stability increment provided by a G·U mismatch in a duplex is large if it is flanked by G·C or C·G pairs, and generally is lower for A·U or U·A pairs, whereas there is only a weakly stabilizing effect if one G·U pair is adjacent to a further G·U mismatch (69, 70). Thus a 5'-UG-3' tandem mismatch still has a small negative Gibbs free-energy increment (and hence is stabilizing the duplex), as is the case for a 5'-GG-3' tandem mismatch, whereas a 5'-GU-3' couple is clearly destabilizing (positive free-energy increment).

Generally, 5'-GU-3' tandem mismatches have the least stability (70). This is also reflected in the imino region of proton NMR spectra of corresponding duplexes. In duplexes containing 5'-GU-3' tandem mismatches, the imino resonances of the mismatch G and U bases are broadened at lower temperatures than are the imino lines of interior Watson–Crick pairs, whereas duplexes with a 5'-UG-3' tandem mismatch display the same temperature-dependent broadening seen with the interior Watson–Crick pairs (69). In addition, a general rule was deduced with respect to the influence of the flanking base pairs on the stability of the tandem mismatch pairs (70): the stability of a given tandem mismatch depends on the flanking base pair in the order GC > CG > UA ≈ AU (with the first base on the 5' side).

Another interesting effect that is possibly associated with particular structural features of G·U mismatches is the specific (weak) binding of manganese ions at or in the vicinity of such mismatch pairs (48, 71, 72). These NMR studies make use of proton NMR line-broadening effects of paramagnetic manganese ions when they are bound close to a proton spin (73). The line broadening grows strongly with decreasing manganese ion–proton distance according to an r^{-6} dependence (r = proton–manganese ion distance).

The addition of small amounts of manganese ions (at a molar ratio of 1:150 to 1:200 to the RNA, in the presence of a 1000-fold excess of magnesium ions over manganese ions) to duplexes derived from the acceptor arm of E. coli tRNA^{Ala} resulted in specific line-broadening effects. This indicated a weak, specific binding of manganese in the major groove of the duplex between the G·U mismatch pair at the third position in the stem (counted from the side of the single-stranded terminus of the tRNA), and the preceding second G·C base pair was found (58). For the acceptor stem duplex of yeast tRNA^{Phe}, where a G·U pair occurs in a very similar context but is shifted to the fourth position in the stem, a fully analogous binding pattern of the

Mn^{2+} ion was observed: from the particularly strong broadening effects on Mn^{2+} addition for the imino protons of the third (G·C) base pair and the U imino proton resonance of the mismatch, a specific manganese binding was determined to occur in the major groove between the third and the fourth (G·U) base-pairs (71). However, broadening effects comparable to those for the original tRNA^{Ala} acceptor stem have also been detected for a variant with a G·C pair in place of the mismatch G·U pair (i.e., particularly strong broadening of the imino resonances of the base pairs two and three) (71); it must be concluded, therefore, that, at least in this case, the G·U pair is not responsible in the first place for the creation of a specific Mn^{2+} binding site but rather it is the structural peculiarities associated with the track of four purines (Gs) in this duplex (71). If the G·U mismatch in the tRNA^{Ala} acceptor stem is replaced by an I·U pair, both the strength and the specificity of Mn^{2+} ion binding are drastically reduced (71).

In a similar case, a specific line broadening was also observed with the acceptor stem of yeast tRNA^{Asp} (71), where the G·U mismatch is accommodated in a different sequence context (the mismatch U flanked by two Gs) and is located at the fifth position in the stem. Here the imino protons of the mismatch U and that of the fourth (G·C) pair become specifically broadened, though the broadening effects are less dramatic than for the tRNA^{Ala} and tRNA^{Phe} acceptor stems (71).

A specific Mn^{2+} binding site has also been found at the G·U mismatch pair in the P1 helix of a group I self-splicing intron (72), where the G·U pair occurs in a sequence similar to that of the G·U pair in yeast tRNA^{Asp} (the mismatch U is flanked by two purines) and accordingly comparable specific line broadening effects are reported (72).

For the two duplexes containing tandem G·U mismatches in opposite sequence polarity (Fig. 2) distinctly different binding specificities for Mn2+ ions were found (S. Limmer and G. Ott, unpublished results). With the 5' → 3' duplex (Fig. 2a) containing a 5'-UG-3' tandem mismatch, the most pronounced line broadening was measured for the U imino proton of the G·U pair; a somewhat weaker (and almost equal) broadening for the imino lines originating from the preceding base pairs five (C·G) and four (U·A) was measured. For the 3' → 5' duplex with the opposite sequence polarity (Fig. 2b), however, the strongest broadening is seen for the G imino proton line of the fifth base preceding the G·U mismatch, and smaller effects are detected for the U imino resonance of the G·U mismatch pair. Here, possibly the previously discussed potential of three (or more) consecutive purines is more effective for the creation of a Mn^{2+} binding site (which is, perhaps, somewhat enhanced by the presence of the G·U pair), whereas in the former case (5' → 3' duplex) apparently the G·U pair is of greater importance for the formation of the Mn^{2+} binding site.

II. G·A Base Pairs

A. General Features and X-Ray Structure Analyses

Intuitively, it might seem that the formation of purine–purine mismatch base pairs is unfavorable for steric and geometric reasons, and would give rise to severe distortions of the regular helical arrangement. However, it turns out that G·A base-pairs can be stably formed in both DNA (74–78) and RNA (79–81); this was directly demonstrated by the X-ray structural analyses.

However, the existence of G·A mismatches had already been deduced from sequence analyses and secondary structure predictions of ribosomal RNAs (27–30, 82, 83, 83a). Next to G·U pairs, G·A mismatches are the most frequently encountered non-Watson–Crick pairs in ribosomal RNAs, with a remarkable preference of 5'-GA-3' tandem mismatches (27–30, 70, 83a). In contrast to the G·U pair, G·A mismatches display a greater conformational variability, i.e., there are different kinds of base pair arrangements involving distinct hydrogen-bonding patterns [(see Fig. 3) (5, 78, 79)], giving rise to considerably varying C-1'–C-1' distances and glycosyl bond angles (74). In particular, in the G(*anti*)–A(*anti*) conformation (Fig. 3a) the C-1'–C-1' distance is increased by almost 2 Å as compared to both regular Watson–Crick pairs as well as other forms of G·A pairs (78, 79).

It should be noted that for DNA the conformation of the G·A mismatch pair is both sequence and pH dependent (84–90): at pH > 7 the G(*anti*)–A(*anti*) arrangement (Fig. 3a) is observed, whereas below pH ≈ 6 (when A becomes protonated at N-1) the conformation of G switches to *syn* while A remains in the *anti* conformation (Fig. 3c). At the same time, the hydrogen bond pattern is altered. Hydrogen bonds are formed between the keto group of the mismatch G and the amino group of A, and another one between the protonated N-1 of A and the N-7 of G (Fig. 3c).

Up to now only one crystal structure analysis of an RNA duplex containing two (isolated) G·A mismatch pairs has been reported (79). The self-complementary sequence 5'-CGCGAAUUAGCG-3' (residues involved in mismatch base pairing are underlined) was analyzed where the two G·A pairs occur, at the fourth and ninth positions of the duplex stem. The duplex forms an A-type double helix with 10 regular Watson–Crick base pairs and two G·A mispairs. The global helical twist of 34.2° (corresponding to 10.5 residues per full helical turn), the average helical rise (2.62 Å), and the sugar-phosphate backbone torsion angles and the average groove widths (as far as measurable) are close to the values expected for a canonical A-RNA helix (12). However, there are distinct local variations of these parameters along the helical stem, in particular at and in the immediate vicinity of the G·A pairs (79). The G·A mispairs are found to adopt the G(*anti*)–A(*anti*) conformation sketched in Fig. 3a. Interestingly, in an analogous DNA structure with the sequence

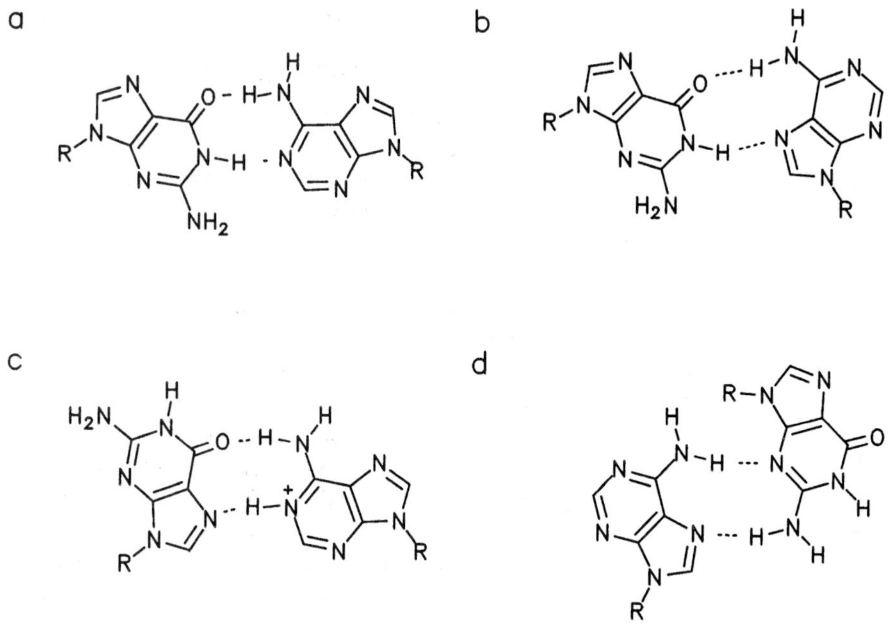

FIG. 3. Base-pairing patterns of G·A mismatch pairs (adapted from Ref. 79).

5'-d(CGCGAATTAGCG-3' (74), which forms a B-DNA helix in the crystal, the G·A pair is in the G(*anti*)–A(*syn*) conformation (Fig. 3b), whereas another B-form decamer, 5'-d(CCAAGATTGG)-3', gives rise to G·A pairs in a G(*anti*)–A(*anti*) arrangement (76) similar to that in the RNA dodecamer duplex (79). The C-1'–C-1' distances 12.4 and 12.9 Å, respectively, for the fourth and ninth G·A pair) are ~2 Å larger than those found for regular Watson–Crick pairs (10.5 Å), as is expected from the geometry of the G(*anti*)–A(*anti*) pairing. The increased C-1'–C-1' distance of the G·A mispair also causes a slight bulging of the duplex (79) in the immediate neighborhood of the mismatches. The phosphodiester backbone of the rest of the duplex, however, is almost unaffected (intrastrand phosphorus–phosphorus distances between 5.3 and 6.3 Å, ribose puckering C-3'-*endo* for all residues). For the G·A pairs, clearly two hydrogen bonds can be deduced from the crystal structure: one between the N-6 amino group of A and the carbonyl O-6 of G, and the second one between the N-1 imino group of G and the N-1 of A. Moreover, the involvement of an unconventional reverse three-center hydrogen bond is discussed, where both the G imino proton and an N-2 amino proton form hydrogen bonds to the adenine N-1 nitrogen, which could lend additional stability to the G·A pairs (79). NMR studies of a DNA duplex containing a G·A mispair (87) seem to corroborate this suggestion.

In the previously discussed duplex structure, four steps between base pairs occur that involve mismatch G·A pairs, namely, steps 3 and 9 of type 5′-CG-3′/5′-AG-3′, and steps 4 and 8 of type 5′-GA-3′/5′-UA-3′. Generally, in all cases pronounced intrastrand purine–purine stacking is found with strong overlap of the five-membered ring of a purine base and the six-membered ring of the adjacent purine (79). In addition, weak intrastrand pyrimidine–purine stacking as well as cross-strand purine–purine stacking are observed. In comparison to the other steps (average twist of 36.4°), the four steps involving the G·A pairs are marked by a reduced twist angle (average of 28.5°) (79). For the amino group of the G in the G·A pair at the ninth position of the duplex, an interaction with a water molecule was observed; this was not the case, however, for the equivalent amino group of the G in the fourth G·A pair (79). Apparently, tight binding of water molecules to the G amino group does not play a comparably important role for stabilizing the G·A mismatch pair, as in the case of certain G·U wobble pairs (36, 37).

Two further X-ray structure analyses of RNA molecules containing G·A mispairs have been published (80, 81). They characterize the three-dimensional structure of a small catalytic RNA molecule, the so-called "hammerhead" ribozyme (91). The hammerhead secondary structure motif is marked by three base-paired stems and a central core of 15 nucleotides, most of which are strictly conserved (87, 88). Most of these conserved nucleotides are located in two single-stranded regions with no obvious base pair formation. These conserved nucleotides are of crucial importance for ribozymatic activity, i.e., cleavage of the phosphodiester bond between two nucleotides at a well-defined position in the secondary structure (80, 81, 91, 92). Catalytic activity requires the presence of divalent metal ions (e.g., Mg^{2+}, Mn^{2+}, and Ca^{2+}).

The crystal structures of the two hammerhead ribozymes, which differ in certain sequence details, revealed very similar global fold reminiscent of a tuning fork or the greek letter γ (80, 81). Moreover, the base pairing and the arrangement of the conserved nucleotides of the central core were almost identical, though in one case (80) the so-called substrate strand, containing the nucleotide at which the phosphodiester bond is cleaved by the ribozyme strand, consists totally of deoxyribose nucleotides, thus preventing spontaneous cleavage. In the other case only the cytosine at the cleavage site was replaced by a 2′-methoxy derivative of this nucleoside (81), which hence lacks the 2′-hydroxy group that is absolutely necessary for scission of the phosphodiester bond at this site.

From crystal structure analysis it became clear that two of the stems

(stems II and III) (*80, 81*) deduced from the secondary structure are coaxially stacked in the tertiary structure, with the catalytic pocket (containing the cleavage site, i.e., the nucleotide with the scissile bond) and the remaining stem I branching away from this axis (*81*).

The nearly collinear arrangement of stems II and III is brought about by two additional G·A mismatch pairs and an unusual (singly hydrogen-bonded) A·U pair (*80, 81*) formed between strictly conserved nucleotides of the two single-stranded regions in the secondary structure. Stem II together with the three additional base pairs was termed "augmented stem II helix" by Scott *et al.* (*81*). It stacks directly upon the stem III helix, thus forming a prolonged, distorted pseudocontinuous A-type helix.

The two consecutive G·A mismatch pairs occur in a (conserved) sequence track 5′-CGAA-3′/5′-UGAG-3′. They form hydrogen bonds between the 2-amino groups of G and N-7 of A, and between the 6-amino group of A and the guanine N-3 (Fig. 3d). Moreover, the second hydrogen atom of the 6-amino group of adenine is involved in hydrogen bonds to O-4′ in the first of the two G·A pairs, and to the 2′-hydroxy group of the second one (*80, 81*). In addition, there are further hydrogen bonds involving the functional groups of the nucleotides in the G·A pairs as well as phosphate groups connecting both nucleotides of the same strand and of the opposite strand. They may help to stabilize this nonstandard structure and are also important for the catalytic activity of the hammerhead ribozyme (*80*).

The three consecutive purines in the track 5′-GAA-3′ continue the stacking of the stems in a roughly parallel manner (*80*). On the opposite strand of this mismatched duplex (nucleotides 3′-AGU-5′), however, the G and U bases appear to be largely destacked, whereas there is a distinct cross-strand base plane overlap between both the two Gs and the two As, respectively. All of the nucleotides are in *anti* conformation, the sugar puckering is always C-3′-*endo*, except for the G of the first G·A pair (in the track 5′-GAA-3′), which is found to adopt a 2′-*endo* ribose pucker (*80*). Moreover, the latter base is significantly tilted.

It has been discussed (*80*) that a sequence track

$$\begin{matrix} 5' \text{ G A R} \\ 3' \text{ R A G} \end{matrix} \quad (R = \text{purine})$$

could give rise to a local structure that brings about an arrangement of functional groups of the bases, as well as a sugar-phosphate backbone conformation that allows specific binding of catalytically important metal ions (*80*).

B. NMR Studies and Thermodynamic Stability

In Fig. 3 different possible base-pairing arrangements for G·A mismatch pairs are shown. In two cases, hydrogen bonds are formed involving the G–

N-1 amino proton (Fig. 3a, b) in addition to the hydrogen bonds between the 6-amino group of adenine and the 2-carbonyl oxygen of guanine. Another base-pairing scheme including an imino hydrogen bond is possible if the adenine base is protonated (at pH < 6) at its N-1 nitrogen, which interacts with the N-7 of the guanine base (Fig. 3c). Such a base pairing has indeed been detected in several DNA sequences containing G·A mispairs (84–90).

With respect to NMR analysis of G·A mismatches, it is important to note that an imino proton involved in a stable hydrogen bond should give rise to a proton resonance signal in the region below (downfield from) ~11 ppm, and moreover should have a linewidth comparable (at least for nonterminal G·A pairs) to that of an ordinary Watson–Crick pair. If the imino resonance, however, is shifted considerably upfield from the usual imino resonance region (above ~11 ppm) and is comparatively broad, this can be taken as an indication of the noninvolvement of the G imino proton in base pair hydrogen bonding. The validity of this rule is corroborated by NMR studies of both DNA (84, 87, 88, 90, 93–95) and RNA (96–100) duplexes.

Interestingly, as was pointed out by SantaLucia et al. (96), relatively narrow and downfield-shifted imino signals are observed for tandem mismatches of the type 5'-AG-3', whereas the altered tandem mispair arrangement 5'-GA-3' usually gives rise to broader, upfield-shifted resonances. However, in the same sequence context, the latter 5'-GA-3' tandem mismatch pair is generally more stable than the former tandem mismatch, which has been shown by extensive thermodynamic investigation of duplex stabilities (70, 98). Clearly this indicates different base pair arrangements and corresponding different hydrogen bonding patterns in the two different G·A tandem mismatches.

The importance of the flanking base pairs for both stability and base pairing is documented by the observation of distinct imino resonance positions found for different duplexes with 5'-GA-3' tandem mismatches. If the preceding base pair X·Y in a sequence context 5'-XGA-3'/5'-GAY-3' is A·U, U·A, C·G, G·U, or U·G, in all cases a relatively broad imino signal around 10 ppm is observed, in agreement with the rule formulated above. However, with a flanking G·C pair the G imino resonance in the spectrum of the corresponding duplex with a 5'-GGA-3'/5'-GAC-3' track [which is the thermodynamically most stable configuration for G·A tandem mismatch pairs (70)], the imino signal appears significantly downfield shifted at ~12 to 13 ppm (98, 100), indicative of an involvement of the G imino proton in the base pair formation.

Accordingly, evaluation of the imino region of the proton NMR spectrum of mismatch duplexes containing G·A pairs can serve to discriminate between the different kinds of base pairing (Fig. 3). Further details of the base pair geometry can then be derived from NMR analyses, including the spectral regions of the nonexchangeable protons as well as information originating from nuclei other than protons (e.g., ^{31}P or ^{13}C), using more sophisticated multidimensional NMR methods (97, 99, 101–104).

Such a structure analysis by NMR has been performed for two synthetic self-complementary duplexes containing a central G·A tandem mismatch with identical flanking base pairs: 5'-CGAG-3'/5'-CGAG-3' (97, 99). In both studies, the G imino resonance of the G·A pairs was fairly broad and resonated at about 10 ppm, as expected for a base-pairing scheme not involving imino proton hydrogen bonds. Structure determination based on data from two-dimensional NMR experiments and restrained molecular dynamics calculations revealed that the nucleosides in the G·A base pairs indeed adopt *anti* conformations and are arranged in a "head-to-head" manner involving hydrogen bonds between the 6-amino group of guanine and the N-7 nitrogen of adenine (Fig. 3d).

The helical twist between the two mismatch G·A pairs is significantly higher (~81°) than the standard twist angle for a regular A-RNA. Correspondingly, the twist angle between the flanking C·G Watson–Crick pair and the adjacent G·A pair is distinctly diminished (~18°) in comparison to a canonical A-RNA helix (97). The "sheared" arrangement of the bases in this mismatch pair gives rise to a comparatively short C-1'–C-1' distance of only ~9.1 Å (97). At the same time, the unusual twist angles between the G·A base pairs, and the G·A pairs and their neighboring base pairs, allow the accommodation of these mismatch pairs in the duplex without gross distortions of the overall A-helical geometry (97).

A remarkable feature of this type of tandem G·A mismatch is the high degree of interstrand base plane overlap between the adenines and the guanines of the two adjacent G·A pairs (97), at the expense of the intrastrand stacking between A and G of the same strand. There, the interstrand stacking between the adenine bases seems to be still more pronounced than that of the guanines. The interstrand stacking of the guanines is actually even fairly poor in the duplex characterized by Katahira *et al.* (99). Both studies (97, 99) agree on the weak stacking between the G·A mispair and the flanking C·G pair.

A quite similar geometry and base pairing has also been reported for a particularly stable hairpin structure with a 5'-GCAA-3' tetraloop (105). Here, the first and the fourth nucleotide of the loop form a G·A mismatch pair that is again flanked by a C·G Watson–Crick pair of the helical stem, as in the duplexes described above. Hydrogen bonds are formed without participation of the G imino proton according to the scheme depicted in Fig. 3d. Unlike in the UUCG tetraloop hairpin (42, 43), all nucleotides, including all loop residues, are in *anti* conformation. Both G and A of the loop G·A mispair stack on the bases of the closing Watson–Crick C·G pair of the stem.

An NMR structural analysis of a variant tetraloop (5′-GAGA-3′) hairpin sequence with a different loop-closing Watson–Crick (A·U) pair arrives essentially at the same conclusion with respect to the arrangement and hydrogen-bonding scheme of the G·A mismatch pair built by the first and last nucleotides of the tetraloop (106).

Very similar "sheared" G·A mismatch pair geometries have also been derived from NMR analyses of hexanucleotide (5′-GUAAUA-3′) hairpin loops found in a highly conserved sequence of the L11 protein-binding region in 23S ribosomal RNAs (106a, 106b). There, the first (G) and the last (A) nucleotides of the loop form a G·A mispair of the type depicted in Fig. 3d, with both nucleosides adopting *anti* conformation. The G·A pair obviously stacks on the closing C·G pair of the stem.

The structural analysis of the highly conserved α-sarcin/ricin hairpin from 28S/23S ribosomal RNAs (101, 102) revealed that the loop, which according to the secondary structure predictions consisted of 17 unpaired nucleotides, actually forms a very compact structure with several mismatch base pairs continuing the helical stem of regular Watson–Crick base pairs. Only the four nucleotides (5′-GAGA-3′) at the top of the former 17-nucleotide loop give rise to a tetraloop, which is closed by a C·G base-pair and hence largely resembles the tetraloops described before. The hydrogen bonding pattern, base arrangement, and stacking of the G·A mismatch pair built up from the first and the fourth nucleotides in this residual tetraloop conform to the G·A mismatch pair structure found in the GNRA (R = purine, N = any nucleotide) tetraloop (105, 106).

A further G·A mismatch pair (101, 102) in this sequence is arranged in a quite similar way. However, its stacking is unique and extraordinary because the G of the mismatch is stacked on the U, and the mismatch A on the A of a preceding reverse-Hoogsteen A·U pair, such that each base of the mismatch overlaps with the base of the reverse-Hoogsteen pair on the opposite strand. In contrast, there is obviously good intrastrand stacking between the G·A mismatch bases and the bases of the C·G pair on the other side of the mispair (101, 102).

Another RNA secondary structure element that consists of an asymmetric purine-rich internal loop flanked by two regularly base-paired stem regions is the Rev responsive element (RRE) from the *env* gene of HIV. It is the binding site for the viral Rev protein. There have been two NMR studies aimed at the structural characterization of the RRE RNA in free form and when bound to peptides derived from the Rev protein (103, 104). It was clearly demonstrated in both analyses that binding of the peptide by the RNA causes a stabilization of the internal loop base pairs that is unambiguously detected by the appearance of additional imino resonances in the proton NMR spectra on peptide binding, indicating the strengthening of hydrogen bonds involving imino protons (or, more accurately, the reduction of the imino proton exchange rates of the concerned bases). Though on the basis of the NMR analysis of the nonexchangeable protons a certain predominant conformation of the internal loop in the free (not peptide bound) RNA can be derived, it must be concluded that there is some exchange between two or more different conformations (conformational averaging) (*103, 104*).

On peptide binding, the structure of the internal loop is stabilized and accordingly better defined. In particular, two nucleotides (one A and one U) become bulged out, and several base pairs (both Watson–Crick and purine–purine mismatch pairs) are stably formed (or at least distinctly stabilized). One of the additionally emerging imino signals (downfield from 12 ppm) is definitely due to the formation of a G(*anti*)–A(*anti*) mismatch base pair involving a hydrogen bond between the G N-1 imino proton and the N-1 of A (Fig. 3a). From the sequential NOE connectivities an intrastrand stacking of the G·A mismatch bases is deduced (*103*) for the mispair in the specific context 5'-GGG-3'/5'-GAC-3', where the formation of the second flanking G·G mismatch pair requires a U in the sequence track 5'-CAUG-3' to be bulged out (*103, 104*).

A mismatch G·A pair formation was also assumed to occur in the loop E of eukaryotic 5S ribosomal RNA (*107*), with a sequence context similar to the one discussed before, 5'-GGA-3'/5'-UAC-3'. Here, a relatively narrow imino resonance at ~11.6 ppm was assigned to the N-1 proton of the mispair G (*107*). However, the more detailed analysis of the resonances of the nonexchangeable protons suggest a G(*anti*)–A(*anti*) base pair arrangement with hydrogen bonds involving the amino group, but not the imino proton (Fig. 3d). Consistent with this is the strong interstrand stacking between the G·A mismatch bases and the bases of the adjacent reverse-Hoogsteen A·U pair. In addition, there is obviously also only poor stacking between the G·A mismatch pair and the Watson–Crick G·C pair on the other side.

It seems interesting to note that the thermodynamically most stable G·A tandem mismatch sequence contexts (5'-GGAC-3'/5'-GGAC-3', 5'-GAGC-3'/5'-GAGC-3', and 5'-CAGG-3'/5'-CAGG-3') display base pair arrangements involving imino proton hydrogen bonds (*70, 96, 98*). G·A tandem mismatches in these contexts are stabilizing the concerned duplexes by Gibbs free-energy increments of about -2.5 kJ mol^{-1} up to almost -12 kJ mol^{-1} (*70*). However, flanking A·U or U·A base pairs in combination with G·A or A·G tandem mismatches destabilize the duplexes [Gibbs free-energy contributions of about $+0.5$ to 3.8 kJ mol^{-1} (*70*)]. Most probably, these different stabilization effects are only indirectly related to the variant hydrogen-bonding patterns, but rather to the distinct base stacking found for G·A pairs in different sequence contexts (compare also the G·A pairs described in Refs. *103* and *104* with those described in Refs. *97, 105* and *106*).

A very interesting phenomenon regarding the influence of isolated G·A mismatch pairs on the thermodynamic stability of self-complementary RNA duplexes has been reported by Morse and Draper (*108*). They find that these G·A mispairs affect the duplex stability to a much different extent, depending on the number of intervening regular G·C Watson–Crick base pairs, even when the immediate neighborhood of the mismatches remains unchanged. If the G·A pairs are separated by four (alternating) G·C pairs, the duplexes are destabilized by about 13 kJ mol^{-1} as compared to a reference duplex with only regular, alternating G·C pairs. For duplexes with G·A pairs separated by only two G·C pairs, a drastic decrease in stability is observed (~29 kJ mol^{-1} for sequence 5'-CAG-3'/5'-CAG-3'), and in the duplex with a sequence track 5'-CGG-3'/5'-CAG-3', no duplex formation, even at pH values around 7.0, could be achieved. However, a stable duplex was formed at pH 5.0 whereas the former one became distinctly more stable (*108*). Very probably, under these conditions a stable G(*syn*)–A$^+$H(*anti*) base pair (Fig. 3c) is formed, as was detected in NMR studies of DNA duplexes at low pH.

It seems noteworthy that in cases in which isolated G·A pairs obviously appear to stabilize (or, at least do not significantly destabilize) the duplex structure, the base pair 5' of the A in the G·A mismatch is *not* a regular Watson–Crick pair but rather a reverse Hoogsteen (*101, 102, 103*), or a further purine–purine (G·G) mismatch (*103, 104*). This also applies to the 5'-GA-3' tandem mismatch, however, not to the 5'-AG-3' tandem mispair, which, in an appropriate sequence context, can stabilize a duplex (*70, 98*) (see also previous discussion).

III. U·U and U·C Base Pairs

A. General Features and Crystal Structures of Mismatch Duplexes

Generally, U·U and U·C mismatch pairs appear to occur much less frequently than G·U and G·A pairs in the secondary structure of naturally occurring RNAs, which is evident in the case of tandem mismatches in ribosomal RNAs (*27, 28, 70*). Nevertheless, both X-ray structure analyses (*36, 37, 40, 41*) and NMR studies (*101, 107, 109*) prove the existence of stable mismatch U·U and U·C base pairs. The base pair patterns that can be expected for such mismatches are compiled in Fig. 4.

It should be noted that the U·U mismatch pair formation involves two imino hydrogen bonds with closely neighboring imino protons, which is, of course, important with respect to the NMR detection of stable U·U mismatch pairs (Fig. 4a, b). A very similar arrangement is also suggested also for the

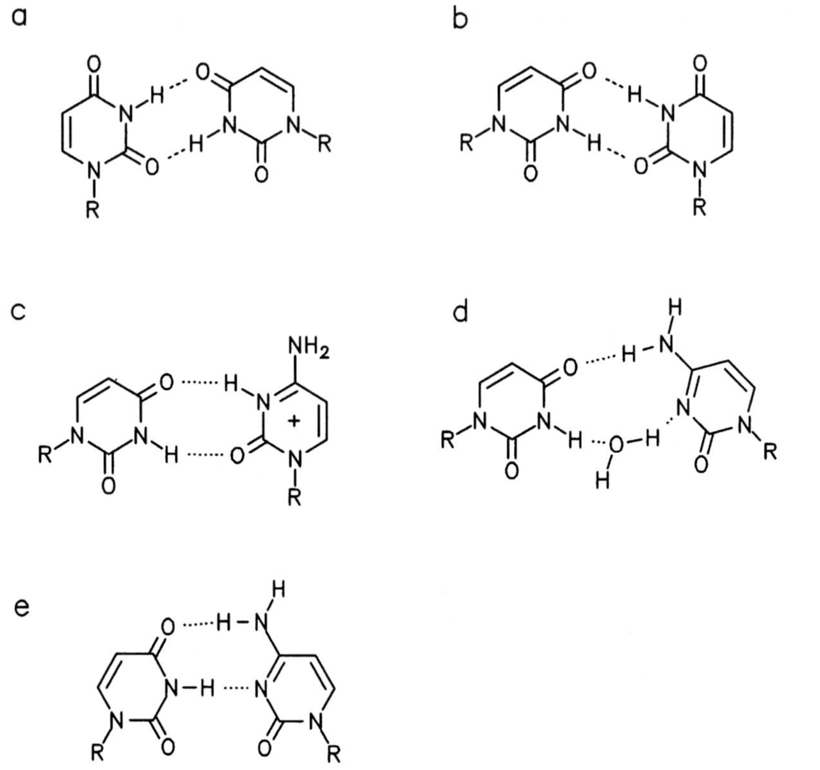

FIG. 4. Base pair arrangements (schematically) for U·U (a, b) and U·C (c, d, e) mismatches, respectively. (c) Base pairing between U and the protonated form of C (C$^+$H), which is stable at low pH.

U·C$^+$ mismatch pair (Fig. 4c), whereas in the "neutral" U·C mispair (Fig. 4d, e) the N-3 imino proton of U forms a water-mediated hydrogen bond to the N-3 of C. The only crystal structure with an internal U·U mismatch (in contrast to those where the mismatch formation is brought about by crystal packing effects; see later) is provided by the self-complementary duplex with the sequence 5'-GGACUUUGGUCC-3' (40); this has been discussed already in connection with the G·U pair conformation (see Section I,A).

In this duplex, two central U·U pairs are formed and flanked by G·U base pairs. The U·U pairs are of the wobble type depicted schematically in Figs. 4a and 4b. Actually, both types of U·U wobble pairs are found in this duplex: in one pair, the O-4 of the uracil in one strand (I) extends into the major groove, while the O-4 of the U in the opposite strand (II) is involved in hydrogen bonding to the N-3H of the base-paired U (Fig. 4a); in the second one the O-4 of the uracil on strand II protrudes into the major groove, and its N-3H makes a hydrogen bond to the O-4 of the U on strand I (Fig. 4b). In the specific crystal structure, one of the U·U pairs is highly twisted (by about 45°) with respect to the relative base plane orientations such that, in effect, only one hydrogen bond (N-3H–O-2) is intact (*40*). By this, the O-4 oxygens of the sequentially adjacent Us on opposite strands approach each other (distance ~3.2 Å), thus permitting the binding of a water molecule that bridges the two O-4 carbonyl oxygens of the adjacent U·U pairs in the major groove.

There is a quite large helical twist of (~55°) between the two U·U pairs; this affects the base stacking within the U·U tandem. Generally, stacking between the bases of the two U·U pairs is rather poor, though it is somewhat better on one strand. Both U·U pairs stack over the guanine bases of the adjacent G·U pairs (*40*).

As mentioned in Section I,A, this duplex gives rise to a distinctly less pronounced A-helix deformation as compared to the structure observed for a similar duplex having, instead of the central U·U tandem mismatch, a U·C tandem (*40*). The presence of numerous bound water molecules in the major groove of the latter duplex, which obviously contributes to the stabilization of the U·C base pairs, has been discussed as a reason for this. Probably, the accommodation of these water molecules requires a widening of the major groove (*36*), whereas the single water molecule bridging the two Us on opposite strands in the tandem U·U mismatch duplex (*40*) can be accommodated without strong distortions of the A-helix geometry just by rotation of the base plane of one of the uracils.

Totally different U·U pair arrangements are found in the crystal structure of the oligoribonucleotide 5′-UUCGCG-3′, which gives rise to the formation of regular Watson–Crick G·C base pairs due to antiparallel association of two of these strands constituting an A-type helical stem (*41*). The two 5′-terminal ("overhang") Us, however, form novel, unusual Hoogsteen-like base pairs with the 5′-UU overhang of an adjacent duplex. With respect to the orientation of their glycosyl bonds (N-1–C-1′), these U·U pairs are in a trans-like conformation (Fig. 5), in contrast to the cis-like arrangement in regular Watson–Crick pairs.

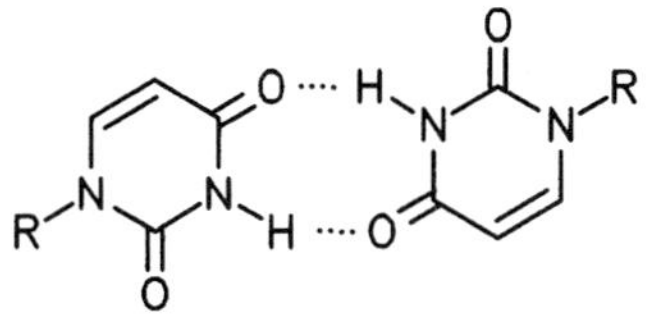

FIG. 5. Base arrangement in the unusual trans-like U·U pair (*38*).

One normal hydrogen bond is found between the N-3 imino group of one U and the carbonyl O-4 of the base-paired U. In addition, a second extraordinary hydrogen bond was seen between the O-4 of one uracil and the C-5H of the U in the opposite strand (*41*). The base pairs are further stabilized by an ordered water molecule bridging the O-4 of the uracil in one strand with the O-2 of the U in the opposite strand. Moreover, further water-mediated hydrogen bonds involving less highly ordered water contribute to the stability of the mismatch U·U pairs (*41*). The nucleotide conformation is *anti* for all residues in the duplex, and all riboses adopt the 3'-*endo* pucker.

The glycosyl bond angles are, correspondingly, drastically altered (about $-9°$ and $30°$, respectively) as compared to the values of Watson–Crick pairs in standard A-helical geometry (*12*). Though the helical twist ($32°$) between the two unusual U·U mispairs hardly deviates from the standard value, it is significantly diminished between the terminal C·G pair of the stem and the (first) U·U mismatch pair ($\sim 22°$). Both of the base plane overlaps, i.e., between the uracils in the two U·U base pairs and the one between the terminal C·G pair of the regular duplex stem and the adjacent U·U pair, are fairly poor, with slightly better stacking between the two U·U pairs (*41*).

Another duplex displaying U·U mismatch formation was observed for the self-complementary duplex 5'-GCUUCGGC-dBrU-3' (where dBrU is 5-bromo-2'-deoxyguanosine) with a 3' dangling end consisting of a modified deoxy nucleotide (*37*). The 3'-dBrU nucleotide of one duplex forms a U·U base pair of the type sketched in Fig. 4b, involving two hydrogen bonds with the 3'-dBrU of another symmetry-related duplex, thus constituting continuous helical stacks in the crystal (*37*). This duplex also contains a central U·C tandem mismatch flanked by G·U pairs (cf. Section I,A). The average twist angle amounts to $37.4°$ (corresponding to ~ 9.6 base pairs per full helix turn), which is somewhat larger than the average helical twist for regular A-RNA helices. The C-1'–C-1' distances (12 Å) in the two U·C pairs are considerably larger than those for regular Watson–Crick pairs. All sugars adopt 3'-*endo* conformation.

The two central U·C mismatch pairs are arranged according to the pattern sketched in Fig. 4c, with one hydrogen bond between the carbonyl O-4 of the uridine and the N-4 amino group of cytidine, and another one mediated by an ordered water molecule bridging the N-3 nitrogens of the two paired bases (*36*, *37*). Thereby the U·C base pair is not only stabilized, but due to the separation of the two O-2 atoms by the inserted water molecule, an orientation of the glycosyl bonds is achieved at the same time, similar to that in regular Watson–Crick pairs (*37*).

In this central mismatch region of the duplex there is only poor base stacking. For the duplex with the same arrangement of the central mismatch core (two U·C mispairs adjacent to U·G mismatches) and flanking C·G Watson–Crick pairs, moreover, a significantly increased width of the major groove [7.5 Å, compared to a standard A-RNA value of 4.2 Å (36)] was reported. This is mainly attributed to the extensive hydration of the major groove around the U·C mismatch region, which is necessary for the stabilization of the U·C mispair (36, 40).

B. NMR Investigations and Thermodynamic Stability

As can be derived from Fig. 4, both U·U mismatch pairs and the protonated U·C$^+$ mispair (at pH < 6) involve (direct) hydrogen bonds with imino protons. These should give rise to downfield-shifted proton resonances in the region between about 14 and 10 ppm if the base pairing is stable enough to prevent rapid imino proton exchange with the solvent (exchange rates $\gtrsim 10^3$ sec^{-1}). Indeed, imino signals were observed for U·U tandem mismatches in different sequence contexts (100, 109, 113), as well as for an isolated U·U pair (107), U·C tandem mismatches (100), and an isolated U·C pair (111, 112).

Comparison of the imino resonance regions of the proton NMR spectra of duplexes with tandem U·U mismatches in different sequence contexts (70, 100, 109) reveals that, independent of the specific nature of the flanking base pairs in the studied self-complementary duplexes, two distinct imino signals were detected between about 11.5 and 10.0 ppm. This is consistent with the formation of two stable imino hydrogen bonds between the uracils of a U·U wobble pair, as sketched in Fig. 4a and b. This conclusion is corroborated by the extremely strong NOEs between the imino resonance assigned to the U·U mispairs (70, 100, 109).

Interestingly, the two U·U imino resonances display a much greater spectral separation if the tandem U·U mismatch is accommodated in a sequence track of the form 5'-YUUR-3'/5'-YUUR-3' (R = purine, Y = pyrimidine), where R and Y form either regular Watson–Crick or G·U wobble pairs (70, 100). In these cases, the more downfield-shifted line appears at about 11.2–11.3 ppm and the upfield signal around 10.4–10.5 ppm (70, 100, 109). In a thorough NMR analysis using ^{15}N- and ^{13}C-labeled RNA in combination with multidimensional homo- and heteronuclear NMR experiments (109), the downfield resonance was uniquely assigned to the U 3' neighbor to the pyrimidine (C) in the Watson–Crick (C·G) pair preceding the U·U pair in the above-mentioned sequence track. However, in a sequence context 5'-RUUY-3'/5'-RUUY-3', both imino resonances appear about 11.1 ppm (70) with a chemical shift difference of ~0.2 ppm or less. For both resonance patterns, beyond the very strong intrapair imino–imino NOEs, there are also NOEs between the mispair U imino peaks and the imino signals of the neighboring Watson–Crick base pair (70, 109). In particular, the variant imino resonance dispersions could be associated with different stacking patterns in the two types of sequence contexts. However, it should also be taken into account that the shielding effect of a 3′ neighboring purine is clearly greater than that of a pyrimidine (62, 63), and hence a more upfield-shifted resonance position can be expected for a U imino proton 5′ of a purine. At the same time, the effect of the base 5′ of the U should—at least in a more or less regular A-RNA helix–have a less pronounced effect on the shielding, and consequently on the resonance position of the U imino line, than the base 3′ of U (62). Nevertheless, differences in base pair stacking will also contribute to the specific imino resonance patterns.

The isolated U·U pair in the loop E sequence of eukaryotic 5S ribosomal RNA (107) is accommodated between a further A·A mismatch pair and a regular A·U pair (context 5′-AUA-3′/5′-UUA-3′) and gives rise to two relatively broad imino peaks at ~10.5 and 10.2 ppm, respectively. Here again, the Us have both purine 3′ neighbors. Though the NMR data in this study do not allow a detailed structural analysis with respect to the geometry of the U·U pair, they permit the conclusion that a stable, closed base pair of the type shown in Fig. 4a and b is formed (107).

Thermodynamic analyses of duplexes containing tandem U·U mismatches (70, 100) showed that they contribute to the overall stability of the duplexes about -2.1 and -0.4 kJ mol^{-1} in sequence contexts 5′-GUUC-3′/5′-GUUC-3′ and 5′-CUUG-3′/5′-CUUG-3′, respectively, whereas they destabilize the duplexes when the tandem U·U pairs are flanked by U·A or A·U pairs, respectively [increase in Gibbs free energy up to ~6 kJ mol^{-1} for the combinations 5′-UUUA-3′/5′-UUUA-3′ (70)]. Interestingly, comparing the effects of flanking A·U and U·A pairs, and of G·C/C·G base pairs among each other, it turns out that the combinations with purines 5′ of the mismatch U are more stable for each of the two Watson–Crick pairs compared to the corresponding combination with a 5′ pyrimidine, i.e., the 5′-G/3′-C flanking base pair lends more stability to the U·U tandem mismatch than the inverse base pair orientation 5′-C/3′-G. The same is valid for the combination with flanking A·U pairs (70).

Relatively few NMR studies of sequences with U·C mismatches have been published so far (70, 100, 111, 112). In the proton NMR spectra of the analyzed self-complementary duplexes, no resonances in the spectral region between about 15 and 9 ppm are observed at pH 6.5 (70, 100), independent of the nature of the Watson–Crick base pairs flanking the U·C tandem mismatch pairs. However, if the pH is lowered to 5.3, for the duplex with a mismatch core 5′-CCUG-3′/5′-CCUG-3′, three additional lines emerge at 11.4, 9.7, and 9.1 ppm (100). This points to the formation of a base pair type sketched in Fig. 4c involving imino hydrogen bonds between the carbonyl oxygens of U and the protonated N-3 of cytosine, as well as between the N-3 imino proton of the neutral U and the O-4 oxygen of C^+, in a manner that was originally suggested for a T·C mismatch in a DNA duplex (*110*).

By contrast, a duplex with a U·C tandem mismatch in the sequence context 5'-CUCG-3'/5'-CUCG-3' shows only little pH dependence (*100*). On lowering the pH, the imino resonances originating from the Watson–Crick pairs of the mismatch-flanking stems, which are already present at pH 6.5, merely become significantly broader, and no new resonances appear (*100*). Though the base-pairing patterns (or at least the base pair stabilities) of the two duplexes previously mentioned display different pH dependences, the overall duplex stability is not significantly affected by the pH variation (*100*), indicating that hydrogen bonding changes have only minor influence on the mismatch pair stability in these cases.

Generally, U·C tandem mismatches (and very probably also isolated U·C mispairs) are destabilizing, with the least destabilization increments found for both 5'-UC-3'/5'-UC-3' and 5'-CU-3'/5'-CU-3' tandems when flanked by G·C Watson–Crick pairs [Gibbs free energy contributions of ~3.4 to 4.4 kJ mol^{-1} (*70*)], whereas flanking A·U pairs give rise to stronger destabilizations [~10 to 12 kJ mol^{-1} (*70*)].

A U·C mismatch base pair was detected by NMR (*111, 112*) in the common arm base-paired segment of wheat germ 5S ribosomal RNA. The U·C mispair there is formed between the two nucleotides of the 12-nucleotide loop that border on the 4-bp stem in the secondary structure model of this 26-nucleotide RNA. Hydrogen bonds are formed between the N-4 amino group of C and the O-4 carbonyl oxygen of U, and between the N-3 imino group of U and the N-3 nitrogen of C, according to the scheme given in Fig. 4e. Both the imino (at ~12.4 ppm) and the amino resonances (at ~9.2 and 7.9 ppm, respectively) of the U·C base pair could be assigned (*111*). Beyond the strong intrapair NOEs between the U imino and C amino signals, additional NOEs were also found to the imino and amino resonances of the preceding A·U pair of the stem (*111*). The formation of the C·U pair, which is stacked above the A·U pair of the stem, was confirmed by analysis of the nonexchangeable proton NMR spectra (*112*). Moreover, an involvement of further nucleotides of the loop in the base stacking continuing the helical structure of the stem was derived from this study. However, no hydrogen bonding was found to occur between the bases facing each other on the opposite strands (*112*). Interestingly, the cytidine of the U·C pair adopts 2'-*endo* conformation. If an A-helical conformation derived by computer modeling is imposed on the fragment including the U·C mispair, the U·C pair must be severely buckled (*112*).

IV. Other Mismatches

Studies of RNA molecules containing mismatches other than G·U, G·A, U·U, and U·C mispairs are rather scarce. All of these mismatches destabilize the duplexes in which they are accommodated (70, 100, 108, 114).

No crystal structures of RNA molecules with those types of mismatches are presently available, in contrast to DNA (13, 78). There are, however, a few NMR studies of oligoribonucleotides with A·C, C·C, G·G, and A·A mismatches (70, 100, 104, 107, 114, 114a). In the Rev-responsive element, a stable G·G mismatch pair is formed, at least on binding of the Rev-derived peptide (103, 104), in a manner schematically depicted in Fig. 6. In particular, two imino proton resonances for the two base-paired guanines are observed, between which strong NOEs are detected.

The G·G mispair is sandwiched between a Watson–Crick C·G pair and another mismatch pair (G·A) [sequence context 5'-GGC-3'/5'-GGA-3', with a bulged-out U nucleotide between A and G not indicated in this sequence track (103, 104)]. The G base in the G·G mismatch 3' of the G in the adjacent G·A mispair is oriented *anti* with respect to its ribose, whereas the other G nucleotide of the G·G pair adopts *syn* conformation (103, 104). The G·G pair seems to stack on both the G·A and G·C pairs adjacent to it.

In general, G·G mismatches destabilize the duplexes in which they occur with the exception of the mismatch in a sequence track 5'-GGC-3'/5'-GGC-3' (108), which duplex has about the same stability as a reference duplex consisting only of C·G pairs. The strong influence of the sequence context, especially for such an isolated G·G mismatch pair, is documented by the destabilization by nearly 17 kJ mol^{-1} measured for a duplex with the G·G mismatch in a different context, 5'-CGG-3'/5'-CGG-3' (108).

On lowering the pH from 7.0 (for which the above-mentioned data apply) to 5.0, the stability of the 5'-GGC-3'/5'-GGC-3' duplex decreased by ~15 kJ mol^{-1}, whereas other purine–purine mismatch combinations were

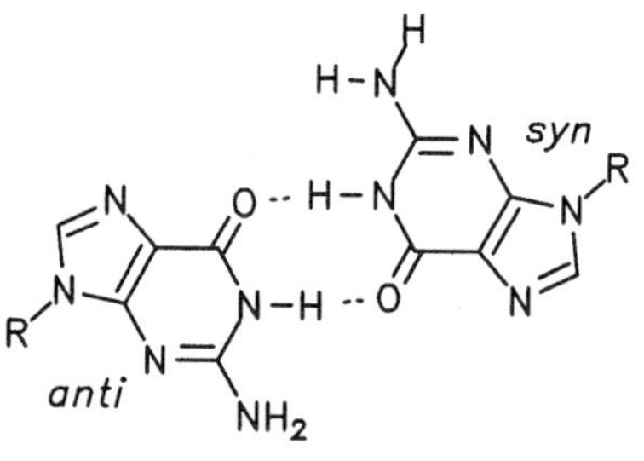

FIG. 6. Schematic drawing of the base pair arrangement of a G·G pair as found in the Rev-responsive element of the HIV *env* gene (103, 104).

much less affected (*108*). Because G nucleotides have p*K* values less than 3, it seems rather more probable that an alternative base pairing in the adjacent C·G Watson–Crick pairs involving protonated C$^+$ bases gives rise to the observed stability increase (*108*).

A significant destabilization, also reported for duplexes with both isolated (*108*) and tandem A·A (*70*) mismatches, is (at least for the tandem mismatches) particularly pronounced when flanked by A·U or U·A base pairs [~13 kJ mol^{-1} (*70*)].

An A·A mismatch was suggested in the NMR study of loop E of eukaryotic 5S rRNA (*107*). The base pair geometry could not be accurately determined, although the structural model based on the NMR data is compatible with the assumption of a symmetric A·A mismatch where both nucleotides adopt *anti* conformation (*107*).

Formation of an A·A mismatch pair is also postulated to occur in the α-sarcin/ricin loop from 28S/23S rRNA (*101, 102*). There it stacks on a further unusual reverse-Hoogsteen A·U pair, whereas on the other side it is adjacent to a C·C combination within which no obvious base pairing is recognizable (*101, 102*). Nevertheless, the adenine bases appear to stack above the Cs. According to the structural model (*101, 102*), the A bases in the mismatch appear to be arranged in a "side-by-side," or sheared, arrangement with (near) *anti* conformation of both nucleosides. However, the precision of the NMR-based structure in this region of the molecule is not high enough to permit more detailed and accurate assertions.

A mismatch whose DNA counterpart has been characterized structurally by X-ray crystallography (*13, 115*) and NMR spectroscopy (*14, 110, 116*) is the A·C mispair. As mentioned before, it is destabilizing in all sequence contexts (*70, 100*), with the least destabilization effect in combination with flanking G·C pairs (*70*). There is a small difference of ~3.8 kJ mol^{-1} in the destabilization increments for tandem A·C mismatches with flanking 5'-C/3'-G pairs, between 5'-CA-3'/5'-CA-3' and 5'-AC-3'/5'-AC-3' tandem mismatches, with the latter being more stable (*70*). For all other variants of flanking base pairs, the nature (orientation) of the tandem mismatch only negligibly affects the overall stability (*70*).

Interestingly, the stability of tandem A·C mismatches increases on lowering the pH value from 7.0 to 5.3. The gain in stability is more pronounced for the 5'-CCAG-3'/5'-CCAG-3' duplex than for the 5'-CACG-3'/5'-CACG-3' duplex (*100*). However, in the proton NMR spectra of duplexes with tandem A·C mismatches, no additional lines appeared on lowering the pH (*100*). Puglisi *et al.* (*114*) characterized the structure of the oligoribonucleotide 5'-GCGAUUUCUGACCGCC-3' in solution by NMR. It forms a hairpin with a 6-bp stem and a trinucleotide (UCU) loop that is closed by a G·C pair. The stem contains an A·C mismatch pair between two Watson–Crick pairs

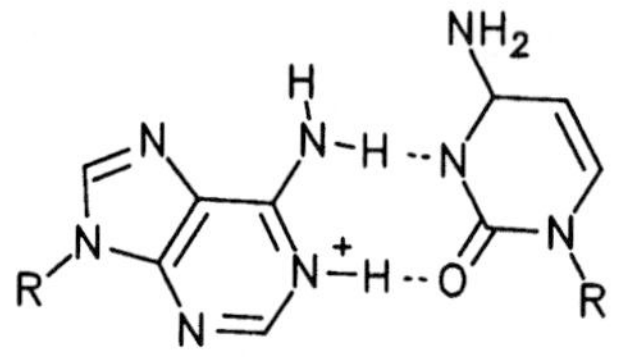

FIG. 7. Scheme of the base pair arrangement in a protonated A$^+$·C mismatch pair (*110, 114, 115*).

in the sequence context 5'-GAU-3'/5'-ACC-3'. From the pH dependence of the NMR spectra, it is concluded that an A$^+$·C base pair is formed with the adenine protonated at its N-1 position. The base-pairing pattern shown schematically in Fig. 7 has been observed before in a B-DNA duplex with an A·C mismatch (*13, 113*) and derived from NMR solution studies of mismatch DNA and RNA duplexes (*14, 110, 114, 114a*). Such an A·C mismatch pair is easily incorporated into an A-form helix without gross distortions. The C-1'–C-1' distance is only slightly altered as compared to a regular helix, and the major difference from standard Watson–Crick geometry is a shift of the adenine toward the minor groove and a shift of the C toward the major groove (*115*).

Under the same buffer conditions, the hairpin stability of the A·C mismatch sequence is distinctly lower as compared to an analogous structure having a regular A·U pair instead of the A·C mispair. The melting temperature is decreased by 11.5 K; the Gibbs free energy of hairpin formation increases by as much as 8.8 kJ mol^{-1} (at pH 6.4, without divalent cations) (*114*).

An A·C mispair has also been found in the internal loop A between helices I and II of the hairpin ribozyme–substrate complex (*114a*). Here, an A·C mismatch pair of the type sketched in Fig. 7 is formed, involving hydrogen bonds between the amino group of A and the N-3 of C, and another one between the protonated N-1 of A and the O-2 of C (*114a*). At pH 6.4, where the NMR studies on which the structural characterization is based have been performed, the adenine is partially protonated due to the unusually high pK of 6.2 (*114a*). The A·C pair stacks on the preceding G·C pair of the helical stem without significant distortion of the helical geometry. Both riboses are in 3'-*endo* conformation. Interestingly, the A·C mismatch is flanked by another A·G mispair of the "sheared" type shown in Fig. 3d (both nucleosides *anti*), with G having a 2'-*endo* ribose pucker (*114a*). The A·G pair continues the base pair stacking.

Finally, mismatches of the C·C type are briefly discussed. They can base pair involving imino hydrogen bonds if one of the cytosines is protonated (at pH < 6) according to the scheme displayed in Fig. 8. Obviously, this arrange-

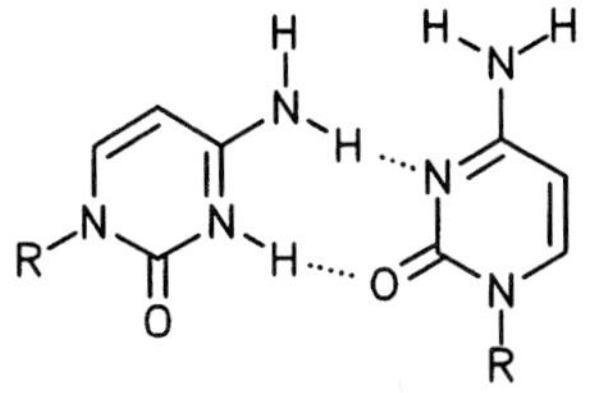

FIG. 8. Base pairing of a $C^+\cdot C$ mismatch pair involving a protonated cytosine (at low pH).

ment is isomorphous with the U·U and U·C$^+$ mismatch pairs discussed previously (Fig. 5). For the self-complementary duplex 5'-CGC<u>CC</u>GCG-3' with a central C·C tandem mismatch, three additional lines appear in the proton NMR spectral region between 15 and 9 ppm (at ~10.85, 10.05, and 9.2 ppm) on lowering the pH from 6.5 to 5.3 (*100*). They can be associated with the formation of protonated C$^+$·C base pairs [pK for C protonation, ~4.2 (*12*)] according to the scheme of Fig. 8, which has been suggested before for both RNA (*117, 118*) and DNA (*119, 120*). Moreover, at low pH (5.5) the duplex becomes distinctly more stable, as indicated by a melting temperature increase by 12 K and a more favorable Gibbs free energy of duplex formation of more than 10 kJ mol^{-1} (*100*), which is consistent with the formation of (additional) hydrogen bonds. The assumption of C$^+$·C base pair formation is further corroborated by the analysis of the corresponding CD spectra.

V. Possible Biological Roles of Mismatch Pairs

It has been discussed in the preceding sections that, in general, mismatch base pairs of different types can be accommodated in RNA double-helical regions without causing very severe distortions of the canonical A-RNA geometry. However, they produce not only subtle local deviations from the regular A-form structure (such as altered C-1'–C-1' distances, specific changes in helical or propeller twists, groove widths, base plane shifts, etc.), which might constitute recognition sites for the correct interaction with specific partners (proteins, further RNAs, metal ions, drugs, etc.). Moreover, due to their nonstandard base arrangement they also display a different pattern of functional groups in both grooves as compared to regular Watson–Crick pairs, which also should be of importance for the specific interaction with ligands, because, by mismatch pairs, functional groups (e.g., hydrogen bond donors or acceptors) are presented in a unique way. Thus an optimal interaction between these functional groups and complementary groups of the ligand is enabled.

A very interesting case exemplifying both of the features of these mismatches with respect to ligand recognition/interaction is provided by *E. coli* tRNA^{Ala}. It has been demonstrated by two different groups (*17, 18*) that a specific G·U pair at the third position of the acceptor stem represents the major recognition site for the cognate aminoacyl-tRNA synthetase. However, both groups interpret the function of the G·U pair in this recognition process in different ways: McClain and co-workers (*17, 19, 121*) conclude from an analysis of several suppressor tRNAs with mutations at and in the vicinity of the G·U pair that the helix geometry modifications introduced by the G·U wobble pair are the main recognition signal for the cognate alanine-specific aminoacyl-tRNA synthetase. Support for this assumption comes from the observation that mutant tRNAs with other mismatch pairs (in place of the G·U pair), such as C·A or G·A mispairs, were also correctly charged with alanine, whereas replacement of the G·U pair by Watson–Crick pairs prevented correct aminoacylation (*17, 19, 121*).

In contrast, Schimmel and Hou (*18*) stress the crucial importance of the 2-amino group of the mismatch G, which is not involved in base pairing and projects into the minor groove, for the specific interaction with the cognate synthetase (*122*). This assumption is corroborated by the total loss of substrate properties of minihelices derived from the acceptor stem of tRNA^{Ala} (*123*) if the mismatch G is replaced by inosine (I). Inosine lacks the 2-amino group of G. Both aspects could be important for optimal interaction between tRNA^{Ala} and its cognate synthetase, possibly by presenting the G amino group in an orientation that is most favorable for making contacts with specific groups of the enzyme and that is ensured by the local structure modifications induced by the G·U mismatch. One can also wonder if the specifically bound divalent metal ions near a G·U pair (*48, 71, 72*) might play a role in this recognition process.

Interestingly, the absolutely conserved G·U pair in the P1 helix of group I self-splicing introns, which is crucially important for substrate recognition, can be replaced only by another A·C mismatch pair without loss of function, but not by G·C or A·U Watson–Crick pairs (*124*). This again stresses the importance of slight structural modifications, produced by the insertion of mismatches, in intermolecular recognition (*35*).

The relatively numerous G·U mispairs found in secondary structure analyses of ribosomal RNAs (*27–31*) could hint at their relevance for the specific interaction of the respective RNA with ribosomal proteins and/or further ribosomal RNAs. This is of major importance in the self-assembly of ribosomes. This, of course, also applies to the other types of mismatches occurring in ribosomal RNAs, in particular G·A or A·A mismatches, some of which have been structurally characterized (*101, 102, 107*).

That the use of mismatch pairs as recognition sites for the intermolecular interaction is not restricted to ribosomal RNAs or tRNAs is made clear by the role played by the asymmetric loop in the HIV-1 Rev-responsive element RNA in the binding of the Rev protein, or, respectively, Rev-derived peptides (*103, 104*). Analysis of the conformational changes of the internal loop in this RNA on peptide binding reveals another feature of mismatched regions in double-stranded RNA that might also play a role in further intermolecular recognition processes involving RNA. As has been shown before, many of the described mismatch base pairs are distinctly less stable than regular Watson–Crick pairs. Accordingly, the mismatch nucleotides may not only present certain functional groups in an unusual, unique manner, but they can additionally offer a higher degree of conformational flexibility, which makes them more readily adaptable to complementary parts of their interaction partners (proteins, RNAs, low-molecular-weight ligands), enabling thus an "induced fit" of the RNA to its ligand. In the case of the peptide binding by the Rev-responsive element RNA (*103, 104*), this was impressively demonstrated by the observation of distinctly better defined and more stable base pairing in the mismatched internal loop on peptide binding.

Further support for the suggested importance of conformational flexibility (and hence adaptability) created by single-stranded or mismatched RNA regions is provided by the observation of RNAs that have been obtained via *in vitro* selection methods (*125, 126*) to bind specifically certain ligands. These sequences invariably contain unpaired regions (loops, bulges) (*127*), which possibly also could give rise to mismatch pairs in view of the many unexpected mispairs found in the α-sarcin/ricin loop of 28S/23S rRNA (*101, 102*).

Acknowledgments

I thank Martin Vogtherr for typing the manuscript and preparing the figures. Thanks are due to Melanie Beikman for help with preparation of the manuscript, and to Mathias Sprinzl for discussions. This work was supported by the Deutsche Forschungsgemeinschaft (Projekt Li-722/1-2).

References

1. M. Delarue and D. Moras, in "Nucleic Acids and Molecular Biology" (F. Eckstein and D. M. J. Lilley, eds.), Vol. 3, p. 182. Springer-Verlag, Berlin and New York, 1989.
2. I. Tinoco, Jr., J. D. Puglisi, and J. R. Wyatt, in "Nucleic Acids and Molecular Biology" (F. Eckstein and D. M. J. Lilley, eds.), Vol. 4, p. 205. Springer-Verlag, Berlin and New York, 1990.
3. M. Chastain and I. Tinoco, Jr., *This Series* **41**, 131 (1990).
4. D. E. Draper, *Acc. Chem. Res.* **25**, 201 (1992).

5. J. R. Wyatt and I. Tinoco, Jr., *in* "The RNA World" (J. F. Atkins and R. F. Gesteland, eds.), p. 465. Cold Spring Harbor Lab., Cold Spring Harbor, New York, 1993.
6. G. Varani and A. Pardi, *in* "RNA–Protein Interactions" (Kiyoshi Nagai and Iain W. Mattaj, eds.), p. 1. IRL Press, Oxford, 1994.
7. D. H. Turner, N. Sugimoto, and S. M. Freier, *Annu. Rev. Biophys. Biophys. Chem.* **17,** 167 (1988).
8. E. Westhof and F. Michel, *in* "RNA–Protein Interactions" (Kiyoshi Nagai and Iain W. Mattaj, eds.), p. 25. IRL Press, Oxford, 1994.
9. J. A. Jaeger, J. SantaLucia, Jr., and I. Tinoco, Jr., *Annu. Rev. Biochem.* **62,** 255 (1993).
10. G. Varani and I. Tinoco, Jr., *Q. Rev. Biophys.* **24,** 479 (1991).
11. P. B. Moore, *Acc. Chem. Res.* **28,** 251 (1995).
12. W. Saenger, "Principles of Nucleic Acid Structure," Springer-Verlag, Berlin and New York, 1984.
13. O. Kennard, *in* "Nucleic Acid Research and Molecular Biology" (F. Eckstein and D. M. J. Lilley, eds.), Vol. 1, p. 25. Springer-Verlag, Berlin and New York, 1987.
13a. W. N. Hunter, *Meth. Enzymol.* **211,** 221 (1992).
14. D. J. Patel, L. Shapiro, and D. Hare, *in* "Nucleic Acid Research and Molecular Biology" (F. Eckstein and D. M. J. Lilley, eds.), Vol. 1, p. 70. Springer-Verlag, Berlin and New York, 1987.
15. F. H. C. Crick, *J. Mol. Biol.* **19,** 548 (1966).
16. M. Sprinzl, C. Steegborn, F. Hübel, and S. Steinberg, *Nucleic Acids Res.* **24,** 68 (1996).
17. W. H. McClain and K. Foss, *Science* **240,** 793 (1988).
18. Y.-M. Hou and P. R. Schimmel, *Nature (London)* **333,** 140 (1988).
19. W. H. McClain, Y. M. Chen, K. Foss, and J. Schneider, *Science* **242,** 1681 (1988).
20. S. H. Kim, J. L. Sussman, F. L. Suddath, G. J. Quigley, A. McPherson, A. H. J. Wang, N. C. Seeman, and A. Rich, *Proc. Natl. Acad. Sci. U.S.A.* **71,** 4970 (1974).
21. J. L. Sussman, S. R. Holbrook, R. W. Warrant, and S. H. Kim, *J. Mol. Biol.* **123,** 607 (1978).
22. A. Klug, J. Ladner, and J. D. Robertus, *J. Mol. Biol.* **89,** 511 (1974).
23. A. Jack, J. E. Ladner, and A. Klug, *J. Mol. Biol.* **108,** 619 (1976).
24. C. D. Stout, H. Mizuno, S. T. Rao, P. Swaminathan, J. Rubin, T. Brennan, and M. Sundaralingam, *Acta Crystallogr.* **B34,** 1529 (1978).
25. D. Moras, M. B. Comarmond, J. Fischer, J. C. Thierry, J. P. Ebel, and R. Giegé, *Nature (London)* **288,** 669 (1980).
26. E. Westhof, P. Dumas, and D. Moras, *J. Mol. Biol.* **184,** 119 (1985).
27. R. R. Gutell, M. N. Schare, and M. W. Gray, *Nucleic Acid Res.* **20** (Suppl.), 2095 (1992).
28. R. R. Gutell, M. W. Gray, and M. N. Schare, *Nucleic Acid Res.* **21,** 3055 (1993).
29. C. R. Woese, R. R. Gutell, R. Gupta, and H. F. Noller, *Microbiol. Rev.* **47,** 621 (1983).
30. R. R. Gutell, N. Larsen, and C. R. Woese, *Microbiol. Rev.* **58,** 10 (1994).
31. D. Gautheret, D. Konings, and R. R. Gutell, *RNA* **1,** 807 (1995).
32. H. Mizuno and M. Sundaralingam, *Nucleic Acids Res.* **5,** 445 (1978).
33. P. H. van Knippenberg, L. J. Formenoy, and H. A. Heus, *Biochim. Biophys. Acta* **1050,** 14 (1990).
34. D. Moras, P. Dumas, and E. Westhof, *in* "Structure and Dynamics of RNA" (P. H. van Knippenberg and C. W. Hilbers, eds.), p. 113. Plenum, New York, 1986.
35. F. H.-T. Allain and G. Varani, *J. Mol. Biol.* **250,** 333 (1995).
36. S. R. Holbrook, C. Cheong, I. Tinoco, Jr., and S.-H. Kim, *Nature (London)* **353,** 579 (1991).
37. W. B. T. Cruse, P. Saludjian, E. Biala, P. Strazewski, T. Prangé, and O. Kennard, *Proc. Natl. Acad. Sci. U.S.A.* **91,** 4160 (1994).
38. C. Betzel, S. Lorenz, J.-P. Fürste, R. Bald, M. Zhang, T. R. Schneider, K. S. Wilson, and V. A. Erdmann, *FEBS Lett.* **351,** 159 (1994).

39. G. A. Leonard, K. E. McAuley-Hecht, S. Ebel, D. M. Lough, T. Brown, and W. N. Hunter, *Structure* **2**, 483 (1994).
40. K. J. Bayens, H. L. DeBondt, and S. R. Holbrook, *Nature Struct. Biol.* **2**, 56 (1995).
41. M. C. Wahl, S. T. Rao, and M. Sundaralingam, *Nature Struct. Biol.* **3**, 24 (1996).
42. C. Cheong, G. Varani, and I. Tinoco, Jr., *Nature (London)* **346**, 680 (1990).
43. G. Varani, C. Cheong, and I. Tinoco, Jr., *Biochemistry* **30**, 3280 (1991).
44. K. Wüthrich, "NMR of Proteins and Nucleic Acids," Wiley, New York, 1986.
45. C. W. Hilbers, in "Biological Applications of Magnetic Resonance" (R. G. Shulman, ed.), p. 1. Academic Press, New York, 1979.
46. J.-L. Leroy, D. Broseta, N. Bolo, and M. Guéron, in "Structure and Dynamics of RNA" (P. H. van Knippenberg and C. W. Hilbers, eds.), p. 31. Plenum, New York, 1986.
47. M. Guéron and J.-L. Leroy, in "Nucleic Acids and Molecular Biology" (F. Eckstein and D. M. J. Lilley, eds.), Vol. 6, p. 1. Springer-Verlag, Berlin and New York, 1992.
48. St. Limmer, H.-P. Hofman, G. Ott, and M. Sprinzl, *Proc. Natl. Acad. Sci. U.S.A.* **90**, 6199 (1993).
49. P. D. Johnston and A. G. Redfield, *Nucleic Acids Res.* **5**, 3913 (1978).
50. P. D. Johnston and A. G. Redfield, *Biochemistry* **20**, 1147 (1981).
51. A. Heerschaap, C. A. G. Haasnoot, and C. W. Hilbers, *Nucleic Acids Res.* **10**, 6981 (1982).
52. R. E. Hurd and B. R. Reid, *Biochemistry* **18**, 4017 (1979).
53. S. Roy and A. G. Redfield, *Nucleic Acids Res.* **9**, 7073 (1981).
54. D. R. Hare and B. R. Reid, *Biochemistry* **21**, 5129 (1982).
55. M. J. Kime and P. B. Moore, *Biochemistry* **22**, 2615 (1983).
56. L.-H. Chang and A. G. Marshall, *Biochemistry* **25**, 3056 (1986).
57. H. A. Heus, J. M. A. van Kimmenade, P. H. van Knippenberg, C. A. G. Haasnoot, S. H. de Bruin, and C. W. Hilbers, *J. Mol. Biol.* **170**, 939 (1983).
58. K. M. Lee and A. G. Marshall, *Biochemistry* **25**, 8245 (1986).
59. D. R. Hare, N. S. Ribeiro, D. E. Wemmer, and B. R. Reid, *Biochemistry* **24**, 4300 (1985).
60. S. A. White, M. Nilges, A. Huang, A. T. Brünger, and P. B. Moore, *Biochemistry* **31**, 1610 (1992).
61. St. Limmer, B. Reif, G. Ott, L. Arnold, and M. Sprinzl, *FEBS Lett.* **385**, 15 (1996).
62. D. B. Arter and P. G. Schmidt, *Nucleic Acids Res.* **3**, 1437 (1976).
63. C. Giessner-Prettre and B. Pullman, *Q. Rev. Biophys.* **20**, 113 (1987).
64. D. W. Hoffmann, R. A. Colvin, M. A. Garcia-Blanco, and S. W. White, *Biochemistry* **32**, 1096 (1993).
65. D. W. Hoffmann and S. W. White, *Nucleic Acids Res.* **23**, 4058 (1995).
66. J. D. Puglisi, J. R. Wyatt, and I. Tinoco, Jr., *Biochemistry* **29**, 4215 (1990).
67. F. M. Jucker and A. Pardi, *Biochemistry* **34**, 14416 (1995).
68. T. Sakata, H. Hiroaki, Y. Oda, T. Tanaka, M. Ikehara, and S. Uesugi, *Nucleic Acids Res.* **18**, 3831 (1990).
69. L. He, R. Kierzek, J. SantaLucia, Jr., A. E. Walter, and D. H. Turner, *Biochemistry* **30**, 11124 (1991).
70. M. Wu, A. McDowell, and D. H. Turner, *Biochemistry* **34**, 3204 (1995).
71. G. Ott, L. Arnold, and St. Limmer, *Nucleic Acids Res.* **21**, 5859 (1993).
72. F. H.-T. Allain and G. Varani, *Nucleic Acids Res.* **23**, 341 (1995).
73. O. Jardetzky and G. C. K. Roberts, "NMR in Molecular Biology," Chap. 3. Academic Press, Orlando, 1981.
74. T. Brown, W. N. Hunter, G. Kneale, and O. Kennard, *Proc. Natl. Acad. Sci. U.S.A.* **83**, 2402 (1986).
75. W. N. Hunter, T. Brown, and O. Kennard, *J. Biomol. Struct. Dynam.* **4**, 173 (1986).

76. G. G. Privé, U. Heinemann, S. Chandrasegaran, L. S. Kan, M. L. Kopka, and R. E. Dickerson, *Science* **238**, 498 (1987).
77. T. Brown, G. A. Leonard, E. D. Booth, and J. Chambers, *J. Mol. Biol.* **207**, 455 (1989).
78. O. Kennard and W. N. Hunter, *Angew. Chem. (Int. Ed. Engl.)* **30**, 1254 (1991).
79. G. A. Leonard, K. E. McAuley-Hecht, S. Ebel, D. M. Lough, T. Brown, and W. N. Hunter, *Structure* **2**, 483 (1994).
80. H. W. Pley, K. M. Flaherty, and D. B. McKay, *Nature (London)* **372**, 68 (1994).
81. W. G. Scott, J. T. Finch, and A. Klug, *Cell* **81**, 991 (1995).
82. W. Traub and J. L. Sussman, *Nucleic Acids Res.* **10**, 2701 (1982).
83. H. F. Noller, *Annu. Rev. Biochem.* **53**, 119 (1984).
83a. D. Gautheret, D. Konings, and R. Gutell, *J. Mol. Biol.* **242**, 1 (1994).
84. X. Gao and D. J. Patel, *J. Am. Chem. Soc.* **110**, 5178 (1988).
85. T. Brown, G. A. Leonard, E. D. Booth, and G. G. Kneale, *J. Mol. Biol.* **212**, 437 (1990).
86. G. A. Leonard, E. D. Booth, and T. Brown, *Nucleic Acids Res.* **18**, 5617 (1990).
87. C. Carbonnaux, G. A. van der Marel, J. H. van Boom, W. Guschlbauer, and G. V. Fazakerley, *Biochemistry* **30**, 5449 (1991).
88. A. N. Lane, T. C. Jenkins, D. J. S. Brown, and T. Brown, *Biochem. J.* **279**, 269 (1991).
89. S. Ebel, A. N. Lane, and T. Brown, *Biochemistry* **31**, 12083 (1992).
90. J. W. Cheng, S. H. Chou, and B. R. Reid, *J. Mol. Biol.* **228**, 1037 (1992).
91. R. H. Symons, *Annu. Rev. Biochem.* **61**, 641 (1992).
92. D. E. Ruffner, G. D. Stormo, and O. C. Uhlenbeck, *Biochemistry* **29**, 10695 (1990).
93. L.-S. Kan, S. Chandrasegaran, S. M. Pulford, and P. S. Miller, *Proc. Natl. Acad. Sci. U.S.A.* **80**, 4263 (1980).
94. L. P. M. Orbons, G. A. van der Marel, J. H. van Boom, and C. Altona, *Eur. J. Biochem.* **170**, 225 (1987).
95. Y. Li, G. Zon, and W. D. Wilson, *Proc. Natl. Acad. Sci. U.S.A.* **88**, 26 (1991).
96. J. SantaLucia, Jr., R. Kierzek, and D. H. Turner, *Biochemistry* **29**, 8813 (1990).
97. J. SantaLucia, Jr. and D. H. Turner, *Biochemistry* **32**, 12612 (1993).
98. A. E. Walter, M. Wu, and D. H. Turner, *Biochemistry* **33**, 11349 (1994).
99. M. Katahira, M. Kanagawa, H. Sato, S. Uesugi, T. Kohno, and T. Maeda, *Nucleic Acids Res.* **22**, 2752 (1994).
100. J. SantaLucia, Jr., R. Kierzek, and D. H. Turner, *Biochemistry* **30**, 8242 (1991).
101. A. A. Szewczak, P. B. Moore, Y.-L. Chan, and I. G. Wool, *Proc. Natl. Acad. Sci. U.S.A.* **90**, 9581 (1993).
102. A. A. Szewczak and P. B. Moore, *J. Mol. Biol.* **247**, 81 (1995).
103. R. D. Peterson, D. P. Bartel, J. W. Szostak, S. J. Horvath, and J. Feigon, *Biochemistry* **33**, 5357 (1994).
104. J. L. Battiste, R. Tan, A. D. Frankel, and J. R. Williamson, *Biochemistry* **33**, 2741 (1994).
105. H. A. Heus and A. Pardi, *Science* **253**, 191 (1991).
106. M. Orita, F. Nishikawa, T. Shimayama, K. Taira, Y. Endo, and S. Nishikawa, *Nucleic Acids Res.* **21**, 5670 (1993).
106a. S. Huang, Y.-X. Wang, and D. E. Draper, *J. Mol. Biol.* **258**, 308 (1996).
106b. M. A. Fountain, M. J. Serra, T. R. Krugh, and D. H. Turner, *Biochemistry* **35**, 6539 (1996).
107. B. Wimberly, G. Varani, and I. Tinoco, Jr., *Biochemistry* **32**, 1078 (1993).
108. S. E. Morse and D. E. Draper, *Nucleic Acids Res.* **23**, 302 (1995).
109. E. P. Nikonowicz and A. Pardi, *J. Mol. Biol.* **232**, 1141 (1993).
110. D. J. Patel, S. A. Kozlowski, S. Ikuta, and K. Itakura, *Fed. Proc., Fed. Am. Soc. Exp. Biol.* **43**, 2663 (1984).
111. J. Wu and A. G. Marshall, *Biochemistry* **29**, 1722 (1990).
112. J. Wu and A. G. Marshall, *Biochemistry* **29**, 1730 (1990).

113. E. P. Nikonowicz, A. Sirr, P. Legault, F. M. Jucker, L. M. Baer, and A. Pardi, *Nucleic Acids Res.* **20,** 4507 (1992).
114. J. D. Puglisi, J. R. Wyatt, and I. Tinoco, Jr., *Biochemistry* **29,** 4215 (1990).
114a. Z. Cai and I. Tinoco, Jr., *Biochemistry* **35,** 6026 (1996).
115. W. N. Hunter, T. Brown, N. N. Anand, and O. Kennard, *Nature (London)* **320,** 552 (1986).
116. D. J. Patel, S. A. Kozlowski, S. Ikuta, and K. Itakura, *Biochemistry* **23,** 3218 (1984).
117. J. Brahms, J. C. Maurizot, and A. M. Michelson, *J. Mol. Biol.* **25,** 465 (1967).
118. W. Guschlbauer, *Nucleic Acids Res.* **2,** 353 (1975).
119. D. Gray, T. Cui, and R. Ratliff, *Nucleic Acids Res.* **12,** 7565 (1984).
120. E. L. Edwards, M. H. Patrick, R. L. Ratliff, and D. M. Gray, *Biochemistry* **29,** 828 (1990).
121. K. Gabriel, J. Schneider, and W. H. McClain, *Science* **271,** 195 (1996).
122. K. Musier-Forsyth, N. Usman, S. Scaringe, J. Doudna, R. Green, and P. Schimmel, *Science* **253,** 784 (1991).
123. C. Francklyn and P. Schimmel, *Nature (London)* **337,** 478 (1989).
124. J. A. Doudna, B. P. Cormack, and J. W. Szostak, *Proc. Natl. Acad. Sci. U.S.A.* **86,** 4702 (1989).
125. C. Tuerk and L. Gold, *Science* **249,** 505 (1990).
126. A. D. Ellington and J. W. Szostak, *Nature (London)* **346,** 818 (1990).
127. D. P. Bartel and J. W. Szostak, *in* "RNA–Protein Interactions" (K. Nagai and I. W. Mattaj, eds.), p. 248. IRL Press, Oxford, 1994.
128. P. Fan, A. K. Suri, R. Fiala, D. Live, and D. J. Patel, *J. Mol. Biol.* **258,** 480 (1996).
129. Y. Yang, M. Kochoyan, P. Burgstaller, E. Westhof, and M. Famulok, *Science* **272,** 1343 (1996).

The Mechanism of 3' Cleavage and Polyadenylation of Eukaryotic Pre-mRNA[1]

ELMAR WAHLE[2] AND UWE KÜHN

Institut für Biochemie
Justus-Liebig-Universität Giessen
35392 Giessen, Germany

I. Pre-mRNA Sequences Directing 3' End Formation	42
A. Polyadenylation Signals in Animal Cells	43
B. Polyadenylation Signals in Yeast	44
C. Polyadenylation Signals in Plants	46
D. Comparison of Polyadenylation Signals in Different Organisms	46
II. Cleavage and Polyadenylation in Animal Cells	48
A. Nuclear 3' Processing	48
B. Cytoplasmic Polyadenylation	54
III. Cleavage and Polyadenylation in Yeast	56
IV. Regulation of Polyadenylation	57
A. Choice between Alternative Sites	58
B. Choice between Splicing and Polyadenylation	60
C. Quantitative Regulation	62
V. Coupling of 3' End Formation to Transcription and Splicing	64
VI. Conclusion	65
References	66

The formation of mature 3' ends by posttranscriptional processing is an obligatory step in the synthesis of eukaryotic mRNAs. The almost universal pathway for 3' end formation consists of two reactions, a specific endonucleolytic cleavage of the mRNA precursor downstream of the coding sequence, followed by template-independent addition of a poly(A) tail to the upstream cleavage product. The only known cellular mRNAs that do not follow this pathway are those encoding the replication-dependent histones in metazoans. These RNAs are also cleaved endonucleolytically, but the reac-

---

[1] Abbreviations: CF_m, mammalian cleavage factor; CF_y, yeast cleavage factor; CPSF, cleavage and polyadenylation specificity factor; CstF, cleavage stimulation factor; PAB, poly(A) binding protein; PF, polyadenylation factor; snRNP, small nuclear ribonucleoprotein particle; snRNA, small nuclear RNA; UTR, untranslated region.

[2] To whom correspondence may be addressed.

tion differs from that of a typical pre-mRNA with respect to both sequences and factors, and no polyadenylation ensues. In yeast and in plants, histone mRNAs are polyadenylated like all others. Viral mRNAs often acquire their poly(A) tails by special mechanisms. Vaccinia virus mRNAs are polyadenylated by a virus-encoded enzyme without undergoing endonucleolytic cleavage (1). The poly(A) tails of polio virus mRNAs are generated by transcription of a poly(U) stretch at the end of the (−) strand RNA (2). Other RNA viruses also synthesize poly(A) tails by transcription and extend them by polymerase "slippage" or "stuttering" on the oligo(U) template (3). Poly(A) tails are also found on mRNAs in bacteria and in mitochondria.

The almost universal conservation of this modification suggests an essential function. Indeed, the genes for several proteins involved in 3′ end formation, including poly(A) polymerase, are essential for viability in yeast (see Section III). Also, mutations in the polyadenylation signal of a gene lead to a strong reduction in the expression of that gene (4–7). It is now well documented that the poly(A) tail plays an essential role in the initiation of translation (8–10) (earlier work is reviewed in Refs. 11 and 12).

The poly(A) tail is also important for the control of mRNA turnover because deadenylation is the first and rate-limiting step in the degradation of many mRNAs (13). Possibly, the tail has an additional function in the export of mRNA from the nucleus to the cytoplasm (14).

This review is devoted to recent progress in the analysis of 3′ cleavage and polyadenylation in metazoan cells and in yeast. Older work is referenced selectively; the reader is referred to the many reviews that have covered this topic previously (15–24). Some of these (18, 21) also contain information on the poly(A) polymerases of *Escherichia coli* and of vaccinia virus, which will not be discussed here. What little is known about 3′ end formation of plant mRNAs has been summarized by Hunt (25) and Rothnie (26), so the subject is not treated in great detail here.

I. Pre-mRNA Sequences Directing 3′ End Formation

In mammalian cells, one of several sequences responsible for cleavage and polyadenylation, the well-known AAUAAA motif, is sufficiently conserved that it was immediately recognized in an early comparison of 3′-terminal sequences from only six mRNAs, before cDNA cloning was practiced (27). In contrast, 3′ end formation signals in plants and in yeast have proved elusive, and the same is true for additional components of mammalian poly(A) sites. Only recently has it become clear that there may be an underlying consensus of the complex 3′ end formation signals in these different organisms.

A. Polyadenylation Signals in Animal Cells

As mentioned, the sequence AAUAAA is the hallmark of poly(A) sites in vertebrate (and probably all metazoan) cells. The sequence is normally located 10–30 nucleotides upstream of the cleavage site, and around 90% of all genes have a perfect copy. The most frequent variant is AUUAAA, which is also processed with reasonable efficiency *in vitro*. Interestingly, other variants do occur. When tested in the form of point mutations in a signal that normally contains AAUAAA, these are essentially not processed *in vitro*. The poly(A) sites that have no discernible AAUAAA sequence at all (*18*) are the most extreme forms. Although a trivial cloning artifact, internal priming by oligo(dT), might account for the apparent absence of AAUAAA, this can clearly be excluded in some cases.

The way in which these strongly aberrant sites are processed has not been investigated. For polyadenylation signals differing from AAUAAA by a single nucleotide, the cooperative nature of poly(A) site recognition may permit other components of the complex signal to compensate (*28*). The efficient use of an unusual CAUAAA signal in a *Xenopus* tubulin gene depended on sequences between 440 and 770 bp downstream of the poly(A) site (*29*). It has been speculated that these sequences might affect processing of the nascent RNA through transcriptional pausing or termination.

The AAUAAA signal alone is not sufficient to induce cleavage and polyadenylation. A second essential element of the polyadenylation signal exists downstream of the cleavage site. The downstream element is poorly conserved, and its essential features have not been identified. Two main types of sequences have been described. One type consists of a run of Us, at least four out of five nucleotides (*30–32*). A second type, usually called a GU-rich element (*30, 31*), may have the consensus YGUGUUYY (*33*). Whereas in some cases just one of these elements was sufficient for poly(A) site function (*30, 32*), two of them worked synergistically in another case (*31*). Interpretation of these results is complicated because the degeneracy of the signals makes it difficult to exclude their presence in surrounding sequences presumed to be inert.

Point mutations in the downstream elements can have measurable but weak effects. Complete inactivation of a poly(A) site requires larger deletions. Although deletions sometimes distinguish two downstream elements (*31*), other downstream signals are redundant or dispersed to such a degree that individual elements can no longer be defined, and only extended deletions affect processing (*34*). The U- or GU-rich sequences are found within 30 nucleotides downstream of the cleavage site. A claim that such a sequence 47–55 nucleotides downstream of the SV40 late cleavage site is required (*35*) has been challenged (*34, 36*) and can probably be explained by the inadvertent introduction of an artificial downstream element (*37*).

A nonessential, stimulatory sequence, apparently distinct from the two sequences just discussed, and located further downstream, has been identified in the SV40 late polyadenylation site (38, 39).

The distance between AAUAAA and the downstream element(s) determines a sequence region within which cleavage can take place. Within this region, a preference for cleavage on the 3' site of an A residue, often a CA dinucleotide, may determine the precise site (37).

Sequences upstream of AAUAAA can enhance the efficiency of 3' processing (17, 18, 21). Although upstream elements were originally identified in viruses, the first example in the poly(A) site of a cellular gene has recently been reported (40). All of these elements are U rich, but a consensus sequence has not emerged so far. It is not even clear whether all of them can be grouped together, because their mechanism of action is still controversial (see Section III,B).

Processing efficiency can also be influenced by sequences immediately surrounding AAUAAA (41–43). Again, favorable sequences have not been clearly defined.

Polyadenylation signals are usually presented as one-dimensional strings of nucleotides. Is this the true structure of 3' processing substrates? Possibly, yes. A thorough study of secondary structure has been performed for the adenovirus L4 site (44). This polyadenylation signal forms several weak, competing stem–loop structures. Exposure of AAUAAA in a loop seems to favor polyadenylation, but is clearly not essential. Computer analysis of a random sample of other polyadenylation sites suggests that about two-thirds can form this type of structure. Although presentation of AAUAAA in a loop has only a minor effect, the single-stranded nature of the signal is essential (44, 45).

However, some poly(A) signals do depend on secondary structure. In the type I human T cell leukemia virus, the AAUAAA signal is separated from the cleavage and polyadenylation site by more than 250 nucleotides. Most of this sequence forms a very stable secondary structure, the binding site for the viral Rex protein. Formation of the structure is essential for 3' end formation, juxtaposing the AAUAAA signal and the downstream element (46, 47). In the same manner, the TAR structure of the human immunodeficiency virus (HIV) is required to bring the upstream polyadenylation element in proximity to the AAUAAA sequence (48). Similar effects have been seen with sequences artificially inserted between AAUAAA and the cleavage site (49).

B. Polyadenylation Signals in Yeast

Poly(A) sites in yeast are much less clearly defined than they are in metazoans. Sequence comparisons have failed to identify conserved motifs comparable to the AAUAAA sequence. Point mutations in yeast poly(A) signals generally have rather weak effects. The most thoroughly investigated poly(A) site, that of the *CYC1* gene, could not even be inactivated by linker scanning mutagenesis (50). The most fruitful approach, therefore, has been to search for sequences that restore a poly(A) site destroyed by an extended deletion (51–53). This type of analysis of *Saccharomyces cerevisiae* poly(A) signals has led to the following model (52, 53). A so-called efficiency element is found at a variable distance upstream of the cleavage site. Sequences of the efficiency element are related to UAUAUA and include the previously described bipartite sequence UAG···UAUGUA (54–56). Cleavage takes place approximately 20 nucleotides downstream of a second sequence, the "positioning element." Interestingly, one of the most efficient versions of the positioning element is AAUAAA. The sequence A_6 is equally efficient, and related motifs are also functional (53, 57). Finally, cleavage usually takes place after the sequence YA_n (58). Although this model is supported by point mutations and sequence comparisons, it is not clear whether it can account for the quite dramatic effect of several point mutations in the *ADH2* gene (59).

An unresolved question is the involvement of sequences downstream of the cleavage site. In the *ADH2* gene, downstream sequences appear to be required for 3′ end formation (59). In other genes, deletion of all or nearly all downstream sequences had weak effects or none at all (54, 60–62). Recognition of a downstream element would allow the cell to distinguish a precursor RNA from one that has already been polyadenylated. Of course, the most rudimentary sequence specificity would suffice to distinguish poly(A) from all other sequences. Such a relaxed sequence specificity cannot be excluded from the data presently available.

An informative experiment would be to incubate a cleaved and polyadenylated RNA in a yeast extract (see Section III) to determine whether it will be cleaved a second time. A different possible mechanism to exclude reprocessing would be if, *in vivo*, only nascent RNAs could serve as substrates for cleavage and polyadenylation, so that any RNA, once cleaved, could not be processed a second time. Although there is currently no evidence for this model, it does not seem entirely unlikely, given the observation that transcription and 3′ processing are coupled (see Section V). Finally, one could imagine that some RNAs are in fact processed a second time, and that only export from the nucleus removes the finished mRNA from the pool of processing substrates. This would waste the effort of the first round of cleavage and polyadenylation, but it would not hurt the RNA.

A large proportion of yeast polyadenylation signals functions in both orientations, but not all of them do (61, 63). Currently, this is unexplained. The kinds of sequences that direct 3′ processing, their degeneracy, and the AU-richness of yeast 3′ untranslated regions (3′ UTRs) suggest the possibility that suitable sequences occur by chance on both strands. A stringent criterion for a truly bidirectional signal would be that all mutations affecting processing have similar effects in either orientation. Application of such a test to the *TRP4* polyadenylation signal gave complex results: One part of the signal appeared to be bidirectional, all mutations having similar effects in both orientations, and another part of the signal was unidirectional (G. Braus, personal communication).

The 3' end formation in yeast also differs from that in metazoan cells in that cleavage sites can be spread out over regions of up to 200 nucleotides (58, 61).

The *CYC1* poly(A) signal from S. *cerevisiae* functions in *Schizosaccharomyces pombe*, and the 3' ends of several S. *pombe* genes are faithfully processed in extracts from S. *cerevisiae* (64). However, a detailed analysis of the polyadenylation signal of the S. *pombe ura4* gene led to a picture somewhat different from that in S. *cerevisiae*. Of two sequences required, an "efficiency element" was located downstream of the cleavage site, but functioned independently of its position. A "site-determining element" was present upstream of the cleavage site (65). With respect to both function and sequence, the relationship of these two sequence motifs to those described for S. *cerevisiae* is not obvious.

C. Polyadenylation Signals in Plants

The formation of 3' ends in plants requires two types of sequence elements. A near-upstream element is present about 20 nucleotides upstream of the cleavage site. This element can have the sequence AAUAAA, but weakly related motifs like AAUGGAAAUG also occur, and AAUAAA is not necessarily the best version. A far-upstream element is also required. It consists of poorly defined, extended U-rich sequence and is located approximately 100 nucleotides upstream of the cleavage site. Cleavage often occurs at a sequence YA, and many genes use multiple cleavage sites. No defined sequences appear to be essential downstream of the cleavage site (26).

D. Comparison of Polyadenylation Signals in Different Organisms

A comparison of 3' end formation signals is presented in Fig. 1. The signals in S. *cerevisiae* and in plants are obviously quite similar. Both require a far-upstream or efficiency element, which is U-rich, and a near-upstream or positioning element, which is related to the mammalian signal AAUAAA. Although the degeneracy and redundancy of these sequences make definitive statements difficult, both types of sequence elements appear to be essential. In S. *cerevisiae* and in plants, cleavage preferentially takes place 3' to a YA sequence. Comparison of cDNA and a genomic sequence does not reveal exactly where cleavage takes place, in other words whether the first A of the poly(A) tail is encoded in the DNA. Precise mapping of the *in vitro* cleavage

3' END PROCESSING OF PRE-mRNA 47

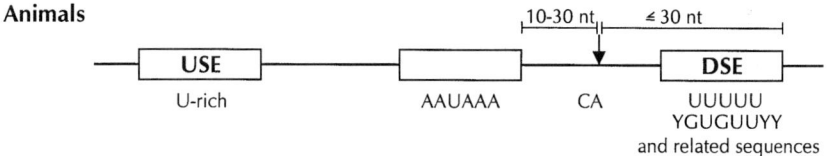

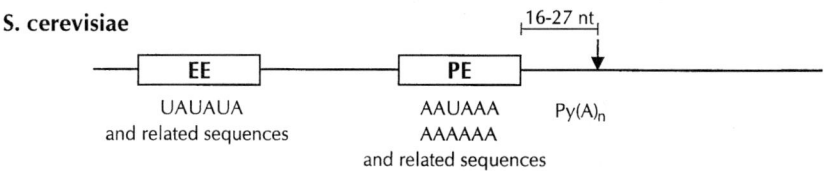

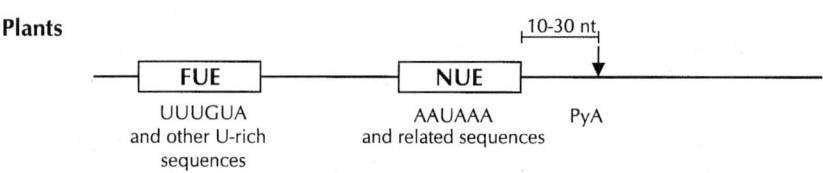

FIG. 1. A comparison of 3' processing signals in animals, yeast, and plants. Names of sequence elements are given in the boxes: USE, upstream element; DSE, downstream element; EE, efficiency element; PE, positioning element; FUE, far-upstream element; NUE, near-upstream element. Possible sequences of these elements are given below the boxes. The AAUAAA sequence in animal polyadenylation signals does not bear a particular name; it probably corresponds to the positioning element in yeast and the near upstream element in plants. The vertical arrows indicate the cleavage sites. See text for details. Reprinted from Ref. 26 by permission of Kluwer Academic Publishers, with modifications.

site has, to our knowledge, not been reported for yeast and is not yet possible for plants, due to the lack of an *in vitro* system. In plant genes and in most genes of S. cerevisiae, sequences downstream of the cleavage site are not known to be required for processing. Although 3' end formation signals in yeast and in plants appear very similar, mutational analysis shows that they are not identical (66).

Mammalian poly(A) signals are more distantly related to those of plants and yeast. The U-rich far-upstream elements are found in many genes, but they are not essential. The near-upstream element is AAUAAA, and, in contrast to the situation in plants and yeast, most point mutations in this sequence inactivate the signal. Cleavage prefers the sequence CA. Cleavage on the 3' side of the A residue has been proved in the two cases analyzed (67, 68). Degenerate downstream elements are essential.

II. Cleavage and Polyadenylation in Animal Cells

A. Nuclear 3′ Processing

Both 3′ cleavage and polyadenylation can be carried out *in vitro*, either in nuclear extracts (69) or by a reconstitution from purified and partially purified proteins. Short RNAs containing just the essential sequences can serve as substrates. These are made *in vitro* by transcription with phage RNA polymerases and permit 3′ processing to be uncoupled from transcription and from other processing steps.

In the *in vitro* reaction, cleavage and polyadenylation are coupled, i.e., cleaved intermediates lacking a poly(A) tail are not detectable under normal reaction conditions. Cleavage depends on the AAUAAA sequence as well as downstream and upstream elements and proceeds in the presence of EDTA, albeit with reduced efficiency. In contrast, polyadenylation is strictly Mg^{2+}-dependent. Thus, in the presence of EDTA, the cleaved intermediate accumulates. A more efficient method for selectively suppressing polyadenylation is the use of a chain-terminating ATP analog, which results in the accumulation of the 5′ cleavage fragment extended by a single adenylate residue. A so-called precleaved RNA, prepared by run-off transcription from a suitably linearized plasmid and corresponding to the cleaved intermediate, can be polyadenylated independently of cleavage. This reaction depends on the AAUAAA motif and, at least in some cases, on upstream sequences.

ATP serves as the precursor for poly(A) synthesis. However, in the absence of ATP, cleavage does not take place either. Analogs with a nonhydrolyzable α/β bond permit cleavage but not polyadenylation. Thus, the reaction has an ATP requirement not just for synthesis of the poly(A) tail but also for cleavage. Hydrolysis of the β/γ bond does not appear to be required. In crude extracts, ATP is required for assembly of the RNA–protein complex catalyzing cleavage (*18*). Its specific role remains to be defined.

Cleavage of the precursor RNA generates an upstream fragment with a 3′-hydroxyl group that can be directly extended by poly(A) polymerase (*67, 68*). The downstream fragment has a 5′-phosphate and is rapidly degraded in extracts. Endonucleases generating 3′-hydroxyls and 5′-phosphates activate water to act as a nucleophile. Typically, a divalent metal ion is involved in this process (*70*). Resistance of the cleavage activity to EDTA is thus unusual.

In nuclear extracts, polyadenylation substrates are assembled into large protein–RNA complexes before any reaction takes place. This is in agreement with reconstitution studies showing that all proteins, including poly(A) polymerase, have to assemble before cleavage occurs.

Chromatographic fractionation of HeLa cell nuclear extracts has resolved six proteins that are required for reconstitution of cleavage and polyadenylation. Whereas both 3' cleavage of histone mRNAs and splicing involve the action of small nuclear ribonucleoprotein particles (snRNPs), 3' processing leading to polyadenylated RNA is almost certainly catalyzed by proteins only.

The AAUAAA sequence is bound by a protein complex termed cleavage and polyadenylation specificity factor (CPSF) (71–73). As the name suggests, CPSF is essential for both cleavage and polyadenylation, as is its binding site. CPSF has at least three subunits with masses of 160, 100, and 70 kDa. A fourth polypeptide of 30 kDa was not present in all preparations (42, 72, 73) and thus is not essential. However, this polypeptide does not just copurify, it is also coprecipitated by antibodies against the larger subunits (74); UV cross-linking studies suggest that it may contribute to RNA binding (71, 74). The largest subunit also binds RNA (71, 74). As an isolated polypeptide, expressed from the cloned cDNA, it has limited specificity for RNAs containing an AAUAAA motif (75). The sequence does not show strong similarities to known RNA binding domains (75, 76). Binding of CPSF to AAUAAA involves contacts to all bases and several ribose residues (77, 78). cDNAs encoding the 100-kDa subunit have also been cloned (74). The function of this subunit, like that of the 70-kDa polypeptide, is unknown.

Binding of CPSF to the substrate RNA is unstable. The complex is stabilized by the addition of a second protein, the cleavage stimulation factor (CstF) (73, 79–83). This factor binds the downstream element of the polyadenylation signal, as demonstrated by functional assays and by UV cross-linking (79, 81–83). CstF has three subunits of 77, 64, and 50 kDa (84). Reconstitution from cDNA clones showed that the three polypeptides are arranged in a linear fashion, with the 77-kDa subunit connecting the two others. All three are essential for cleavage (85). The 64-kDa subunit has a clear match to the RNP-type RNA binding domain (86) (the RNA binding domain is reviewed in Ref. 87) and can be cross-linked to substrate RNAs by UV irradiation (82, 84, 86, 88). The 77-kDa subunit binds the 160-kDa subunit of CPSF (75). Presumably, this interaction is responsible for the cooperativity of RNA binding. A homolog of the 77-kDa subunit of CstF in *Drosophila* is *suppressor of forked* (85). Mutations in this gene modify the effects of transposon insertions, probably through changes in 3' end formation. The function of the smallest subunit of CstF is unknown. Its sequence shows similarities to the transducin repeats found in trimeric G-proteins (89).

Two additional factors required for cleavage are cleavage factors I_m and II_m (CF I_m and II_m) (90). A preparation of CF I_m contained three polypeptides of 68, 59, and 25 kDa (91). It remains to be seen whether all of them are true subunits of the protein. All three polypeptides can be UV cross-linked to substrate RNA. An apparent preference for polyadenylation substrates has not yet been examined in detail. CF I_m weakly binds RNA in gel shift experiments and, like CstF, stabilizes a CPSF–RNA complex (91). Its function in 3′ cleavage remains to be uncovered.

CF II_m has been separated from the other components of the processing system but not yet purified to homogeneity. Again, its function is unknown.

The proteins described so far are essential for cleavage of all poly(A) sites and sufficient for cleavage of at least one poly(A) site, the SV40 late site (90). Which of these polypeptides carries the endonuclease activity is unknown. In addition, cleavage at most sites examined depends on poly(A) polymerase, and the enzyme also stimulates cleavage at the SV40 late site (18, 21). The catalytic activity of this protein is responsible for the polymerization of the poly(A) tail. Its function in cleavage is not known. However, one may presume that the ability of poly(A) polymerase to stabilize a CPSF–RNA complex (73, 92) is essential to its role in cleavage. Interaction between CPSF and poly(A) polymerase again involves the largest subunit of CPSF (75). A possible structure of the cleavage complex is given in Fig. 2A.

Poly(A) polymerase is a single polypeptide with a molecular mass of either 77 or 83 kDa, depending on the splicing pattern (93, 94). Under physiological conditions, the protein on its own has an extremely feeble polymerase activity (95). The activity is stimulated by CPSF and is therefore directed toward those RNAs that have a binding site for CPSF (72, 95–97). Polymerase activity can also be strongly stimulated in a nonspecific fashion by Mn^{2+} (95–98). Independently of the conditions, the enzyme has no specificity with respect to the RNA substrate; specificity for AAUAAA is acquired exclusively through the interaction with CPSF. However, poly(A) polymerase is specific for the utilization of ATP (95, 99). Although polymerization is template independent, it is a conventional polymerase reaction in all other respects: AMP incorporation is primer dependent (95), the second product is pyrophosphate (99), and incorporation of AMP proceeds with inversion of configuration at the α-phosphate (T. Wittmann and E. Wahle, unpublished data). The reaction mechanism thus probably involves an in-line attack without covalent intermediate.

Sequence comparison (100, 101) and site-directed mutagenesis (101) support the membership of poly(A) polymerase in a family of proteins containing the two other known template-independent polymerases (terminal transferase and tRNA nucleotidyl transferase), as well as DNA polymerase β and additional nucleotidyl transferases. Probably all of these enzymes, like a larger group involving all known polymerases, use three (sometimes two) acidic amino acid residues to bind two divalent metal ions, which coordinate the incoming nucleoside triphosphate, stabilize the deprotonated state of the attacking 3′-hydroxyl of the primer RNA, and stabilize the pentacoordinated transition state (102). Based on mutagenesis experiments, the three acidic

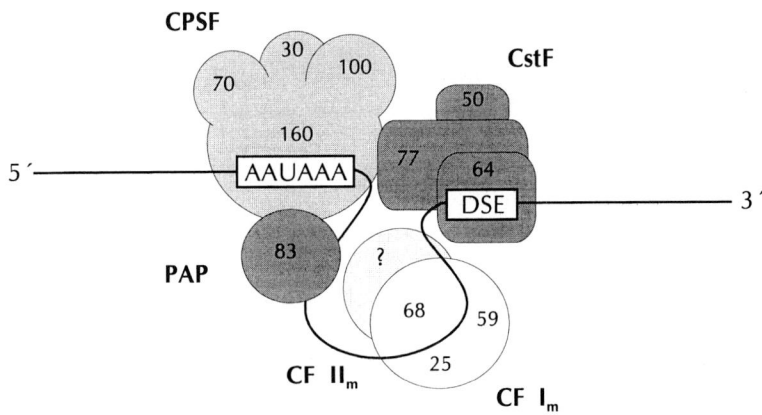

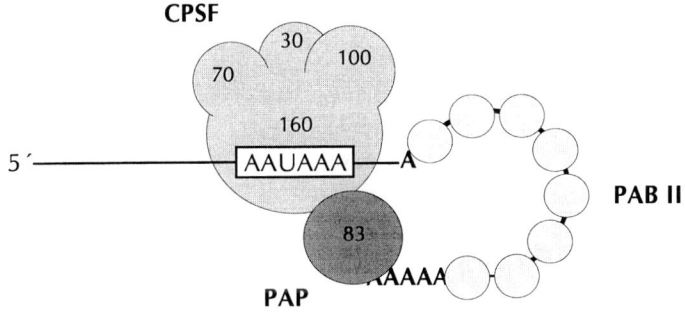

FIG. 2. Models of the cleavage and polyadenylation complexes. (A) The cleavage complex. The arrangement of subunits within CPSF and CF I_m is unknown, as is the positioning of CF I_m and CF II_m. A contact between poly(A) polymerase and the RNA before cleavage is also hypothetical. The other interactions shown are supported by experimental evidence. (B) The elongation complex. Poly(A) polymerase might interact directly with PAB II, but this has not been investigated. PAP, poly(A) polymerase; DSE, downstream element; all other abbreviations as in the text.

amino acids are Asp-113, Asp-115, and Asp-167 in the case of bovine poly(A) polymerase (101). The stereochemistry of the enzyme is consistent with this mechanism.

Interaction of poly(A) polymerase with the primer RNA is weak, as shown by direct binding experiments, by a high apparent K_M in the nonspecific reaction and by the distributive reaction mechanism (95). Although Mn^{2+} decreases the apparent K_M to about $\frac{1}{100}$th, it does not change the distributive mechanism, and its effect remains unexplained (95). Weak binding to RNA is probably responsible for the dependence of polymerase activity on CPSF and/or a functionally similar factor, poly(A) binding protein II. These proteins apparently act by recruiting the polymerase to the primer RNA. A region at the N terminus of poly(A) polymerase with some sequence similarity to an RNP domain was initially proposed to be responsible for RNA binding (93, 103). However, more recently it has been suggested that a region surrounding one of the nuclear localization signals (103) binds the primer (101, 104).

Mammalian poly(A) polymerase contains a large C-terminal domain rich in serine and threonine residues and containing a second nuclear localization signal (103). This domain is not essential for any of the known functions of the enzyme (93, 95, 101, 105). A regulatory role is suggested by multiple phosphorylations and differential splicing in this region (93, 103, 105).

The proteins enumerated so far catalyze cleavage and subsequent poly(A) addition. Although all of them are involved in cleavage, CPSF and poly(A) polymerase suffice for poly(A) addition to a precleaved RNA. However, polyadenylation is slow. One additional protein, poly(A) binding protein II (PAB II) is required for rapid elongation (106). Like the other proteins involved in 3' cleavage and polyadenylation, PAB II is localized in the cell nucleus (107). The protein has a molecular mass of 33 kDa. It contains a single RNP domain in the middle, preceded by a very acidic domain and followed by a very basic domain. The protein tends to form oligomers in the absence of RNA (108). It binds specifically to poly(A) and poly(G) with a site size of approximately 9 nucleotides and a packing density of around 15 nucleotides (108, 109). A *Drosophila* homolog is encoded by the *rox2* gene (110).

Like CPSF, PAB II appears to act by tethering poly(A) polymerase to the RNA, as shown by the following observations: Stimulation of poly(A) extension is dependent on the presence of at least 10 adenylate residues at the 3' end of the RNA; these residues provide a binding site for PAB II. PAB II can act on poly(A) polymerase independently of CPSF or any other protein, using simple homopolymeric poly(A) as a substrate. Both CPSF and PAB II individually stimulate polyadenylation to about the same extent and lead to a very small increase in processivity. However, only the simultaneous presence of both proteins effects the highest rate of poly(A) elongation (around 25 nucleotides/sec) and a fully processive reaction (92, 106, 111).

Once a tail of approximately 250 adenylate residues has been generated, elongation switches to a distributive mode and therefore becomes slow again (111). *In vivo*, 250 nucleotides corresponds to the average length of a newly made poly(A) tail. Thus, the *in vitro* reaction shows proper length control.

This phenomenon is seen only in the presence of PAB II. The basis of length control is a true measurement of poly(A) tail length, not a kinetic mechanism. PAB II may be responsible for the length measurement through stoichiometric binding to the poly(A) tail.

The characteristics of the polyadenylation of a precleaved RNA as just described are in complete agreement with the properties of this reaction in crude nuclear extracts (*112*). A possible structure of the polyadenylation complex is shown in Fig. 2B.

A recurring theme in the reactions described so far is cooperativity. Cooperative interactions have been shown through binding assays and/or functional studies for CPSF–CstF, CPSF–CF I_m, CPSF–poly(A) polymerase, and poly(A) polymerase–PAB II. The data suggest that CPSF, CstF, CF I_m, and poly(A) polymerase are components of the processing complexes observed in nuclear extracts. PAB II probably does not bind until an oligo(A) tail has been generated. Whether CF II_m is part of the complex is not yet known. Cooperativity affords precision in the recognition of poly(A) sites while allowing degeneracy in individual sequence elements of these sites. Cooperative complex formation also ensures that cleavage and polyadenylation are coupled, so that a 3′ end lacking the protective poly(A) tail is not exposed to the action of nucleases. Similarly, the simultaneous action of CPSF and PAB II guarantees that poly(A) polymerase extends the poly(A) tail to its full length before falling off.

There is one obvious gap in this scheme: Binding of CstF requires the downstream sequence element that is removed by cleavage, and CstF, CF I_m, and CF II_m have no effect on the polyadenylation of a precleaved substrate. Consequently, addition of the first 10 nucleotides to a precleaved RNA, which is carried out by CPSF and poly(A) polymerase without assistance by PAB II, is essentially distributive. In other words, the relatively stable cleavage complex would appear to be converted into an unstable (largely distributive) polyadenylation complex, only to be stabilized again by the subsequent binding of PAB II. Why would the cleavage complex be designed to position poly(A) polymerase immediately on the newly created 3′ end, only for the enzyme to fall off repeatedly during the poorly processive addition of the first 10 nucleotides? Currently there is no answer to this question. However, in crude extracts the 3′ processing complex becomes sensitive to competition by the polyanion heparin, not after cleavage but after the addition of the first 10 adenylate residues. Also, specific cross-linking of a protein, almost certainly the 64-kDa subunit of CstF, is lost at the same time (*113*). Because this is the polypeptide binding the downstream sequence element, a reasonable explanation of these data is that the cleavage complex disintegrates not after cleavage, but after polyadenylation has switched to the processive elongation mode through binding of PAB II. Thus, addition of the first 10 A residues is probably distributive only when polyadenylation is uncoupled from cleavage.

B. Cytoplasmic Polyadenylation

The reactions described above are obligatory for almost all mRNAs and take place in the cell nucleus. However, the poly(A) tails of some mRNAs are further processed in the cytoplasm. This is most prominent during oocyte maturation and early embryogenesis (*114–116*). Oocytes store large amounts of mRNAs accumulated during oogenesis. Gene expression during oocyte maturation and the cleavage divisions is regulated at the level of translation; transcription in the developing embryo does not start until later. Because the poly(A) tail is essential for translation, changes in the polyadenylation status of cytoplasmic mRNAs can be used to regulate translation. This type of regulation is essential for development (*117–119*).

Usually, mRNAs translated in the growing oocyte have full-length poly(A) tails and lose them at the same time that their translation is shut down. Other mRNAs, stored in an inactive form in the oocyte, have exceptionally short oligo(A) tails. These oligoadenylated inactive mRNAs are generated by deadenylation of RNA that initially received a full-length poly(A) tail by conventional processing in the nucleus (*120*). Cytoplasmic elongation of the oligo(A) tail is associated with recruitment of the mRNA for translation during the course of development. Studies in mouse, *Xenopus*, and *Drosophila* have shown that *in vitro* addition of a poly(A) tail to a mRNA before injection into the oocyte can be sufficient to induce translation of this mRNA at a wrong developmental stage (*121–124*). In other cases, the presence of the poly(A) tail was not sufficient, and it was suggested that the act of poly(A) elongation in the oocyte was essential to the activation of mRNA (*125, 126*). The connection between poly(A) elongation and translation may be mediated by cap ribose methylation (*127*).

How is cytoplasmic poly(A) extension catalyzed and regulated? There can be little doubt that the basic machinery is identical with or at least very similar to the one operating in the cell nucleus. Poly(A) polymerase differing in no obvious way from the mammalian nuclear enzyme has been identified in *Xenopus* oocytes by partial purification, by cDNA cloning, and by specific antibodies (*128–130*). A fraction of the enzyme is localized in the oocyte cytoplasm and is subject to phosphorylation during development (*130*). Developmentally regulated cytoplasmic polyadenylation depends on the same sequence that is required for nuclear polyadenylation, AAUAAA (*125, 131, 132*). Thus, CPSF or a closely related factor is very likely involved as well. This is also supported by biochemical evidence.

In addition to AAUAAA, poly(A) extension requires upstream sequences, so-called cytoplasmic polyadenylation elements (CPEs). In *Xenopus*, some

CPEs with sequences such as U_5AU, U_5AAU, or $U_6AUAAAG$ cause polyadenylation during oocyte maturation (*122, 125, 131, 132*). Very similar sequences also function as CPEs in mouse oocytes (*120, 133*). Others, for example U_{12} or U_{18}, are responsible for polyadenylation during *Xenopus* embryogenesis (*126, 134*). The spacing between the CPE and AAUAAA influences the timing of polyadenylation. CPEs may well be equivalent to the upstream elements of nuclear poly(A) signals described above. Therefore, the question of how CPEs act is discussed here together with the more general question of how upstream elements act.

One view is that separate factors, distinct from the general polyadenylation machinery, bind upstream elements, including CPEs. Several proteins binding to CPE-containing RNAs have been identified by UV cross-linking. One of them has been purified from *Xenopus* oocytes, and cDNAs have been obtained (*135*). Immunodepletion of this CPE-binding protein (CPEB) from oocyte extracts inhibits CPE-dependent polyadenylation. The activity can be partially restored by readdition of recombinant protein. CPEB is phosphorylated during oocyte maturation, in agreement with the fact that a *cdc2*-like kinase is essential for cytoplasmic polyadenylation and maturation (*132, 135, 136*).

In nuclear polyadenylation in mammalian cells, the A protein of the U1 snRNP has been suggested to bind the upstream element of the SV40 late polyadenylation site by means of its C-terminal RNP domain (*137*). (The N-terminal RNP domain binds the U1 RNA; see Section IV,C.) Others have not confirmed binding of U1A protein to the SV40 late site (*138*), and the C-terminal RNP domain may not bind RNA at all (*139*). *In vitro*, relatively high concentrations of the U1A protein stimulate polyadenylation of a precleaved SV40 late RNA about threefold with respect to poly(A) length and proportion of substrate elongated (*140*). However, it appears that the upstream element of the polyadenylation signal is not required for this effect (*140*). Thus, the stimulation may depend on nonspecific binding to RNA.

An alternative view is that upstream elements are not recognized by separate factors but by the same protein that binds AAUAAA, namely, CPSF. A partial purification of factors involved in CPE-dependent polyadenylation in *Xenopus* oocyte extracts detected only a single RNA-binding activity with specificity for both the CPE and AAUAAA (*128*). Specificities for the two sequences could not be separated in gel retardation assays or by chromatographic fractionation. In fact, CPE-dependent polyadenylation of *Xenopus* RNAs was reconstituted from recombinant mammalian poly(A) polymerase and purified mammalian CPSF (*141*). CPSF binds preferentially to RNAs containing a CPE in addition to AAUAAA. With respect to the reaction in the cell nucleus, *in vitro* polyadenylation of an HIV substrate with recombinant poly(A) polymerase and purified CPSF had the same requirement for an upstream element seen *in vivo,* and the 160-kDa subunit of CPSF could be UV cross-linked to the upstream element (*42*). These data also suggest that no additional protein is needed.

A possible problem with the idea that CPE effects are mediated by CPSF lies in the high degree of fine tuning: All the many RNAs that undergo cytoplasmic polyadenylation differ with respect to the timing and also the length of poly(A) added (*123, 142*). Thus it is conceivable that the activity of CPSF is modified by factors such as CPEB. Assays of this protein in a reconstitution system are eagerly awaited. Recently, a genetic analysis of developmentally regulated polyadenylation has been initiated in *Drosophila*. Two maternal-effect mutations, *grauzone* and *cortex*, were found to be deficient in the polyadenylation of developmentally important mRNAs in the oocyte, including *bicoid, Toll,* and *torso* (*124*). The function of the *grauzone* and *cortex* gene products is still unknown. One may anticipate that genetic tests of the biochemical models discussed herein will be possible in the future.

III. Cleavage and Polyadenylation in Yeast

Yeast mRNAs, like those in mammals, receive their mature 3' ends by cleavage of an extended precursor, followed by the addition of a poly(A) tail (*143–146*). In a crude *in vitro* system (*143, 147*), 3' cleavage is at least partially dependent on a hydrolyzable nucleoside triphosphate, not necessarily ATP. As in mammalian extract, low concentrations of EDTA inhibit polyadenylation but permit cleavage. However, in yeast, the cleaved intermediate is abundant even under normal reaction conditions. Poly(A) tails added *in vitro* have the same length as they do *in vivo*, around 90 nucleotides.

Fractionation of whole cell extracts has identified four different processing factors (*147*). Two of these, cleavage factors I_y and II_y (CF I_y and II_y), are required for specific 3' cleavage. (Note that the names CF I_y and CF II_y do not imply a functional equivalence to the mammalian factors CF I_m and CF II_m.) The lack of coupling between cleavage and polyadenylation is reflected in the fact that poly(A) polymerase is not required for cleavage *in vitro*. However, mutations in the gene encoding poly(A) polymerase affect the choice of the cleavage site, suggesting the possibility that an interaction of poly(A) polymerase with the cleavage complex exists *in vivo* (*148*). CF I_y, polyadenylation factor I (PF I), and poly(A) polymerase are required for poly(A) addition to a precleaved substrate RNA. Polyadenylation of a precleaved RNA also happens *in vivo* when this RNA is generated by the action of a ribozyme built into the primary transcript (*149*).

Poly(A) polymerase of *S. cerevisiae* is a 64-kDa polypeptide and is similar to the previously described mammalian enzyme (*150, 151*). The polymerase is encoded by an essential gene, PAP1 (151, 152). The products of two other essential genes, RNA14 and RNA15 (153), are subunits of CF I$_y$ with molecular masses of 75 and 33 kDa, respectively (154). CF I$_y$ appears to contain two additional polypeptides (L. Minvielle-Sebastia and W. Keller, personal communication). The essential FIP1 gene encodes a subunit (36 kDa) of PF I (155). Purification of PF I suggests a complex of at least seven polypeptides (P. Preker and W. Keller, personal communication). Nothing is known about the identity of CF II$_y$. CF I$_y$ is thought to bind the substrate RNA via its RNA15 subunit (153). The sequence specificity of this interaction remains to be determined. The FIP1 gene product interacts with poly(A) polymerase as well as with RNA14 protein. Thus, PF I may mediate an interaction between the RNA-bound CF I$_y$ and poly(A) polymerase (155). An interaction between poly(A) polymerase and CF I$_y$, direct or indirect, has also been proposed on the basis of codepletion with antibodies directed against the polymerase (156).

Mutations in the nonessential REF2 gene cause moderate defects in 3' end cleavage (157). *In vitro* complementation experiments demonstrate a direct participation of REF2 protein in the reaction. The protein binds RNA, but its exact role in cleavage is unknown.

Additional genetic data are difficult to interpret at the moment. PRP20 is the yeast homolog of the mammalian RCC1 gene, which encodes a guanine nucleotide exchange factor for the Ras-related nuclear G-protein Ran. Mutations in prp20 have a very pleiotropic phenotype, one aspect of which is a defect in 3' end formation (158). Deletion of the nonessential SSM4 gene suppresses certain rna14 alleles (159). The rna14 mutations are also suppressed by overproduction of SSM1 protein, the yeast counterpart of ribosomal protein L1 (160). It is possible that the RNA14 protein has a second function in addition to 3' processing (161). Finally, RNA14 is among the suppressors of mutations in the NPL3 gene, encoding a protein with similarity to mammalian hnRNP proteins (162).

IV. Regulation of Polyadenylation

Formation of 3' ends is essential for the production of translatable mRNA (4–7). Moreover, the inherent "strength" of a poly(A) site can determine the steady-state level of mRNA (36, 163). The limited data available suggest that the strength of a poly(A) site is reflected in the stability of the processing complex that it forms *in vitro* (41, 48, 81, 164). As pointed out before (21), this is not trivial and implies that the dissociation rate of the complex is high compared to the forward rate on the reaction pathway.

The fact that a weak poly(A) site leads to low steady-state levels of mRNA suggests that a fraction of the pre-mRNA fails to be processed and is degraded, although this has not been investigated in detail. Extended RNA molecules can often be detected when a polyadenylation signal is inactivated. The extended chains may represent less efficient cleavage and polyadenylation at a downstream site (6), or they may be unprocessed precursors waiting to be destroyed. Mutation of the polyadenylation signal in the human gene encoding arylsulfatase A leads to the accumulation of large quantities of extended RNAs lacking poly(A). The abundance of these molecules suggests that they are stable. However, the arylsulfatase A gene is unusual in that extended poly(A)$^-$ transcripts are abundant even in wild-type individuals. Inefficient polyadenylation may be due to the unusual polyadenylation signal AAUAAC (7).

As an essential step in gene expression, 3' end formation can be subject to regulation. Different types of regulation are discussed in the next three sections.

A. Choice between Alternative Sites

A given gene can have more than one poly(A) site. In the case of a competition between equivalent sites, a preferred use of the one located upstream suggests a first-come, first-served mechanism (165, 166). However, a strong downstream site can outcompete a weak upstream site (28, 167). Thus, mature mRNA can be generated with a complete, unused poly(A) site in it. However, from the analysis of mature mRNA one cannot decide whether reprocessing at the upstream site is strictly forbidden or happens at a low frequency.

It has been reported that the presence of multiple weak polyadenylation sites does not increase mRNA output compared to output from a single weak site (166). This puzzling observation was explained by a speculative model in which the 3' processing complex gains access to the site from a downstream position, e.g., the advancing RNA polymerase (166). According to the model, the complex binds the site located most closely to the 3' end and either executes the reaction or dissociates, so that only the most distal poly(A) site in the nascent transcript is functional at any time. A specific prediction of the model would be that a weak poly(A) site downstream of a stronger one should be able to reduce the production of mature RNA. We are not aware of a systematic test of this prediction.

Often, the choice of the poly(A) site has no obvious consequences for gene expression. However, in a few cases, selection of the poly(A) site is important. In the mRNA for amyloid precursor protein, a particular 3' UTR sequence, which is present in the mature RNA only if the more distal of two poly(A) sites is used, enhances translation of the RNA (168). The yeast *CBP1* gene is regulated according to the requirement for mitochondrial function by alternative 3' processing with one site located within the open reading frame (169, 170).

The adenovirus major late transcription unit makes use of five different cleavage/polyadenylation sites, and different open reading frames are expressed depending on poly(A) site choice. Early in infection, the first three polyadenylation sites are present in the pre-mRNA, and the L1 site is used most frequently. At later times, all five sites are present, and L1 is used less frequently. Preferential use of L1 compared to L3 can be reproduced in minigene constructs, even in the absence of adenovirus infection (171). When these minigenes are inserted into viable virus, switching to equal use of L1 and L3 occurs late in infection. Switching depends on properties of the L1 site: when it is debilitated by the deletion of sequences flanking the core polyadenylation signal, preferential use in early infection is abolished and, consequently, no switching occurs. Switching does not occur either when the L1 site is replaced by a different polyadenylation signal. In other words, switching is due to a neglect of L1 at the late stage of infection, not to upregulation of L3 (172, 173).

The L1 site is weak; presumably its use early in infection is due to its proximal position (41, 164, 174). The activity of the cleavage factor CstF, as assayed by partial purification, declines during adenovirus infection. Limiting activity of this protein (or any other processing factor) is expected to favor the use of stronger polyadenylation sites, such as L3 (174). However, when cells at a late stage of infection are superinfected with a distinguishable virus, the superinfecting virus shows the early pattern of poly(A) site choice (171). This cis effect on poly(A) site choice is not easily explained by changes in the abundance or activity of a processing protein.

Another prominent case of alternative poly(A) sites is that of retroviruses, pararetroviruses, and retrotransposons. Due to their terminally redundant sequences, most retroviruses have poly(A) signals both at their 5' and their 3' ends. Synthesis of full-length RNA from the DNA requires that the poly(A) site located close to the promoter be ignored and that the poly(A) site at the other end of the transcription unit be used. Choice between the two sites is not known to be regulated in response to cellular or extracellular events. Nevertheless, a discussion of mechanisms contributing to the exclusive use of the downstream poly(A) site is most appropriate in this section.

The promoter-proximal site is functional when placed at the 3' end of a reporter gene. This has led directly to the idea that processing at the first site is inhibited by the proximity of the promoter (175–178). In addition, the U3 sequences, present in the transcript upstream of the promoter-distal but not the promoter-proximal poly(A) site, usually contain an upstream element favoring the use of the distal site (28, 48, 179–185). However, whereas one study concluded that this upstream element is essential for poly(A) site choice

(*183*), others concluded that it was merely of secondary importance (*182, 186*).

Studies aiming to decide whether inhibition by promoter proximity or stimulation by U3 sequences is more important have not only come to different conclusions but, ironically, have introduced additional explanations for poly(A) site choice. One is that the short sequence between the promoter and proximal poly(A) site does not contain an intron. Because splicing and polyadenylation are thought to be coupled (see Section V), the poly(A) site may not be recognized due to the absence of upstream splice sites. In fact, introduction of a small intron upstream of the promoter-proximal poly(A) site of an HIV-derived construct led to cleavage and polyadenylation at this site in transient transfections (*183*). Another explanation is inhibition of the proximal poly(A) site of HIV by the major viral 5' splice site located just downstream. Destruction of this site within a complete, replication-competent HIV provirus indeed led to activation of the promoter-proximal site, even though the short transcript contained no intron (*186*).

Thus, convincing evidence has been presented both for inhibition by the absence of an upstream intron (in chimeric constructs) and for inhibition by the downstream intron (in viable virus). Although it is not clear how these data can be reconciled, it is possible that the idea of inhibition by promoter proximity may have been misleading. There is an additional complication in that only retroviral poly(A) sites show proper site choice, whereas other poly(A) sites are functional when cloned in the promoter-proximal position (*175, 177, 178, 186*). Finally, the dependence of poly(A) site choice in HIV on promoter activation by the viral TAT protein (*187*) also remains to be explained. Because TAT influences the processivity or elongation rate of RNA polymerase, the timing of poly(A) site recognition and splice site recognition may be relevant.

B. Choice between Splicing and Polyadenylation

In some genes, a regulated poly(A) site is located in an intron. Either the poly(A) site can be used so that the open reading frame is terminated, or the intron containing the poly(A) site is spliced out, and the downstream poly(A) site is used. A well-studied example of this kind of regulation is provided by the immunoglobulin genes.

Membrane-bound and secreted immunoglobulins are derived from the same gene by competing processing pathways of the primary transcript. The secreted form is generated by cleavage and polyadenylation at the upstream of two poly(A) sites, located in an intron. Splicing of this intron removes the upstream poly(A) site, leads to the inclusion of two downstream exons and the use of the downstream poly(A) site, and generates the membrane-bound immunoglobulin (Fig. 3A). During B cell development, the ratio of mem-

3' END PROCESSING OF PRE-mRNA

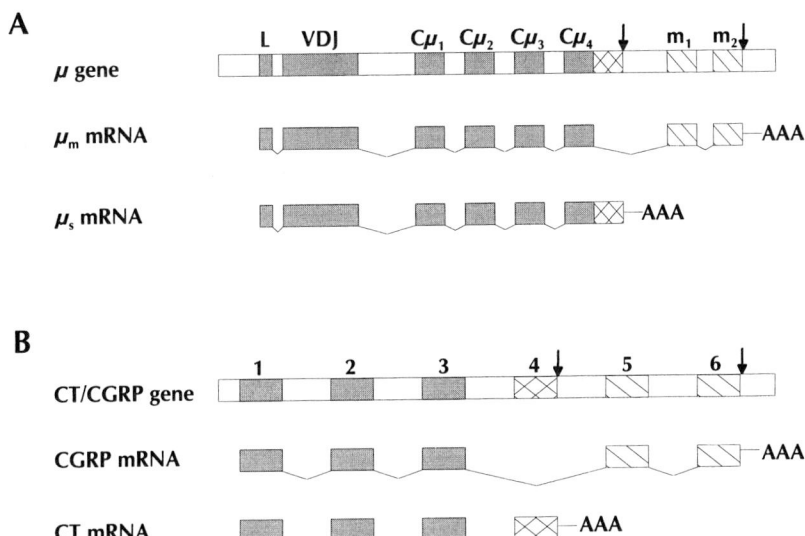

FIG. 3. Two cases of alternative polyadenylation. Grey boxes indicate constitutive exons, hatched and cross-hatched boxes indicate alternative exons present in one or the other of two alternative products. Vertical arrows indicate the cleavage/polyadenylation sites. (A) The immunoglobulin μ chain gene and its two processing products. The μ_m mRNA encodes the membrane-bound form of the protein, whereas the μ_s mRNA encodes the secreted form. Use of the distal polyadenylation site to generate μ_m mRNA is associated with recognition of a 5' splice site within the $C\mu_4$ exon. (B) The gene for calcitonin (CT) and the calcitonin gene-related peptide (CGRP) and its two processing products. The CGRP mRNA encodes the calcitonin gene-related peptide; the CT mRNA encodes calcitonin. Use of the distal poly(A) site generating the CGRP RNA is associated with skipping of exon 4.

brane-bound to secreted immunoglobulin changes: Early B cells and memory cells prefer the splicing event that removes the upstream poly(A) site, whereas plasma cells almost exclusively use the upstream poly(A) site, generating secreted antibodies (188, 189).

The developmental switch operates only when the two alternatively used poly(A) sites are present in the same transcript (190, 191). Moreover, the same regulation of poly(A) site use can be obtained when completely unrelated sequences are used to build a similar transcription unit: an upstream poly(A) site located within an intron and a downstream poly(A) site (192, 193). These two straightforward experiments strongly suggest that there is no specific sequence and no gene-specific factor involved in regulation. Instead, it appears that early B cells have a relatively low 3' end processing activity, thus favoring the competing splicing reaction and use of the stronger downstream poly(A) site. Plasma cells are more proficient in cleavage and polyadenylation such that the weaker upstream poly(A) site is used preferentially (194–196).

What is the cause of this difference? The activity of the cleavage factor CstF, as measured by UV cross-linking to cleavage substrates, has been reported to rise during B cell stimulation (197). However, because binding of CstF to these RNAs is cooperative with the binding of other proteins, one cannot tell whether the difference in CstF activity is cause or effect. An inhibitor of 3' end processing has been identified in nuclear extracts from B cells (198). Its mode of action and role for differential processing remain to be characterized.

A similar pattern of alternative splicing and polyadenylation is found in the calcitonin/calcitonin gene-related peptide transcription unit (Fig. 3B). Mutagenesis experiments indicated that splice-site choice is the dominant factor in alternative processing (199). However, recent evidence supports the idea that poly(A) site selection may also play a role. A sequence located 168 nucleotides downstream of the first poly(A) site is required for use of this site and inclusion of exon 4. The sequence stimulates cleavage and polyadenylation at the exon 4 poly(A) site *in vitro*. Interestingly, this 'polyadenylation enhancer' looks like a 3' splice site immediately followed by a 5' splice site. Binding of several components of the splicing machinery, including the U1 snRNP, appears to be essential for the stimulation of polyadenylation. Probably, use of the poly(A) site at the end of exon 4 favors recognition of the 3' splice site at the upstream end of this exon (200) (see also Section V).

C. Quantitative Regulation

In this section, we discuss RNAs in which either the extent of utilization of a single poly(A) site or the length of the poly(A) tail is regulated.

Expression of the gene encoding the A protein of the U1 snRNP is adjusted in a negative feedback loop by regulation of polyadenylation. By means of the first of its two RNP domains, the U1 A protein binds a stem–loop in U1 small nuclear RNA (snRNA). Two copies of the same 7-nucleotide binding site are also found in the 3' UTR of the A protein mRNA, forming a phylogenetically conserved structure (201). Two molecules of A protein bind to this structure to inhibit 3' end formation *in vivo* and *in vitro* (201, 202). Although the A protein binding site is only 19 nucleotides upstream of the polyadenylation signal (AUUAAA in this case), A protein does not interfere with binding of CPSF, nor does it prevent cleavage. Instead, polyadenylation is inhibited. Remarkably, inhibition is also evident in the Mn^{2+}-dependent nonspecific reaction, implying a direct effect of U1 A protein on poly(A) polymerase. Such a direct interaction has indeed been demonstrated (203). The apparent contradiction between the inhibitory effect of U1A discussed here and the stimulation discussed above may be a matter of protein concentrations and binding sites. With the SV40 poly(A) site, 0.6 μM U1A was stimulatory, whereas 1.2 μM was inhibitory (*140*). The poly(A) site of the U1 A pre-mRNA supplies a tandem binding site for U1 A, and binding of two molecules is essential for inhibition of polyadenylation (*201*). Inhibition was almost complete at 0.1 μM (*202, 203*).

In papilloma viruses, use of the late polyadenylation site is inhibited by sequences resembling 5' splice sites, located in the 3' UTR (*204*). Although U1–snRNP binding is essential for the inhibitory effect, simple steric interference with binding of CPSF or other polyadenylation proteins is not a likely explanation. It is possible that the U1 A protein again acts as an inhibitor. Alternatively, a 5' splice site may interfere with interactions between the 3' splice site and the poly(A) site that are proposed to define the 3'-terminal exon (see Section V).

Nuclear extracts from herpes simplex virus-infected cells efficiently process late viral poly(A) sites, which are weakly processed in control extracts (*205, 206*). The viral gene product IE63 (or ICP27) is required for this activity (*207, 208*). IE63 has pleiotropic effects on RNA metabolism, inducing a relocalization of splicing factors (*209*), inhibiting pre-mRNA splicing (*208, 210*), and possibly stabilizing certain mRNAs by binding to the 3' UTR (*211*). Its role in 3' processing has not been defined.

Cleavage and polyadenylation of the antigenomic RNA of hepatitis delta virus are inhibited by the only viral gene product, the delta antigen, to allow synthesis of full-length antigenomic RNA (*212, 213*). The mechanism of inhibition has not been studied.

Pulse-labeling studies indicate that newly synthesized bulk poly(A) in mammalian cells has a length of approximately 250 nucleotides and is gradually shortened in the cytoplasm. Although poly(A) tail lengths of individual mRNA species can easily be analyzed with great precision, it is not straightforward to distinguish newly made tails from those of aging mRNA molecules, except in bulk RNA. Consequently, it is not usually known whether the poly(A) tail length of a particular mRNA species reflects cytoplasmic or nuclear events.

Several *Xenopus* liver mRNAs encoding serum proteins, including transferrin, have extremely short poly(A) tails, which may play a role in their estrogen-induced degradation. In contrast, transferrin mRNA in the oviduct is induced by the same hormone, and the mRNAs made after hormone induction have long poly(A) tails (*214–216*). When mammalian growth hormone mRNA is induced by dexamethasone, it has longer poly(A) tails than the uninduced mRNA and is more stable (*217*). The same is true when the synthesis of growth hormone is repressed by depletion of thyroid hormone. A nuclear process appears to be responsible (*218*). Longer poly(A) tails are also associated with induction of rat insulin mRNA by glucose (*219*), of chicken very-low-density apolipoprotein II mRNA by estrogen (*220*), and of rat α_1-acid glycoprotein mRNA by inflammation (*221*). In the two latter studies, long poly(A) tails appeared transiently after induction; tail length returned to normal before the end of induction. The explanation that suggests itself is that this reflects a pulse of new mRNA with long poly(A) tails, which are subsequently shortened, as is the case for bulk poly(A). However, in both studies arguments were presented that this was not the case, and that the time courses reflected true changes in poly(A) synthesis or processing. In the case of α_1-acid glycoprotein, the longer poly(A) was seen in mature nuclear RNA.

Poly(A) tail length of rat vasopressin mRNA increases in response to a hyperosmotic stimulus from a normal length of 200 to 400 nucleotides (*222*). Although coupling between longer poly(A) tails and increased mRNA abundance is not obligatory (*223, 224*), longer tails are associated with increased transcription. Thus, as above, one might argue that the apparent increase in tail length simply reflects the higher abundance of young mRNAs that have not yet undergone tail shortening. However, patterns of poly(A) length are not consistent with this explanation (*222*). A nuclear regulatory mechanism has been proposed based on indirect evidence (*224*). Surprisingly, mouse vasopressin mRNA does not show regulation of poly(A) tail length. Vasopressin mRNA in the rat also undergoes a circadian variation of poly(A) tail length (*225*).

In summary, changes in poly(A) tail length may not be uncommon, in particular for hormone-regulated mRNAs. Neither their mechanisms nor their physiological importance have been established. In most examples just described, long poly(A) tails were associated with increased gene expression. However, one case has been reported in which a long tail is associated with a repression of translation of interferon-β mRNA (*226*).

V. Coupling of 3' End Formation to Transcription and Splicing

The study of 3' cleavage and polyadenylation as an isolated event, independent of transcription and splicing, is technically convenient. However, *in vivo*, the substrate for 3' end formation is a nascent transcript, which at the same time also undergoes splicing. Both transcription and splicing appear to be linked to 3' end formation.

A number of studies concluded that initiation from a mRNA-type RNA polymerase II promoter is essential for proper 3' end formation; transcripts initiated from an snRNA-type RNA polymerase II promoter or synthesized by other RNA polymerases were not processed. A second link between transcription and 3' end formation exists: mutations that interfere with 3' processing also impair termination (18, 21).

The exon-definition model of splicing (227) proposes that splicing involves an initial recognition of exons rather than introns. In other words, an interaction of splicing factors across the exon between the upstream 3' splice site and the downstream 5' splice site is postulated. A similar interaction between the 3' splice site preceding the 3'-terminal exon and the poly(A) site is suggested as a solution to the dilemma that the 3'-terminal exon is not followed by a 5' splice site. In fact, reciprocal stimulation between 3' cleavage/polyadenylation and splicing of the 3'-terminal intron has been reported *in vivo* and *in vitro*, in support of the model (21). Inhibition of cleavage and polyadenylation by an upstream 5' splice site (21, 204) can be interpreted as a disruption of the definition of the 3'-terminal exon. However, additional interactions between the splicing apparatus and the cleavage/polyadenylation complex have been discussed. The promoter-proximal poly(A) site of HIV is inhibited by a downstream 5' splice site (186), whereas the alternative poly(A) site located in an intron of the calcitonin/CGRP gene is stimulated by downstream splicing signals (200). The effects of U1 A protein on polyadenylation previously discussed may be reflections of a role normally played by the U1 snRNP. These complications are not explained by the simple exon-definition model. An investigation of these interactions not just in nuclear extract but in reconstituted systems would be helpful.

VI. Conclusion

In the near future, we can anticipate completion of the inventory of 3' processing factors in mammals and in yeast. Further progress will depend on the efficient overproduction of the polypeptides involved to facilitate mechanistic and structural studies. It is hoped that studies of the coordination of 3' processing with transcription and splicing will progress from cells and crude extracts to purified proteins and will also benefit from the opportunities that yeast genetics has to offer.

ACKNOWLEDGMENTS

We are grateful to Gerhard Braus, Barklie Clements, Walter Keller, Lionel Minvielle-Sebastia, Pascal Preker, Nick Proudfoot, and Rozanne Sandri-Goldin for advice and for communicating unpublished information, and to Nick Proudfoot, Helen Rothnie, Ursula Rüegsegger, and Ursel Selent for critical review of the manuscript. Work in our laboratory is supported by the Deutsche Forschungsgemeinschaft.

References

1. G. Rohrmann, L. Yuen, and B. Moss, *Cell* **46,** 1029 (1986).
2. O. C. Richards and E. Ehrenfeld, *Curr. Top. Microbiol. Immun.* **161,** 89 (1990).
3. G. Luo and P. Palese, *Curr. Biol.* **2,** 77 (1992).
4. K. S. Zaret and F. Sherman, *Cell* **28,** 563 (1982).
5. D. R. Higgs, S. E. Y. Goodbourn, J. Lamb, D. S. Weatherall, and N. J. Proudfoot, *Nature (London)* **306,** 398 (1983).
6. S. H. Orkin, T.-C. Cheng, S. E. Antonarakis, and H. H. Kazazian, *EMBO J.* **4,** 453 (1985).
7. V. Gieselmann, A. Polten, J. Kreysing, and K. von Figura, *Proc. Natl. Acad. Sci. U.S.A.* **86,** 9436 (1989).
8. D. R. Gallie, *Genes Dev.* **5,** 2108 (1991).
9. N. Iizuka, L. Najita, A. Franzusoff, and P. Sarnow, *Mol. Cell. Biol.* **14,** 7322 (1994).
10. S. Tarun and A. B. Sachs, *Genes Dev.* **9,** 2997 (1995).
11. A. B. Sachs and R. W. Davis, *Cell* **58,** 857 (1989).
12. R. J. Jackson and N. Standart, *Cell* **62,** 15 (1990).
13. C. A. Beelman and R. Parker, *Cell* **81,** 179 (1995).
14. R. Eckner, W. Ellmeier, and M. L. Birnstiel, *EMBO J.* **10,** 3513 (1991).
15. J. L. Manley, *Biochim. Biophys. Acta* **950,** 1 (1988).
16. M. Wickens, *Trends Biochem. Sci.* **15,** 277 (1990).
17. N. Proudfoot, *Cell* **64,** 671 (1991).
18. E. Wahle and W. Keller, *Annu. Rev. Biochem.* **61,** 419 (1992).
19. E. Wahle, *BioEssays* **14,** 113 (1992).
20. A. Sachs and E. Wahle, *J. Biol. Chem.* **31,** 2295 (1993).
21. E. Wahle, *Biochim. Biophys. Acta* **1261,** 183 (1995).
22. J. L. Manley, *Curr. Opin. Gen. Dev.* **5,** 222 (1995).
23. W. Keller, *Cell* **81,** 829 (1995).
24. W. Keller, in "Pre-mRNA Processing" (A. I. Lamond, ed.), p. 113. R. G. Landes Comp., Austin, 1995.
25. A. G. Hunt, *Annu. Rev. Plant Physiol. Plant Mol. Biol.* **45,** 47 (1994).
26. H. M. Rothnie, *Plant Mol. Biol.* **32,** 43 (1996).
27. N. J. Proudfoot and G. G. Brownlee, *Nature (London)* **263,** 211 (1976).
28. R. Russnak and D. Ganem, *Genes Dev.* **4,** 764 (1990).
29. K. G. Rabbitts and G. T. Morgan, *Nucleic Acids Res.* **20,** 2947 (1992).
30. M. A. McDevitt, R. P. Hart, W. W. Wong, and J. R. Nevins, *EMBO J.* **5,** 2907 (1986).
31. A. Gil and N. J. Proudfoot, *Cell* **49,** 399 (1987).
32. Z.-F. Chou, F. Chen, and J. Wilusz, *Nucleic Acids Res.* **22,** 2525 (1994).
33. J. McLauchlan, D. Gaffney, J. L. Whitton, and J. B. Clements, *Nucleic Acids Res.* **13,** 1347 (1985).
34. D. Zarkower and M. Wickens, *J. Biol. Chem.* **263,** 5780 (1988).
35. M. Sadofsky, S. Connelly, J. L. Manley, and J. C. Alwine, *Mol. Cell. Biol.* **5,** 2713 (1985).
36. E. R. Gimmi, K. J. Soprano, M. Rosenberg, and M. E. Reff, *Nucleic Acids Res.* **16,** 8977 (1988).
37. F. Chen, C. C. MacDonald, and J. Wilusz, *Nucleic Acids Res.* **23,** 2614 (1995).
38. Z. Quian and J. Wilusz, *Mol. Cell. Biol.* **11,** 5312 (1991).
39. P. S. Bagga, L. P. Ford, F. Chen, and J. Wilusz, *Nucleic Acids Res.* **23,** 1625 (1995).
40. A. Moreira, M. Wollerton, J. Monks, and N. J. Proudfoot, *EMBO J.* **14,** 3809 (1995).
41. J. Prescott and E. Falck-Pedersen, *Mol. Cell. Biol.* **14,** 4682 (1994).
42. G. M. Gilmartin, E. S. Fleming, J. Oetjen, and B. R. Graveley, *Genes Dev.* **9,** 72 (1995).
43. D. B. Batt and G. G. Carmichael, *Mol. Cell. Biol.* **15,** 4783 (1995).

44. A. Sittler, H. Gallinaro, and M. Jacob, *J. Mol. Biol.* **248,** 525 (1995).
45. E. R. Gimmi, M. E. Reff, and I. C. Deckmann, *Nucleic Acids Res.* **17,** 6983 (1989).
46. Y. F. Ahmed, G. M. Gilmartin, S. M. Hanly, J. R. Nevins, and W. C. Greene, *Cell* **64,** 727 (1991).
47. A. Bar-Shira, A. Panet, and A. Honigman, *J. Virol.* **65,** 5165 (1991).
48. G. M. Gilmartin, E. S. Fleming, and J. Oetjen, *EMBO J.* **11,** 4419 (1992).
49. P. H. Brown, L. S. Tiley, and B. R. Cullen, *Genes Dev.* **5,** 1277 (1991).
50. B. I. Osborne and L. Guarente, *Proc. Natl. Acad. Sci. U.S.A.* **86,** 4097 (1989).
51. P. Russo, W.-Z. Li, D. M. Hampsey, K. S. Zaret, and F. Sherman, *EMBO J.* **10,** 563 (1991).
52. P. Russo, W.-Z. Li, Z. Guo, and F. Sherman, *Mol. Cell. Biol.* **13,** 7836 (1993).
53. Z. Guo and F. Sherman, *Mol. Cell. Biol.* **15,** 5983 (1995).
54. W. Hou, R. Russnak, and T. Platt, *EMBO J.* **13,** 446 (1994).
55. S. Irniger and G. H. Braus, *Proc. Natl. Acad. Sci. U.S.A.* **91,** 257 (1994).
56. Z. Guo, P. Russo, D.-F. Yun, J. S. Butler, and F. Sherman, *Proc. Natl. Acad. Sci. U.S.A.* **92,** 4211 (1995).
57. S. Heidmann, C. Schindewolf, G. Stumpf, and H. Domdey, *Mol. Cell. Biol.* **14,** 4633 (1994).
58. S. Heidmann, B. Obermaier, K. Vogel, and H. Domdey, *Mol. Cell. Biol.* **12,** 4215 (1992).
59. L. E. Hyman, S. H. Seiler, J. Whoriskey, and C. L. Moore, *Mol. Cell. Biol.* **11,** 2004 (1991).
60. P. P. Sadhale and T. Platt, *Mol. Cell. Biol.* **12,** 4262 (1992).
61. J. A. Peterson and A. M. Myers, *Nucleic Acids Res.* **21,** 5500 (1993).
62. C. M. Egli, C. Springer, and G. H. Braus, *Mol. Cell. Biol.* **15,** 2466 (1995).
63. S. Irniger, C. M. Egli, and G. H. Braus, *Mol. Cell. Biol.* **11,** 3060 (1991).
64. T. Humphrey, P. Sadhale, T. Platt, and N. Proudfoot, *EMBO J.* **10,** 3503 (1991).
65. T. Humphrey, C. E. Birse, and N. J. Proudfoot, *EMBO J.* **13,** 2441 (1994).
66. S. Irniger, H. Sanfacon, C. M. Egli, and G. Braus, *Mol. Cell. Biol.* **12,** 2322 (1992).
67. C. L. Moore, H. Skolnik-David, and P. A. Sharp, *EMBO J.* **5,** 1929 (1986).
68. M. D. Sheets, P. Stephenson, and M. P. Wickens, *Mol. Cell. Biol.* **7,** 1518 (1987).
69. C. L. Moore and P. A. Sharp, *Cell* **41,** 845 (1985).
70. W. Saenger, *Curr. Opin. Struct. Biol.* **1,** 130 (1991).
71. W. Keller, S. Bienroth, K. M. Lang, and G. Christofori, *EMBO J.* **10,** 4241 (1991).
72. S. Bienroth, E. Wahle, C. Suter-Crazzolara, and W. Keller, *J. Biol. Chem.* **266,** 19768 (1991).
73. K. G. K. Murthy and J. L. Manley, *J. Biol. Chem.* **267,** 14804 (1992).
74. A. Jenny, H.-P. Hauri, and W. Keller, *Mol. Cell. Biol.* **14,** 8183 (1994).
75. K. G. K. Murthy and J. L. Manley, *Genes Dev.* **9,** 2672 (1995).
76. A. Jenny and W. Keller, *Nucleic Acids Res.* **23,** 2629 (1995).
77. P. L. Wigley, M. D. Sheets, D. A. Zarkower, M. E. Whitmer, and M. Wickens, *Mol. Cell. Biol.* **10,** 1705 (1990).
78. V. J. Bardwell, M. Wickens, S. Bienroth, W. Keller, B. S. Sproat, and A. I. Lamond, *Cell* **65,** 125 (1991).
79. G. M. Gilmartin and J. R. Nevins, *Genes Dev.* **3,** 2180 (1989).
80. J. Wilusz, T. Shenk, Y. Takagaki, and J. L. Manley, *Mol. Cell. Biol.* **10,** 1244 (1990).
81. E. A. Weiss, G. M. Gilmartin, and J. R. Nevins, *EMBO J.* **10,** 215 (1991).
82. G. M. Gilmartin and J. R. Nevins, *Mol. Cell. Biol.* **11,** 2432 (1991).
83. C. C. MacDonald, J. Wilusz, and T. Shenk, *Mol. Cell. Biol.* **14,** 6647 (1994).
84. Y. Takagaki, J. L. Manley, C. C. MacDonald, J. Wilusz, and T. Shenk, *Genes Dev.* **4,** 2112 (1990).
85. Y. Takagaki and J. L. Manley, *Nature (London)* **372,** 471 (1994).
86. Y. Takagaki and J. L. Manley, *Proc. Natl. Acad. Sci. U.S.A.* **89,** 1403 (1992).
87. C. G. Burd and G. Dreyfuss, *Science* **265,** 615 (1994).
88. J. Wilusz and T. Shenk, *Cell* **52,** 221 (1988).
89. Y. Takagaki and J. L. Manley, *J. Biol. Chem.* **267,** 23471 (1992).

90. Y. Takagaki, L. C. Ryner, and J. L. Manley, *Genes Dev.* **3,** 1711 (1989).
91. U. Rüegsegger, K. Beyer, and W. Keller, *J. Biol. Chem.* **271,** 6107 (1996).
92. S. Bienroth, W. Keller, and E. Wahle, *EMBO J.* **12,** 585 (1993).
93. T. Raabe, F. J. Bollum, and J. L. Manley, *Nature (London)* **353,** 229 (1991).
94. E. Wahle, G. Martin, E. Schiltz, and W. Keller, *EMBO J.* **10,** 4251 (1991).
95. E. Wahle, *J. Biol. Chem.* **226,** 3131 (1991).
96. Y. Takagaki, L. C. Ryner, and J. L. Manley, *Cell* **52,** 731 (1988).
97. G. Christofori and W. Keller, *Cell* **54,** 875 (1988).
98. G. Christofori and W. Keller, *Mol. Cell. Biol.* **9,** 193 (1989).
99. C. M. Tsiapalis, J. W. Dorson, and F. J. Bollum, *J. Biol. Chem.* **250,** 4486 (1975).
100. L. Holm and C. Sander, *Trends Biochem. Sci.* **20,** 345 (1995).
101. G. Martin and W. Keller, *EMBO J.* **15,** 2593 (1996).
102. H. Pelletier, M. R. Sawaya, A. Kumar, S. H. Wilson, and J. Kraut, *Science* **264,** 1891 (1994).
103. T. Raabe, K. G. K. Murthy, and J. L. Manley, *Mol. Cell. Biol.* **14,** 2946 (1994).
104. A. M. Zhelkovsky, M. M. Kessler, and C. L. Moore, *J. Biol. Chem.* **270,** 26715 (1995).
105. A.-C. Thuresson, J. Aström, A. Aström, K.-O. Grönvik, and A. Virtanen, *Proc. Natl. Acad. Sci. U.S.A.* **91,** 979 (1994).
106. E. Wahle, *Cell* **66,** 759 (1991).
107. S. Krause, S. Fakan, K. Weis, and E. Wahle, *Exp. Cell Res.* **214,** 75 (1994).
108. A. Nemeth, S. Krause, D. Blank, A. Jenny, P. Jenö, A. Lustig, and E. Wahle, *Nucleic Acids Res.* **23,** 4034 (1995).
109. E. Wahle, A. Lustig, P. Jenö, and P. Maurer, *J. Biol. Chem.* **268,** 2937 (1993).
110. S. F. Brand, S. Pichoff, S. Noselli, and H.-M. Bourbon, *Gene* **154,** 187 (1995).
111. E. Wahle, *J. Biol. Chem.* **270,** 2800 (1995).
112. M. D. Sheets and M. Wickens, *Genes Dev.* **3,** 1401 (1989).
113. V. J. Bardwell and M. Wickens, *Mol. Cell. Biol.* **10,** 295 (1990).
114. M. Wickens, *Trends Biochem. Sci.* **15,** 320 (1990).
115. M. Wickens, *Dev. Biol.* **3,** 399 (1992).
116. J. D. Richter, *in* "Translational Control" (J. Hershey, N. Sonenberg, and M. Mathews, eds.). Cold Spring Harbor Lab., Cold Spring Harbor, New York, 1995.
117. F. Gebauer, W. Xu, G. M. Cooper, and J. D. Richter, *EMBO J.* **13,** 5712 (1994).
118. F. J. Salles, M. E. Lieberfarb, C. Wreden, J. P. Gergen, and S. Strickland, *Science* **266,** 1996 (1994).
119. M. D. Sheets, M. Wu, and M. Wickens, *Nature (London)* **374,** 511 (1995).
120. J. Huarte, A. Stutz, M. L. O'Connell, P. Gubler, D. Belin, A. L. Darrow, S. Strickland, and J.-D. Vassalli, *Cell* **69,** 1021 (1992).
121. J.-D. Vassalli, J. Huarte, D. Belin, P. Gubler, A. Vassalli, M. L. O'Connell, L. A. Parton, R. J. Rickles, and S. Strickland, *Genes Dev.* **3,** 2163 (1989).
122. J. Paris and J. D. Richter, *Mol. Cell. Biol.* **10,** 5634 (1990).
123. M. D. Sheets, C. A. Fox, T. Hunt, G. Vande Woude, and M. Wickens, *Genes Dev.* **8,** 926 (1994).
124. M. E. Lieberfarb, T. Chu, C. Wreden, W. Theurkanf, J. P. Gergen, and S. Strickland, *Development* **122,** 579 (1996).
125. L. L. McGrew, E. Dworkin-Rastl, M. B. Dworkin, and J. D. Richter, *Genes Dev.* **3,** 803 (1989).
126. R. Simon, J.-P. Tassan, and J. D. Richter, *Genes Dev.* **6,** 2580 (1992).
127. H. Kuge and J. D. Richter, *EMBO J.* **14,** 6301 (1995).
128. C. A. Fox, M. D. Sheets, E. Wahle, and M. Wickens, *EMBO J.* **11,** 5021 (1992).
129. F. Gebauer and J. D. Richter, *Mol. Cell. Biol.* **15,** 1422 (1995).
130. S. Ballantyne, A. Bilger, A. Astrom, A. Virtanen, and M. Wickens, *RNA* **1,** 64 (1995).
131. C. A. Fox, M. D. Sheets, and M. P. Wickens, *Genes Dev.* **3,** 2151 (1989).

132. L. L. McGrew and J. D. Richter, *EMBO J.* **9**, 3743 (1990).
133. F. J. Salles, A. L. Darrow, M. L. O'Connell, and S. Strickland, *Genes Dev.* **6**, 1202 (1992).
134. R. Simon and J. D. Richter, *Mol. Cell. Biol.* **14**, 7867 (1994).
135. L. E. Hake and J. D. Richter, *Cell* **79**, 617 (1994).
136. J. Paris, K. Swenson, H. Piwnica-Worms, and J. D. Richter, *Genes Dev.* **5**, 1697 (1991).
137. C. S. Lutz and J. C. Alwine, *Genes Dev.* **8**, 576 (1994).
138. G. G. Simpson, G. P. Clark, H. M. Rothnie, W. Boelens, W. van Venrooij, and J. W. S. Brown, *EMBO J.* **14**, 4540 (1995).
139. J. Lu and K. B. Hall, *J. Mol. Biol.* **247**, 739 (1995).
140. C. S. Lutz, K. G. K. Murthy, N. Schek, J. P. O'Connor, J. L. Manley, and J. C. Alwine, *Genes Dev.* **10**, 325 (1996).
141. A. Bilger, C. A. Fox, E. Wahle, and M. Wickens, *Genes Dev.* **8**, 1106 (1994).
142. B. Stebbins-Boaz and J. D. Richter, *Mol. Cell. Biol.* **14**, 5870 (1994).
143. J. S. Butler and T. Platt, *Science* **242**, 1270 (1988).
144. J. S. Butler, P. P. Sadhale, and T. Platt, *Mol. Cell. Biol.* **10**, 2599 (1990).
145. P. P. Sadhale, R. Sapolsky, R. W. Davis, J. S. Butler, and T. Platt, *Nucleic Acids Res.* **19**, 3683 (1991).
146. L. E. Hyman and C. L. Moore, *Mol. Cell. Biol.* **13**, 5159 (1993).
147. J. Chen and C. Moore, *Mol. Cell. Biol.* **12**, 3470 (19920).
148. E. Mandart and R. Parker, *Mol. Cell. Biol.* **15**, 6979 (1995).
149. C. M. Egli and G. H. Braus, *J. Biol. Chem.* **269**, 27378 (1994).
150. J. Lingner, I. Radtke, E. Wahle, and W. Keller, *J. Biol. Chem.* **266**, 8741 (1991).
151. J. Lingner, J. Kellermann, and W. Keller, *Nature (London)* **354**, 496 (1991).
152. D. Patel and J. S. Butler, *Mol. Cell. Biol.* **12**, 3297 (1992).
153. L. Minvielle-Sebastia, B. Winsor, N. Bonneaud, and F. Lacroute, *Mol. Cell. Biol.* **11**, 3075 (1991).
154. L. Minvielle-Sebastia, P. J. Preker, and W. Keller, *Science* **266**, 1702 (1994).
155. P. J. Preker, J. Lingner, L. Minvielle-Sebastia, and W. Keller, *Cell* **81**, 379 (1995).
156. M. Kessler, A. M. Zhelkovsky, A. Skvorak, and C. L. Moore, *Biochem.* **34**, 1750 (1995).
157. R. Russnak, K. W. Nehrke, and T. Platt, *Mol. Cell. Biol.* **15**, 1689 (1995).
158. W. Forrester, F. Stutz, M. Rosbash, and M. Wickens, *Genes Dev.* **6**, 1914 (1992).
159. E. Mandart, M.-E. Dufour, and F. Lacroute, *Mol. Gen. Genet.* **245**, 323 (1994).
160. A. Petitjean, N. Bonneaud, and F. Lacroute, *Mol. Cell. Biol.* **15**, 5071 (1995).
161. N. Bonneaud, L. Minvielle-Sebastia, C. Culin, and F. Lacroute, *J. Cell Sci.* **107**, 913 (1994).
162. M. Henry, C. Z. Borland, M. Bossie, and P. A. Silver, *Genetics* **142**, 103 (1996).
163. S. Carswell and J. C. Alwine, *Mol. Cell. Biol.* **9**, 4248 (1989).
164. J. C. Prescott and E. Falck-Pedersen, *J. Biol. Chem.* **267**, 8175 (1992).
165. Y. Luo and G. G. Carmichael, *Mol. Cell. Biol.* **11**, 5291 (1991).
166. R. M. Denome and C. N. Cole, *Mol. Cell. Biol.* **8**, 4829 (1988).
167. N. Levitt, D. Briggs, A. Gil, and N. J. Proudfoot, *Genes Dev.* **3**, 1019 (1989).
168. F. de Sauvage, V. Kruys, O. Marinx, G. Huez, and J. N. Octave, *EMBO J.* **11**, 3099 (1992).
169. S. A. Mayer and C. L. Dieckmann, *Mol. Cell. Biol.* **9**, 4161 (1989).
170. S. A. Mayer and C. L. Dieckmann, *Mol. Cell. Biol.* **11**, 813 (1991).
171. E. Falck-Pedersen and J. Logan, *J. Virol.* **63**, 532 (1989).
172. J. D. DeZazzo, E. Falck-Pedersen, and M. J. Imperiale, *Mol. Cell. Biol.* **11**, 5977 (1991).
173. J. D. DeZazzo and M. J. Imperiale, *Mol. Cell. Biol.* **9**, 4951 (1989).
174. K. P. Mann, E. A. Weiss, and J. R. Nevins, *Mol. Cell. Biol.* **13**, 2411 (1993).
175. C. Weichs an der Glon, J. Monks, and N. J. Proudfoot, *Genes Dev.* **5**, 244 (1991).
176. H. Sanfacon and T. Hohn, *Nature (London)* **346**, 81 (1990).
177. K. Iwasaki and H. M. Temin, *Gene Express.* **2**, 7 (1992).

178. K. Iwasaki and H. M. Temin, *Genes Dev.* **4**, 2299 (1990).
179. A. Valsamakis, S. Zeichner, S. Carswell, and J. C. Alwine, *Proc. Natl. Acad. Sci. U.S.A* **88**, 2108 (1991).
180. R. H. Russnak, *Nucleic Acid Res.* **19**, 6449 (1991).
181. A. Valsamakis, N. Schek, and J. C. Alwine, *Mol. Cell. Biol.* **12**, 3699 (1992).
182. J. Cherrington and D. Ganem, *EMBO J.* **11**, 1513 (1992).
183. J. D. DeZazzo, J. M. Scott, and M. J. Imperiale, *Mol. Cell. Biol.* **12**, 5555 (1992).
184. J. D. DeZazzo, J. E. Kilpatrick, and M. J. Imperiale, *Mol. Cell. Biol.* **11**, 1624 (1991).
185. P. Brown, L. S. Tiley, and B. R. Cullen, *J. Virol.* **65**, 3340 (1991).
186. M. P. Ashe, P. Griffin, W. James, and N. J. Proudfoot, *Genes Dev.* **9**, 3008 (1995).
187. C. Weichs an der Glon, M. Ashe, J. Eggermont, and N. J. Proudfoot, *EMBO J.* **12**, 2119 (1993).
188. P. Early, J. Rogers, M. Davis, K. Calame, M. Bond, R. Wall, and L. Hood, *Cell* **20**, 313 (1980).
189. F. W. Alt, A. L. Bothwell, M. Knapp, E. Siden, E. Mather, M. Koshland, and D. Baltimore, *Cell* **20**, 293 (1980).
190. G. Galli, J. W. Guise, M. A. McDevitt, P. W. Tucker, and J. R. Nevins, *Genes Dev.* **1**, 471 (1987).
191. M. L. Peterson and R. P. Perry, *Proc. Natl. Acad. Sci. U.S.A.* **83**, 8883 (1986).
192. M. L. Peterson, *Mol. Cell. Biol.* **14**, 7891 (1994).
193. M. L. Peterson, *Gene Express.* **2**, 319 (1992).
194. M. L. Peterson and R. P. Perry, *Mol. Cell. Biol.* **9**, 726 (1989).
195. M. L. Peterson, E. R. Gimmi, and R. P. Perry, *Mol. Cell. Biol.* **11**, 2324 (1991).
196. M. L. Peterson, M. B. Bryman, M. Peiter, and C. Cowan, *Mol. Cell. Biol.* **14**, 77 (1994).
197. G. Edwards-Gilbert and C. Milcarek, *Mol. Cell. Biol.* **15**, 6420 (1995).
198. D.-H. Yan, E. A. Weiss, and J. R. Nevins, *Mol. Cell. Biol.* **15**, 1901 (1995).
199. S. E. Leff, R. M. Evans, and M. G. Rosenfeld, *Cell* **48**, 517 (1987).
200. H. Lou, R. F. Gagel, and S. M. Berget, *Genes Dev.* **10**, 208 (1996).
201. C. W. G. van Gelder, S. I. Gunderson, E. J. R. Jansen, W. C. Boelens, M. Polycarpou-Schwarz, I. W. Mattaj, and W. J. van Venrooij, *EMBO J.* **12**, 5191 (1993).
202. W. C. Boelens, E. J. R. Jansen, W. J. van Venrooij, R. Stripecke, I. W. Mattaj, and S. I. Gunderson, *Cell* **72**, 881 (1993).
203. S. J. Gunderson, K. Beyer, G. Martin, W. Keller, W. C. Boelens, and I. W. Mattaj, *Cell* **76**, 531 (1994).
204. P. A. Furth, W.-T. Choe, J. H. Rex, J. C. Byrne, and C. C. Baker, *Mol. Cell. Biol.* **14**, 5278 (1994).
205. J. McLauchlan, S. Simpson, and J. B. Clements, *Cell* **59**, 1093 (1989).
206. F. McGregor, A. Phelan, J. Dunlop, and J. B. Clements, *J. Virol.* **70**, 1931 (1996).
207. J. McLauchlan, A. Phelan, C. Loney, R. M. Sandri-Goldin, and J. B. Clements, *J. Virol.* **66**, 6939 (1992).
208. R. M. Sandri-Goldin and G. E. Mendoza, *Genes Dev.* **6**, 848 (1992).
209. A. Phelan, M. Carmo-Fonseca, J. McLauchlan, A. I. Lamond, and J. B. Clements, *Proc. Natl. Acad. Sci. U.S.A.* **90**, 9056 (1993).
210. M. A. Hardwicke and R. M. Sandri-Goldin, *J. Virol.* **68**, 4797 (1994).
211. C. R. Brown, M. S. Nakamura, J. D. Mosca, G. S. Hayward, S. E. Straus, and L. P. Perera, *J. Virol.* **69**, 7187 (1995).
212. S.-Y. Hsieh, P.-Y. Yang, J. T. Ou, C.-M. Chu, and Y.-F. Liaw, *Nucleic Acids Res.* **22**, 391 (1994).
213. S.-Y. Hsieh and J. Taylor, *J. Virol.* **65**, 6438 (1991).
214. R. L. Pastori, J. E. Moskaitis, S. W. Buzek, and D. R. Schoenberg, *J. Steroid Biochem. Mol. Biol.* **42**, 649 (1992).

215. R. L. Pastori, J. E. Moskaitis, S. W. Buzek, and D. R. Schoenberg, *Mol. Endocrinol.* **5,** 461 (1991).
216. D. R. Schoenberg, J. E. Moskaitis, L. H. Smith, and R. L. Pastori, *Mol. Endocrinol.* **3,** 805 (1989).
217. I. Paek and R. Axel, *Mol. Cell. Biol.* **7,** 1496 (1987).
218. D. Murphy, K. Pardy, V. Seah, and D. Carter, *Mol. Cell. Biol.* **12,** 2624 (1992).
219. R. Muschel, G. Khoury, and L. M. Reid, *Mol. Cell. Biol.* **6,** 337 (1986).
220. A. W. Cochrane and R. G. Deeley, *J. Mol. Biol.* **203,** 555 (1988).
221. B. R. Shiels, W. Northemann, M. R. Gehring, and G. H. Fey, *J. Biol. Chem.* **262,** 12826 (1987).
222. E. J. Carrazana, K. B. Pasieka, and J. A. Majzoub, *Mol. Cell. Biol.* **8,** 2267 (1988).
223. D. A. Carter and D. Murphy, *J. Biol. Chem.* **264,** 6601 (1989).
224. D. Murphy and D. Carter, *Mol. Endocrinol.* **4,** 1051 (1990).
225. B. G. Robinson, D. M. Firm, W. J. Schwartz, and J. A. Majzoub, *Science* **241,** 342 (1988).
226. E. Dehlin, A. von Gabain, G. Alm, R. Dingelmaier, and O. Resnekov, *Mol. Cell. Biol.* **16,** 468 (1996).
227. B. L. Robberson, G. Cote, and S. M. Berget, *Mol. Cell. Biol.* **10,** 84 (1990).

Stimulation of Kinase Cascades by Growth Hormone: A Paradigm for Cytokine Signaling

TIMOTHY J. J. WOOD[*,‡] LARS-ARNE HALDOSÉN,[‡] DANIEL SLIVA,[‡] MICHAEL SUNDSTRÖM,[†] AND GUNNAR NORSTEDT[*,‡]

Departments of *Cell Biology and
†Structural Biology
Pharmacia and Upjohn
112 87 Stockholm, Sweden
‡Department of Medical Nutrition
Karolinska Institute
141 86 Huddinge, Sweden

I. The Cytokines	74
A. General Features	74
B. Expression and Effects of Growth Hormone	74
II. The Hematopoietic Cytokine Receptor Superfamily	76
A. General Features	76
B. Structure of the Growth Hormone Receptor	78
III. Growth Hormone Receptor Signal Transduction via Phosphorylation Cascades	80
A. Cellular Phosphorylation Pattern	80
B. Signaling via STATs	81
C. MAP Kinases	86
IV. The SPI 2.1 Promoter as a Model System for Cytokine Signaling	87
V. Conclusion	89
Note Added in Proof	91
References	91

Cells signal to each other via direct physical contact using so-called adhesion factors or via soluble signaling molecules, such as small peptides or peptide derivatives, polypeptides, or steroids. These may be produced either in endocrine glands and then secreted into the general circulation, or in small quantities close to the site of action. The message carried by a particular signaling molecule is usually transduced into the cell by a unique receptor molecule. On ligand binding, the receptor initiates a series of events leading to the stimulation or inhibition of cellular events such as enzyme catalysis, structural rearrangement, and expression of specific genes. Receptor molecules may be dissolved in the cyto- or nucleoplasm or bound to the cell membrane. Membrane-bound receptors often stimulate the activation of a signaling cascade of accessory transduction molecules, resulting in stimulation or inhibition of gene expression. This activation process commonly involves the phosphorylation of specific serine, threonine, or tyrosine residues of intracellular proteins. In the following discussion, we will attempt to give an overview of the kinase-dependent signaling pathways used by the cytokine family of molecules, using growth hormone signaling as a paradigm.

I. The Cytokines

A. General Features

The cytokine family of signaling molecules has grown in recent years to embrace a variety of molecules involved in regulation of cell differentiation, tissue development, and homeostasis (1). The family includes several interleukins and growth- and colony-stimulating factors, erythropoietin, the interferons, and several tissue necrosis factors, as well as prolactin and growth hormone. As a consequence of their role as major regulators of localized response to immunological stress, the effects elicited by cytokines are generally restricted to the site of production. Prolactin and growth hormone are exceptions to this rule, because they have both systemic and local effects.

The recent discovery that leptin shares homology with the cytokines adds a new dimension to the control of body homeostasis by this family of molecules. Most of the cytokines are pleiotropic, having multiple targets of action. In addition, many of these targets are shared between several cytokines, indicating a certain redundancy in cytokine signaling. Whether the cytokines have evolved from a single primordial molecule or have converged from several sources is not clear. It seems certain, however, that the prolactin–growth hormone family of molecules is derived from a common ancestor that existed at least 500 million years ago (2).

B. Expression and Effects of Growth Hormone

Growth hormone is an approximately 22-kDa single-domain protein consisting of four long α helices and three minihelices in the connecting loops. The two first helices are parallel to each other and antiparallel to the last two. The main site of growth hormone production is the pituitary somatotroph, which synthesizes and releases growth hormone in secretory pulses principally under the control of growth-hormone-releasing factor and somatostatin

(3). A variety of other factors, for example, the sex steroids, also influence the amplitude and frequency of growth hormone secretory pulses. This gives rise to a male and a female pattern of growth hormone release in a variety of species. Several independently regulated extrapituitary sources of growth hormone also exist. These include lymphocytes (4), placenta (5), and breast (6). Thus growth hormone can be classed as both a hormone having a central site of production and systemic effects and a cytokine having a number of dispersed sites of production and local effects.

Growth hormone is the major endocrine factor in the control of longitudinal growth. The effects of growth hormone can be categorized according to a number of criteria. Viewed from an endocrinological perspective, the effects may be either direct or mediated via the stimulation of production of somatomedins such as the insulin-like growth factors (IGFs). At the level of individual cells, growth hormone stimulation can be considered either to result in mitogenesis or cellular differentiation, or to cause changes in cell metabolism. At the level of regulation of gene transcription, the effects of growth hormone can be further subcategorized into immediate and delayed effects, depending on the time frame of activation and the requirement for ongoing protein synthesis.

Salmon and Daughaday suggested as early as 1957 that the effects of growth hormone on somatic growth are indirect and dependent on the secretion of so-called sulfation factors (IGF-I and IGF-II) (7). This theory subsequently evolved into the somatomedin or dual-effector theory (8). In its most modern incarnation, the dual-effector theory of growth hormone action has been modified to take account of the observation that IGF-I is produced locally as well as in the liver, as originally suggested, and that growth hormone also has direct effects on bone growth (9, 10).

Growth hormone stimulates DNA synthesis and mitogenesis in a variety of cell types *in vivo*, including skeletal muscle, liver, kidney, heart, and adipose tissue (11), and in the epiphysial growth plate, which in turn leads to bone growth (9). It also stimulates mitogenesis in cultured pancreatic beta cells (12), chondrocytes (13), and osteoblasts (14). The direct and indirect IGF-I-mediated effects of growth hormone on differentiation have been studied chiefly using adipocyte models, specifically the 3T3 preadipocyte cell line (8). Growth hormone induces the differentiation of preadipocytes to adipocytes, prechondrocytes to chondrocytes, and myoblasts to mulitnucleated muscle cells (8, 15). It also promotes differentiation in the rat mammary gland (16).

The metabolic effects of growth hormone can be divided into insulin-like and anti-insulin-like (11). Administration of growth hormone to hypophysectomised rats elicits an insulin-like effect: increased glucose transport and oxidation, and acceleration of glucose conversion to carbon dioxide, free fatty acid, and total lipid. After some hours, the animals become refractory to growth hormone, which reflects the normal state in which cells are continuously exposed to serum growth hormone. Administration of pharmacological doses of growth hormone to normal animals elicits an anti-insulin-like effect: decreased carbohydrate tolerance and hyperglycemia. The mechanism underlying the switch from insulin-like to anti-insulin-like effects is not clear, although it appears to operate postreceptor binding (17). Nor is it clear whether the insulin-like effects have any physiological relevance, although the pulsatile nature of growth hormone secretion suggests that this should not be ruled out.

Growth hormone regulates the transcription of a large number of genes, including those coding for lipoprotein lipase (18), the CCAAT/enhancer-binding protein α and δ transcription factors (19), the AP-1 site binding proteins c-fos and c-jun (20, 21), IGF-I (22), the prolactin receptor (23), a number of cytochrome P-450 enzymes (24), and two members of the serine protease inhibitor family (25, 26). As previously mentioned, the effects of growth hormone on gene expression can be divided into immediate and delayed, depending on the time frame of activation and the requirement for ongoing protein synthesis. The messenger RNA levels of most of the genes listed above increase within a few minutes of growth hormone treatment and are not affected by translation inhibitors such as cycloheximide. Such responses can be classified as immediate.

Using the 3T3-F442A preadipocyte cell line as a model for differentiation, it has been shown that growth hormone enhances the expression of the CCAAT/enhancer-binding protein (C/EBP) β transcription factor without altering messenger RNA levels by a mechanism involving mitogen-activated protein (MAP) kinase and Janus kinase (JAK) 2 (19). The growth-hormone-dependent activation of gene expression via C/EBP β belongs to the delayed group of responses because it requires protein synthesis and a period of time ranging from 30 min to several hours to reach a maximum. C/EBP binding sites are found in several growth-hormone-regulated gene promoters, including the serine protease inhibitor 2.1 gene promoter (26). The growth-hormone-dependent regulation of many genes *in vivo* combines both immediate and delayed responses.

II. The Hematopoietic Cytokine Receptor Superfamily

A. General Features

Considerable progress has been made in recent years toward understanding the molecular mechanisms used by the cytokines for signal transduction. The receptors for a large number of cytokines have been cloned and shown to be membrane-spanning glycoproteins with their N termini in the extracellular space. On the basis of their possession of certain structural similarities they can be grouped together into several families. Most of the cytokines belonging to the four-helix bundle family (including growth hormone) bind to receptors from the so-called hematopoietic cytokine/interferon receptor family (27–29). Several other families of cytokine receptors can also be hypothesized on the basis of sequence homologies. These include the kinase receptors typified by the epidermal growth factor receptor, the tissue necrosis factor-type receptors, and the interleukin-1-type receptors (1). As is the case for their ligands, it is unclear whether the cytokine receptors stem from a primitive ancestor, although this seems likely for the growth hormone and prolactin receptors.

The hematopoietic cytokine receptors are transmembrane glycoproteins made up of an extracellular N-terminal ligand-binding domain, a short transmembrane region, and a C-terminal intracellular domain. The extracellular domains of the hematopoietic cytokine/interferon receptor family contain two conserved domains, each of approximately 100 amino acid residues termed SD-100 domains (Fig. 1) (27). These SD-100 domains are preceded by proline residues and contain a conserved series of hydrophobic and hydrophilic residues with a topology best known as the immunoglobulin fold. Variants of this fold are found in a wide range of proteins, including blood clotting factors, extracellular matrix proteins, cell surface adhesion molecules, intracellular proteins containing src homology (SH) type 2 and type 3 domains, and various DNA binding motifs. The N-terminal SD-100 domain contains a pair of conserved cysteines. The hematopoietic cytokine receptors also possess a pair of conserved cysteines and a conserved WSXWS amino acid motif in the C-terminal SD-100 domain that are lacking in the interferon receptors.

The homologies between the intracellular domains of the hematopoietic cytokine receptors are much harder to define. However, two common features are a proline-rich "box 1" region, and a "box 2" region that does not contain any known consensus sequence but that plays a crucial role in signal transduction (30, 31). The receptors in this family have no intrinsic tyrosine kinase activity. In addition to binding to receptors immobilized in membranes, most cytokines also associate with soluble molecules called binding proteins (32). These are usually identical in sequence to the extracellular domain of the membrane-bound receptor and are derived either from alternative messenger RNA splicing or proteolytic cleavage. Their physiological role is unclear. In most cases they act as specific antagonists in cell culture studies but may enhance bioactivity in intact animals by prolonging the half-life of their ligand.

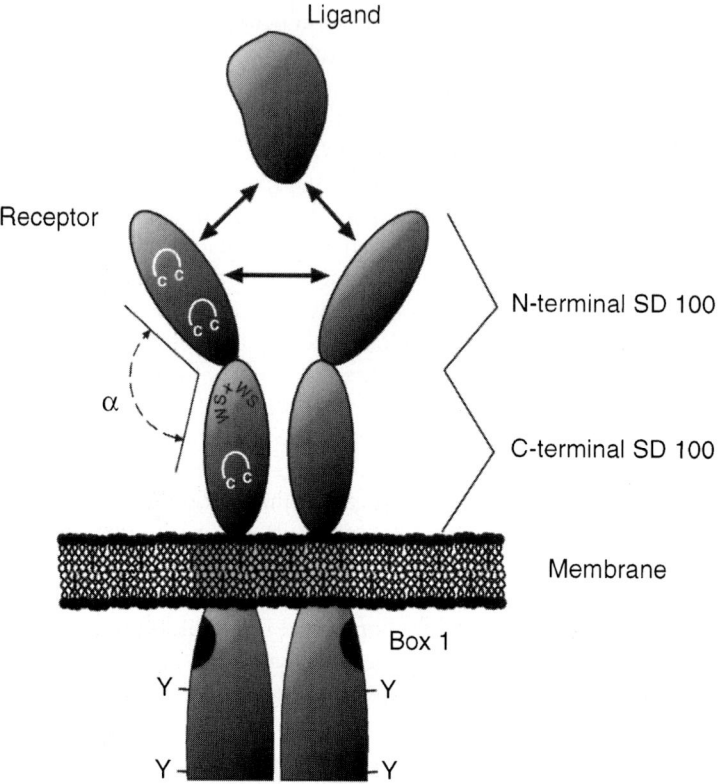

FIG. 1. A cytokine receptor model applicable to the growth hormone receptor. Model of the growth hormone receptor extracellular domain showing SD-100 domains, cysteine bridges, and the WSXWS domain. Binding of hormone is suggested to stabilize hinge angle α.

B. Structure of the Growth Hormone Receptor

The structure of the intracellular domain of the growth hormone receptor as a hormone-bound complex containing two receptor extracellular domains and one growth hormone molecule has been elucidated by X-ray crystallography (Fig. 2) (33). The receptor can be seen to contain two mutually perpendicular domains of approximately 100 amino acids. These are equivalent to the SD-100 motifs modeled by Bazan (27). Both the C-terminal and N-terminal SD-100 domains contain seven β strands that form a sandwich of two antiparallel β sheets, one with four strands and one with three. The two distinct domains are linked by a short four-residue segment. The N-terminal domain contains three disulfide bonds, one of which cross-links the two β sheets.

The crystal structure also shows that growth hormone possesses two unique receptor-binding sites. Site 1 has much higher affinity than site 2 for the receptor. Virtually the same receptor residues are involved in binding both site 1 and site 2. However, there is a clear element of asymmetry because the two receptor molecules are shifted from perfect twofold symmetry by about 10 Å with respect to each other. The area buried between growth hormone site 1 and the receptor (2700 Å^2) is considerably larger than that observed in the growth hormone site 2–receptor interaction (1500 Å^2). This difference is compensated for by the additional surface that is buried between the two C-terminal domains of the receptors (900 Å^2). Interestingly, the interaction between the receptors at the dimer interface is much more extensive than expected. Such close interaction (2–3 Å) is not observed in the interferon-γ receptor crystal structure (34). Receptor–receptor interaction appears to contribute significantly to the overall stability of the receptor extracellular domain–growth hormone complex.

Although strict sequence alignment with other receptors with WSXWS motifs indicates that the human growth hormone receptor WSXWS ranges from Tyr-222 to Ser-226 (YGEFS), structural studies performed in our laboratory suggest a more significant role for residues Lys-179, Trp-186, Arg-211, Phe-225, Arg-213, Tyr-222, and Lys-215. This motif might serve to stabilize the C-terminal SD-100 domain, which lacks intramolecular disulfide bridges. Because the receptor binding sites lie close to the "hinge" region between the two SD-100 domains, it is tempting to speculate that stabilization of the angle between the two domains may be one of the events leading to receptor activation (Fig. 1). Unfortunately, the structure of the free, unbound receptor is not yet available. Thus it is not known whether the receptor conformation changes following growth hormone binding (see Note Added in Proof).

Several lines of evidence, in addition to the growth hormone receptor extracellular domain crystal structure, indicate that ligand-driven receptor dimerization is the key event leading to transduction of cytokine signals across the cell membrane. A 2:1 complex between the human growth hormone extracellular domain and human growth-hormone is observed both in solution and following crystallization (35). A human growth hormone single-site mutant Gly-120 to Arg (G120R), deficient in receptor binding site 2, inhibits receptor dimerization *in vitro* and inhibits signal transduction *in vivo* (36–38). G120R human growth hormone crystallizes as a 1:1 complex with the extracellular domain of the human growth hormone receptor (38a).

The response to growth hormone in several model systems shows biphasic bell-shaped kinetics. Moderate levels of hormone stimulate signaling by inducing the formation of the active 2:1 receptor hormone complex whereas an excess of hormone leads to the formation of 1:1 receptor complexes, which inhibits signaling (39). Prolactin receptor antibodies mimic the action of prolactin on Nb2 pre-T lymphocytes by stimulating the formation of receptor dimers (40). Finally, insertion of a cysteine in the membrane-proximal region of the extracellular region of the erythropoietin receptor creates a constitutively dimerized and hence active receptor (41). The distribution of the growth hormone receptor on the cell surface is not clear, although there is some evidence for the presence of other membrane-bound receptors in specialized regions of the cell membrane. Thus it is a matter of speculation as to whether receptor monomers are brought together following ligand binding to form dimers by lateral diffusion, or whether preformed dimers are stabilized as a result of hormone binding. There is evidence that a cysteine bridge between receptors may stabilize the growth hormone receptor dimer following ligand binding (42).

The intracellular domain of the growth hormone receptor is still virtually a black box. Nothing is known about its three-dimensional structure. The only region with a defined role in signaling is the proline-rich box 1 region, which binds the cytoplasmic tyrosine kinase Janus kinase 2 (JAK 2). Several other molecules involved in cytokine signaling also bind to the growth hormone receptor. These include insulin receptor substrate 1 (IRS-1), Grb-2, and SHC (43, 44). The exact location of the binding sites for these proteins remains to be established.

The intracellular domain of the growth hormone receptor also contains 10 tyrosine residues. A role in activation of protein synthesis and lipogenesis has been ascribed to the two tyrosines closest to the membrane (45). However, the phosphorylation status of the other tyrosines following ligand binding to the receptor is unclear. Indeed, it is still not known which tyrosines are exposed to the cytoplasm. A potential role of the other tyrosines in activation of gene transcription is discussed in Section III,B.

III. Growth Hormone Receptor Signal Transduction via Phosphorylation Cascades

A. Cellular Phosphorylation Pattern

Ligand binding to cytokine receptors often induces the phosphorylation of intracellular proteins. The growth hormone receptor is no exception. Thus, following growth hormone treatment of cells, phosphorylated proteins with molecular masses 36, 40, 42, 120, and 95 kDa can be separated on an SDS-polyacrylamide gel. Experiments performed in several laboratories showed that these species correspond to the MAP kinases (36, 40, and 42 kDa) (46, 47), the growth hormone receptor itself, JAK 2 (both 120 kDa) (48–50), and members of the signal transducer and activator of transcription (STAT) family of transcription factors (95 kDa) (51–54). Curiously, the separation of the 120-kDa tyrosine-phosphorylated SDS-PAGE band into two components, the receptor and a kinase, was a result of the expression of an aberrant but functional growth hormone receptor in Chinese hamster ovary cells. This receptor has a molecular mass of only 84 kDa (55).

Because the growth hormone receptor does not possess intrinsic tyrosine kinase activity, it is logical to hypothesize the existence of an associated tyrosine kinase molecule. This kinase has been identified as JAK 2. Five unique Janus kinases have been identified to date; Janus kinases 1–3, tyk 2, and the *Drosophila hopscotch* gene product (56). Each of these contains a tyrosine kinase domain and a tyrosine kinase-like domain but all lack SH2 or SH3 domains, which are often found in tyrosine kinases. Mutational studies indicate that JAK 2 binds to the proline-rich box 1 region of the receptor (57, 58). By analogy with the tyrosine kinase receptor family, one may speculate that ligand-driven receptor dimerization brings two JAK 2 molecules into mutual proximity, thereby allowing trans-phosphorylation to occur. JAK 2 also phosphorylates a number of tyrosine residues in the receptor intracellular domain, paving the way for the binding of additional signaling molecules such as IRS-1, Grb-2, and SHC to the receptor (43, 44).

B. Signaling via STATs

Concurrent with studies aimed at identifying the proteins phosphorylated following growth hormone treatment of cells, other laboratories, using a combination of biochemical and genetic techniques, performed experiments aimed at identifying the key signaling components in the interferon-α, -β, and -γ signaling pathways. The discoveries made as a result of this work have been important in understanding signaling by the growth hormone receptor and its relatives.

The following model for the interferon-dependent transcriptional activation of a number of genes has been hypothesized (59). Hormone binding leads to activation of members of the Janus kinase family of receptor-associated protein tyrosine kinases (60, 61). This is followed by phosphorylation of both the kinase and the associated receptor. Activation of the Janus kinase molecule is followed by further tyrosine phosphorylation of cytosolic proteins, including members of the STAT family of DNA-binding transcription enhancers (62–68). Following their phosphorylation on tyrosine residues, STATs form homo- or heterodimers (69). STAT dimerization is accompanied by nuclear transport (70) and activation of gene transcription. This is mediated by the binding of the STAT dimer to palindromic response elements with the consensus TTCCCCCAA. These have been termed γ-activated sequences (GASs).

The discovery of the interferon–JAK–STAT signaling pathways led the way for an intense period of research, during which a large number of cytokines were shown to utilize similar pathways. Much of this was the result either of the observation of GAS-like elements in the promoters of cytokine-activated genes or the discovery of cytokine-activated phosphoproteins showing immunological similarities with STAT 1. Six unique STATs have been cloned: p91 and its splice variant p84 (termed STAT 1α and 1β, respectively) (62); p113 (STAT 2) (62); epidermal growth factor and interleukin-6-activated STAT 3 (acute-phase response factor) (71); STAT 4 (71); STAT 5 (mammary gland factor) (72); and interleukin-4-activated STAT 6 (IL-4 STAT) (73). We have also observed STAT-like DNA-binding activity in sf9 insect cells. The STATs contain both SH2 and SH3 domains in addition to their DNA-binding domains. STAT dimerization appears to be the result of SH2 domain–phosphotyrosine binding.

The first indication of a close relationship between the signaling pathways of growth hormone and other cytokines came from the identification of a 9-bp

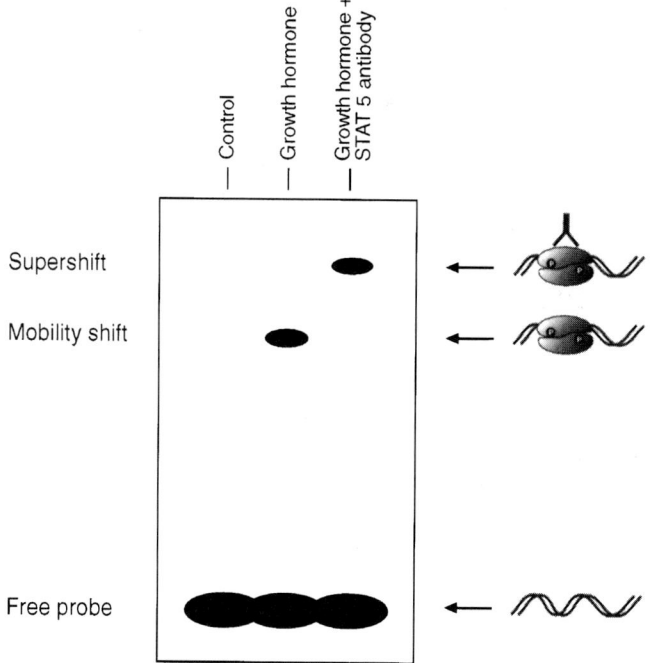

FIG. 3. Gel mobility-shift assay. Following serum starvation, cells are hormone treated and then protein extracts are prepared by salt extraction of nuclei, isolated by osmotic shock and differential centrifugation. Nuclear proteins are then incubated with radioactively labeled oligonucleotide DNA and separated by gel electrophoresis. Binding of nuclear proteins decreases the mobility (mobility shift) of the labeled oligo. This mobility may be further decreased (supershift) if specific antibodies for DNA-binding proteins are included in the incubation.

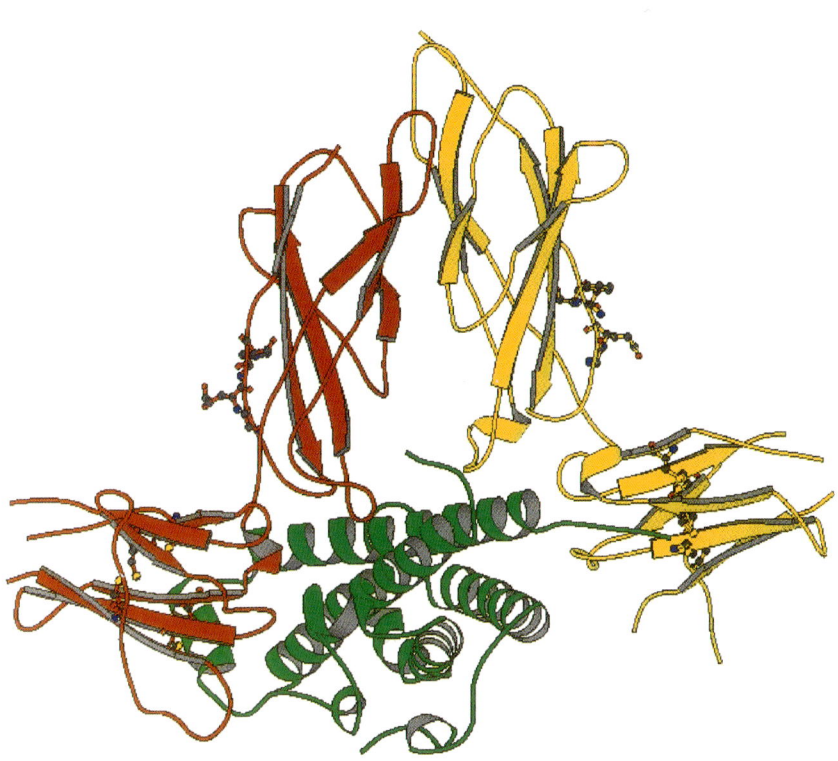

Fig. 2. Structure of the growth hormone receptor. Ribbon diagram of the extracellular domain of the human growth hormone receptor/human growth hormone 2:1 complex. The "WSXWS" box and cysteine bridges are shown in ball and stick format. (Data derived from Ref. *38a*.)

growth-hormone-responsive DNA element in the serine protease inhibitor 2.1 and 2.2 promoters (74). Using the gel electrophoresis mobility-shift assay (Fig. 3), this element was found to bind nuclear proteins present in extracts prepared from growth-hormone-treated cultured cells. It was also found to confer growth hormone responsiveness on the expression of adjacent cDNAs in reporter gene experiments (Fig. 4). Because the element has significant homology with members of the GAS family of STAT-binding DNA elements, it

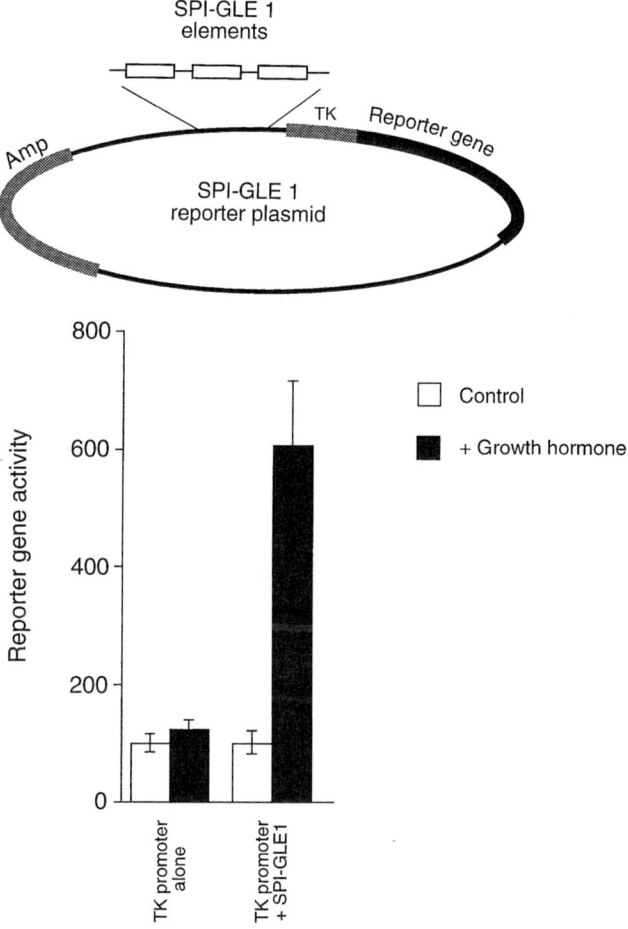

FIG. 4. Reporter gene assay. Cells expressing appropriate combinations of signaling molecules are transfected with a reporter gene containing a reporter cDNA, such as the luciferase or chloramphenicol-acetyltransferase cDNA, downstream of the weak thymidine kinase (TK) promoter, with or without the SPI-GLE 1 enhancer element. Following growth hormone treatment, extracts are prepared by detergent lysis and reporter activity is measured.

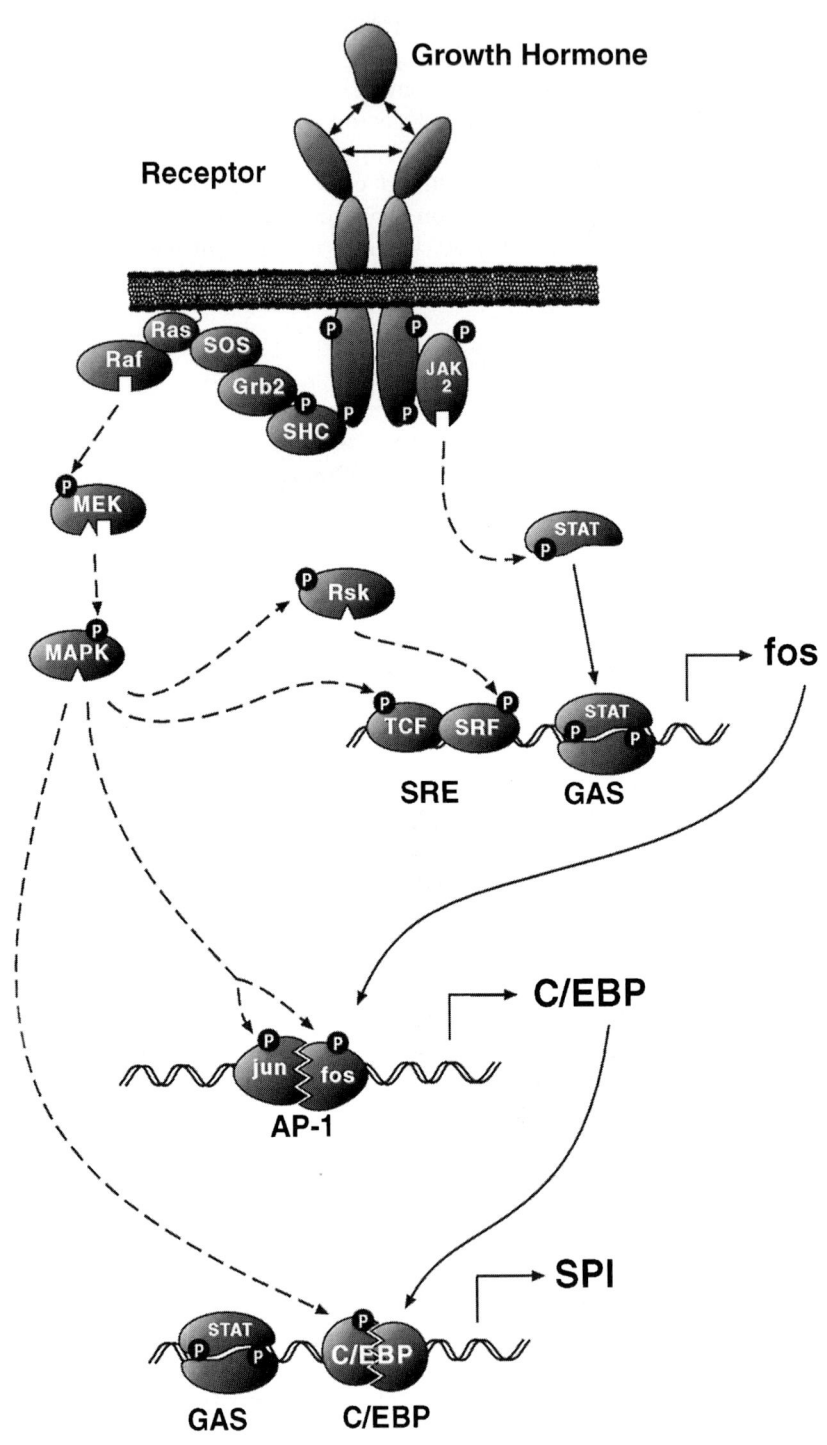

was designated the serine protease inhibitor (SPI) GAS-like element 1 (SPI-GLE 1).

The time frame of appearance and the resistance to inhibition by cycloheximide treatment of the SPI-GLE 1 DNA-binding activity were consistent with the activation of a preexisting but latent transcription factor. In subsequent studies the SPI-GLE 1 binding complex was shown to contain STAT 5, which had recently been purified and cloned on the basis of its binding to DNA elements in the prolactin-responsive β-casein promoter (*51, 72*). The current state of knowledge concerning the transduction of growth hormone signals from the cell surface to the SPI promoter is summarized in Fig. 5. Binding of growth hormone to its receptor results in the formation of a receptor dimer. This causes the activation of JAK 2, which catalyzes the phosphorylation of STAT 5 on tyrosine residue 694 (*75*). The phosphorylated STAT 5 then forms a dimer and migrates to the nucleus, where it binds specific DNA response elements, thereby stimulating gene transcription.

There are a number of gaps in our understanding of STAT-mediated cytokine signaling. Most of these relate either to the mechanisms determining signaling specificity or the mechanism by which activated STAT signals are transported to the nucleus. That a large family of cytokines signal via a small family of STATs and an even smaller family of JAKs is paradoxical. However, several layers of specificity determination can be envisaged. Clearly, tissue specificity in the expression of members of the cytokine receptor and the STAT gene families could lead to tissue-specific responses. In addition, direct interactions between specific phosphorylated receptor tyrosines and STATs have been described. There is some controversy as to whether this applies to STAT 5. A wide variety of receptors (*76–80*), including severely truncated growth hormone receptors lacking tyrosines, can activate STAT 5 (*81*), although the presence of phosphorylated receptor tyrosines appears to en-

---

Fig. 5. Growth hormone signaling via phosphorylation cascades. Binding of the growth hormone to its receptor results in the formation of a receptor dimer. This causes the activation of JAK 2, which catalyzes the phosphorylation of STATs. The phosphorylated STAT then forms a dimer and migrates to the nucleus, where it binds GAS-like response elements, thereby stimulating transcription of the SPI 2.1/2.2 and c-*fos* genes. Growth hormone induces phosphorylation of members of the ERK subtype of MAP threonine/serine kinases via the formation of a complex between growth hormone, JAK 2, SHC, and Grb-2, and probably Raf, ras, SOS, and MEK. This eventually leads to stimulation of a complex between ternary complex factor (TCF) and serum response factor (SRF) on the c-*fos* promoter serum response element. This enhances expression of c-*fos*. The substrates of the MAP kinases also include the transcription factors c-jun (JNK/SAPK), c-fos (ERK), and C/EBP. c-jun and c-fos can bind as a heterodimer to the AP-1 site of the C/EBP promoter, thereby stimulating its transcription. C/EBP in turn stimulates transcription of the SPI 2.1/2.2 genes. Stimulation of phosphorylation is denoted with dashed lines and stimulation of expression or movement of molecules is denoted with solid lines. Square "kinase active sites" are tyrosine specific; triangular "active sites" are serine/threonine specific.

hance the efficiency of STAT activation (*82–84*). We have been unable to show a direct association between the ligand-bound growth hormone receptor and STAT 5 in extracts from cultured cells, although an association between STAT 5 and activated JAK 2 can easily be demonstrated. Phosphorylated tyrosines may play an accessory role in recruiting STAT 5 to the membrane but not be absolutely required for JAK 2-catalyzed phosphorylation of STAT 5.

The role of cross-talk with other signaling pathways, notably those leading to the serine/threonine phosphorylation of STATs, has recently emerged (*68, 85, 86*). For example, STATS 1 and 3 have been shown to contain potential targets for MAP serine/threonine kinases (*86*). Finally, a number of candidate molecules capable of associating with STATs in the transcription complex have been described. These include YY 1 (*87, 88*), Jun (*89*), and Sp 1 (*90*). The mechanism by which STATs migrate to the nucleus is also unknown, although the rapidity with which this occurs suggests a specific mechanism. Because the growth hormone receptor moves rapidly to the nucleus following ligand binding (*91*), it is tempting to speculate that this mechanism involves the receptor.

C. MAP Kinases

The mitogen-activated protein kinases comprise a large family of serine/threonine kinases expressed in yeast, insect, and mammalian cells and are characterized by their activation by dual-specificity kinases on conserved TXY amino acid motifs. The MAP kinases can be divided into three subtypes on the basis of their TXY motifs: the extracellular signal-regulated kinase (ERK) subtype, which contains TEY motifs; the c-Jun NH_2-terminal kinase/serum response factor-associated protein kinase (JNK/SAPK) subtype, which contains TPY motifs; and the p38 kinase subtype, which contains TGY motifs. The MAP kinases are downstream of a cascade of kinases activated by mitogens such as phorbol esters, epidermal growth factor, and growth hormone. This cascade includes the MAP kinase/ERK kinase kinases (MEKs), and a variety of MAP kinase kinase kinases, such as Raf.

The substrates of the MAP kinases include the S6 kinase and p90^{rsk}, PLA_2, and the transcription factors c-jun (JNK/SAPK), c-fos (ERK), and C/EBP. Growth hormone has been shown to induce phosphorylation of members of the ERK subtype of MAP kinases (*46, 47, 92*). This seems to occur via the formation of a complex between growth hormone receptor, JAK 2, SHC, and Grb-2 (*44*) and probably Raf, ras, and SOS. ERK activation eventually leads to stimulation of a ternary complex on the c-fos promoter serum response element, thereby enhancing expression of c-fos. Growth hormone has been shown to induce phosphorylation of IRS-1 (*93*), which lies on another route potentially leading to MAP kinase activation. A direct role for JAK 2 in activation of MAP kinases has been described (94).

IV. The SPI 2.1 Promoter as a Model System for Cytokine Signaling

The serine protease inhibitors comprise a large family of molecules involved in inflammatory responses, blood clotting, and complement activation. The gene expression of two of the serine protease inhibitors, SPI 2.1 and 2.2, is tightly controlled by growth hormone in rat liver. In the normal animal both inhibitors are constitutively expressed. However, expression is repressed by hypophysectomy and inflammation. Expression levels in the hypophysectomized animal can be partially restored by a combination of growth hormone and dexamethasone treatment. A related SPI, SPI 2.3, is poorly expressed in normal animals and is induced by inflammation. The SPI 2.1 promoter contains two independent growth-hormone-responsive regions (Fig. 6): one proximal to the transcription start site between positions -41 and $+8$ (growth-hormone-responsive element I, GHRE I) and another more distal between positions -175 and -114 (growth-hormone-responsive element II, GHRE II) (26, 95). A glucocorticoid-responsive element is located between GHREs I and II.

The mechanism of growth hormone regulation of SPI 2.1 expression by the proximal promoter region is unknown, although the identification of several C/EBP binding sites suggests that this transcription factor family may be involved (26, 96). C/EBP gene expression is regulated by growth hormone via a mechanism involving JAK 2 and MAP kinase (19). As described previously, the distal growth-hormone-responsive promoter region contains a STAT 5 binding motif, SPI-GLE 1. A simplified model of regulation of the SPI 2.1 promoter by growth hormone is presented in Fig. 5.

Visual examination of the sequence surrounding SPI-GLE 1 and comparison with the SPI 2.2 and 2.3 promoter reveal a number of interesting features (Fig. 6). Immediately upstream of the SPI-GLE 1 lies a second GAS-like element termed SPI-GLE 2 (74). A sequence similar to the YY1 transcription factor binding consensus, CCATnT, separates the two GAS-like elements. Neither the YY1 element or SPI-GLE 2 binds nuclear proteins from extracts of cultured cells (74), nor does SPI-GLE 2 seem to be required for the stimulation of transcription of plasmid reporter genes mediated by the SPI GHRE II. However, it has been shown that SPI-GLE 2 weakly binds protein from rat liver (97), and the fact that this element is disrupted by a 42-bp insertion in the GH-refractory SPI 2.3 promoter indicates that it may have a

```
                                                                        -194
SPI 2.3  CACTTTCCACATTGACTCTGAACCATTTGGTAAACCAC.GACATGCCCAGGAG
SPI 2.2  CACTTTCCACATTGACTTTGAGCCACTCAATAAATAAAAGATGTGTTCAAGAG
SPI 2.1  CACTTTCCTCATTGACTTTGA.CCACTCAATAAATAAAAGGTGTGCTCAGGAG

                                  SPI 2.3 insertion              -143
SPI 2.3  ATCGGTAGGCTTGTGTAGAGCAGAGATGGGAATTTTCCCATCAGAAAGTAGGG
SPI 2.2  ATCAGTACGCTT........................................
SPI 2.1  ATCAGTACGCTT........................................

         SPI-GLE 2       SPI-GLE 1                                -92
SPI 2.3  ACTATGAGTCCATGTTCCCAGAAATCATCCCGTCTGCCCATTTCCAGTCTAAA
SPI 2.2  .CTACTAATCCATGTTCTGAGAAATCATCCCGTCTGCCCATCTGTAATCTGAA
SPI 2.1  .CTACTAATCCATGTTCTGAGAAATCATCCAGTCTGCCCATATGTAATCTGAA
                             GHRE-II
                                                                 -41
SPI 2.3  CACAGAGCACAGCTTGTCCATGCCAGCATTTCCTAAGAGGAGGGAGGAGCCCT
SPI 2.2  CACAAAGCACAG.TTATCCAAGGCAGTATTTCCTAAGAGGAGGGAGGAGCCCT
SPI 2.1  CACAAAGCACAGGTTGTCCGAGGCAACATTTCCTAAGAGGAGGGAGGAGCCTT

                 GRE                                         +1
SPI 2.3  TGGTGGGGTTAAATAGGCATCCCTGTGCACTGGGGATAACA
SPI 2.2  TGGTGGGGATAAATAGCCATCGCTGTGCACAAGAGAGCAAA
SPI 2.1  TGGTGTAGATAAATAGGCATCCCTGTGCACTGGGGATAACA
                            GHRE-I
```

FIG. 6. Lineup of SPI 2.1, 2.2, and 2.3 promoters. C/EBP sites [consensus T(T/G)XXG-NAA/T/G] are shown underlined. SPI-GAS-like elements are shown in black boxes. Growth-hormone-responsive elements I and II and glucocorticoid response element are shown in grey boxes.

role *in vivo* in growth hormone regulation. This is interesting in light of the recent observation that both the prolactin-regulated β-casein (87, 88) and the interferon-γ-regulated *mig* gene promoters (98) contain tandem repeated GAS-like DNA elements that mediate regulation of transcription.

The two GAS-like elements in the β-casein promoter are separated by a binding site for the transcription factor YY1 (87, 88). Binding of STAT 5 appears to enhance transcription of the β-casein gene at least in part by displacing YY1. In the SPI 2.3 promoter SPI-GLE 1 is mutated from TTCT-GAGAA to TTCCCAGAA. We have shown that the former mediates trans-activation by STAT 5 but not by STATs 1 and 3. The SPI 2.3 GAS-like element, which is more similar to the GAS consensus, should respond to a broader range of STATs compared to SPI-GLE 1. This is in fact the case (99) and explains the activation of SPI 2.3 expression by interleukin-6 and interferon-γ (100). It does not, however, explain why growth hormone does not enhance SPI 2.3 expression.

V. Conclusion

This review has been concerned mainly with growth-hormone-dependent regulation of gene expression via phosphorylation cascades and the properties of the growth hormone receptor molecule that facilitate signaling via such pathways. Growth hormone signals are also transduced by other pathways involving calcium ions (84, 101), phospholipase C (102, 103), phospholipase A_2 (104), G-proteins (103), and protein kinase C (21). These are likely to integrate together with and modulate the JAK 2/STAT and MAP kinase pathways.

Obviously, regulation of receptor expression and hormone presentation is the first level at which specific modulation of growth hormone signaling can take place. Tissue-specific and temporal patterns of receptor expression determine the ability of a cell to respond to a particular cytokine whereas the levels of soluble cytokine-binding protein determine the effective serum ligand concentration. Regulation of receptor internalization may also be important. JAK 2 appears to be expressed in most cell types. However, it is possible that it is regulated by posttranslational events such as phosphorylation.

One can envisage three levels of modulation of STAT activation. Again, tissue-specific and temporal patterns of expression may play a role. In addition, posttranslational modifications such as serine phosphorylation may modulate STAT activity. The role of MAP kinase in modulation of STATs has already been mentioned in this context.

Finally, interaction with other proteins in the transcription complex may play an important role in STAT signaling. Interactions with YY1, Jun (89), and Sp 1 have already been described. Interactions with other transcription factors such as the steroid receptors may also exist. To date little is known about the mechanisms that bring about the deactivation of STATs. Thus it is not clear whether specific phosphatases catalyze the rapid dephosphorylation of STATs observed *in vivo* or whether this is the result of "housekeeping" phosphatase activity. The kinetics of STAT activation and the concomitant stimulation of gene expression by STATs are also poorly described, although it has been reported that interleukin-6 and interferon-γ activate DNA binding of a complex containing STAT 3 with different kinetics (105).

In spite of, or perhaps because of, the intensity with which laboratories around the world have struggled to unravel the molecular mechanisms facilitating transduction of cytokine signals from the membrane to the nucleus, relatively few STAT-activated genes have been identified. This is best exemplified by the fact that the most fully understood growth-hormone-regulated promoter is present in the SPI 2.1 gene, which codes for a protein of unknown physiological function.

A role for STAT 5 in sex-specific regulation of hepatic genes has been suggested following the observation that hepatic STAT 5 is activated by intermittent but not continuous growth hormone treatment of hypophysectomized rats (106). The regulation of hepatic expression of the cytochrome P-450 enzymes in rats is dependent on the sex-specific pattern of growth hormone secretion. Thus the pulsatile male-specific pattern of secretion stimulates expression of cytochrome P-450 2C11, whereas the continuous, female-specific pattern stimulates expression of cytochrome P-450s 2C12 and 2C7 and represses the expression of cytochrome P-450s 2C11 and 2C13 (24). Because stimulation of cytochrome P-450 2C11 expression in hepatocytes usually requires 1 to 2 days of exposure to hormone, whereas STAT 5 is rapidly and transiently activated, it seems likely that the effect of STAT 5 on P-450 2C11 expression may be indirect.

Another well-characterized growth-hormone-regulated gene containing a STAT response element is the c-fos transcription factor gene (107). The identification of additional growth hormone/STAT-regulated genes is eagerly awaited.

Most of the cytokines have been shown to signal via members of the STAT transcription factor family. In addition, most of the cytokines also activate MAP kinases. Thus despite its historical position as a classical systemic hormone, growth hormone shares many properties with the locally acting cytokines, both in terms of its nonpituitary secretory profile and, as has been described here, its signal transduction. As previously mentioned it is not clear whether the cytokines and their receptors have converged or diverged through evolution. Nor is it certain that the cytokines have acquired the ability to activate shared signal transduction pathways or whether this stems from a primeval signaling system. The signaling cascade initiated by the growth hormone receptor is currently one of the most fully understood of the cytokine signaling systems and thus seems likely to lead our way to a greater understanding of this large and complex branch of biology.

Acknowledgments

This work was funded by the Swedish Medical Research Council (Grant Number 8556), the Swedish Cancer Society (Grant Number 3020-B94-02XBB), the Swedish Medical Association, the Czech Grant Agency (Grant Number 505/95/1134), and Pharmacia and Upjohn.

NOTE ADDED IN PROOF: The structure of the free receptor extracellular region has now been solved by X-ray crystallography and as predicted is significantly different to the ligand bound receptor extracellular region (M. Sundström, B. Nilsson, M. Hartmanis, and G. Norstedt, *Nature Structural Biology*, in press).

REFERENCES

1. N. A. Nicola, *in* "Guidebook to Cytokines and their Receptors" (N. A. Nicola, ed.), p. 1. Sambrook and Tooze, Oxford, 1994.
2. C. S. Nicoll and R. A. Baldocchi, *in* "Progress in Endocrinology" (H. Imura, ed.), p. 1379. Elsevier, Amsterdam, 1988.
3. J. Devesa, L. Lima, and J. A. F. Tresguerres, *Trends Endocrinol. Metab.* **3,** 175 (1992).
4. D. A. Weigent, J. D. Baxter, W. E. Wear, L. R. Smith, K. L. Bost, and J. E. Blalock, *FASEB J.* **2,** 2812 (1988).
5. F. Frankenne, J. Closset, F. Gomez, M. L. Scippo, J. Smal, and G. Hennen, *J. Clin. Endocrinol. Metab.* **66,** 1171 (1988).
6. P. J. Selman, J. A. Mol, G. R. Rutteman, E. Van Garderen, and A. Rijnberk, *Endocrinology* **134,** 287 (1994).
7. W. D. Salmon and W. H. Daughaday, *J. Lab. Clin. Med.* **49,** 825 (1957).
8. H. Green, M. Morikawa, and T. A. Nixon, *Differentiation* **29,** 195 (1985).
9. O. G. P. Isaksson, J.-O. Jansson, and I. A. M. Gause, *Science* **216,** 1237 (1982).
10. O. G. Isaksson, C. Ohlsson, A. Nilsson, J. Isgaard, and A. Lindahl, *Pediatr. Nephrol.* **5,** 451 (1991).
11. O. G. Isaksson, S. Eden, and J. O. Jansson, *Annu. Rev. Physiol.* **47,** 483 (1985).
12. N. Billestrup and J. H. Nielsen, *Endocrinology* **129,** 883 (1991).
13. K. Madsen, U. Friberg, P. Roos, S. Eden, and O. Isaksson, *Nature (London)* **304,** 545 (1983).
14. R. Barnard, K. W. Ng, T. J. Martin, and M. J. Waters, *Endocrinology* **128,** 1459 (1991).
15. H. Green and B. T. Nixon, *Proc. Natl. Acad. Sci. U.S.A.* **81,** 3429 (1984).
16. M. Feldman, W. Ruan, B. C. Cunningham, J. A. Wells, and D. L. Kleinberg, *Endocrinology* **133,** 1602 (1993).
17. G. Grichting, L. K. Levy, and H. M. Goodman, *Endocrinology* **113,** 1111 (1983).
18. S. M. Francis, S. Enerback, C. Moller, B. Enberg, and G. Norstedt, *Mol. Endocrinol.* **7,** 972 (1993).
19. R. W. Clarkson, C. M. Chen, S. Harrison, C. Wells, G. E. Muscat, and M. J. Waters, *Mol. Endocrinol.* **9,** 108 (1995).
20. V. N. Sumantran, M. L. Tsai, and J. Schwartz, *Endocrinology* **130,** 2016 (1992).
21. M. C. Slootweg, R. P. de-Groot, M. P. Herrmann-Erlee, I. Koornneef, W. Kruijer, and Y. M. Kramer, *J. Mol. Endocrinol.* **6,** 179 (1991).
22. J. Isgaard, C. Moller, O. G. Isaksson, A. Nilsson, L. S. Mathews, and G. Norstedt, *Endocrinology* **122,** 1515 (1988).
23. J. A. Robertson, L. A. Haldosen, T. J. Wood, M. K. Steed, and J. A. Gustafsson, *Mol. Endocrinol.* **4,** 1235 (1990).
24. A. Mode, *J. Reprod. Fertil. (Suppl.)* **46,** 77 (1993).
25. A. Le-Cam, G. Pages, P. Auberger, G. Le-Cam, P. Leopold, R. Benarous, and N. Glaichenhaus, *EMBO J.* **6,** 1225 (1987).
26. L. Paquereau, M. J. Vilarem, V. Rossi, J. F. Rouayrenc, and A. Le-Cam, *Eur. J. Biochem.* **209,** 1053 (1992).
27. J. F. Bazan, *Proc. Natl. Acad. Sci. U.S.A.* **87,** 6934 (1990).

28. D. Cosman, S. D. Lyman, R. L. Idzerda, M. P. Beckmann, L. S. Park, R. G. Goodwin, and C. J. March, *Trends Biochem. Sci.* **15,** 265 (1990).
29. E. Thoreau, B. Petridou, P. A. Kelly, J. Djiane, and J. P. Mornon, *FEBS Lett.* **282,** 26 (1991).
30. M. Murakami, M. Narazaki, M. Hibi, H. Yawata, K. Yasukawa, M. Hamaguchi, T. Taga, and T. Kishimoto, *Proc. Natl. Acad. Sci. U.S.A.* **88,** 11349 (1991).
31. R. Fukunaga, E. Ishizaka-Ikeda, C. X. Pan, Y. Seto, and S. Nagata, *EMBO J.* **10,** 2855 (1991).
32. S. Rose-John and P. C. Heinrich, *Biochem. J.* **300,** 281 (1994).
33. A. M. de-Vos, M. Ultsch, and A. A. Kossiakoff, *Science* **255,** 306 (1992).
34. M. R. Walter, W. T. Windsor, T. L. Nagabhushan, D. L. Lundell, C. A. Lunn, P. J. Zauodny, and S. K. Narula, *Nature (London)* **376,** 230 (1995).
35. B. C. Cunningham, M. Ultsch, A. M. De-Vos, M. G. Mulkerrin, K. R. Clauser, and J. A. Wells, *Science* **254,** 821 (1991).
36. M. Ultsch and A. M. de-Vos, *J. Mol. Biol.* **231,** 1133 (1993).
37. M. T. Dattani, P. C. Hindmarsh, C. G. Brook, I. C. Robinson, J. J. Kopchick, and N. J. Marshall, *J. Biol. Chem.* **270,** 9222 (1995).
38. W. Y. Chen, N. Y. Chen, J. Yun, T. E. Wagner, and J. J. Kopchick, *J. Biol. Chem.* **269,** 15892 (1994).
38a. M. Sundström, T. Lundquist, J. Rödin, L. B. Siebel, D. Millisan, and S. Norstedt, *J. Biol. Chem.* (in press) 1996.
39. M. M. Ilondo, A. B. Damholt, B. A. Cunningham, J. A. Wells, P. De-Meyts, and R. M. Shymko, *Endocrinology* **134,** 2397 (1994).
40. H. Rui, J. J. Lebrun, R. A. Kirken, P. A. Kelly, and W. L. Farrar, *Endocrinology* **135,** 1299 (1994).
41. S. S. Watowich, D. J. Hilton, and H. F. Lodish, *Mol. Cell. Biol.* **14,** 3535 (1994).
42. S. J. Frank, G. Gilliland, and C. Van-Epps, *Endocrinology* **135,** 148 (1994).
43. L. S. Argetsinger, G. W. Hsu, M. Myers, Jr., N. Billestrup, M. F. White, and C. Carter-Su, *J. Biol. Chem.* **270,** 14685 (1995).
44. J. VanderKuur, G. Allevato, N. Billestrup, G. Norstedt, and C. Carter-Su, *J. Biol. Chem.* **270,** 7587 (1995).
45. P. E. Lobie, G. Allevato, J. H. Nielsen, G. Norstedt, and N. Billestrup, *J. Biol. Chem.* **270,** 21745 (1995).
46. C. Moller, A. Hansson, B. Enberg, P. E. Lobie, and G. Norstedt, *J. Biol. Chem.* **267,** 23403 (1992).
47. G. S. Campbell, L. Pang, T. Miyasaka, A. R. Saltiel, and C. Carter-Su, *J. Biol. Chem.* **267,** 6074 (1992).
48. C. M. Foster, J. A. Shafer, F. W. Rozsa, X. Y. Wang, S. D. Lewis, D. A. Renken, J. E. Natale, J. Schwartz, and C. Carter-Su, *Biochemistry* **27,** 326 (1988).
49. G. S. Campbell, L. J. Christian, and C. Carter-Su, *J. Biol. Chem.* **268,** 7427 (1993).
50. L. S. Argetsinger, G. S. Campbell, X. Yang, B. A. Witthuhn, O. Silvennoinen, J. N. Ihle, and C. Carter-Su, *Cell* **74,** 237 (1993).
51. T. J. Wood, D. Sliva, P. E. Lobie, T. J. Pircher, F. Gouilleux, H. Wakao, J. A. Gustafsson, B. Groner, G. Norstedt, and L. A. Haldosen, *J. Biol. Chem.* **270,** 9448 (1995).
52. G. S. Campbell, D. J. Meyer, R. Raz, D. E. Levy, J. Schwartz, and C. Carter-Su, *J. Biol. Chem.* **270,** 3974 (1995).
53. X. Wang, B. Xu, S. C. Souza, and J. J. Kopchick, *Proc. Natl. Acad. Sci. U.S.A.* **91,** 1391 (1994).
54. A. M. Gronowski, Z. Zhong, Z. Wen, M. J. Thomas, J. Darnell, Jr., and P. Rotwein, *Mol. Endocrinol.* **9,** 171 (1995).
55. X. Wang, C. Moller, G. Norstedt, and C. Carter-Su, *J. Biol. Chem.* **268,** 3573 (1993).
56. J. N. Ihle, B. A. Witthuhn, F. W. Quelle, K. Yamamoto, W. E. Thierfelder, B. Kreider, and O. Silvennoinen, *Trends Biochem. Sci.* **19,** 222 (1994).

57. S. J. Frank, G. Gilliland, A. S. Kraft, and C. S. Arnold, *Endocrinology* **135,** 2228 (1994).
58. A. Sotiropoulos, M. Perrot-Applanat, H. Dinerstein, A. Pallier, M. C. Postel-Vinay, J. Finidori, and P. A. Kelly, *Endocrinology* **135,** 1292 (1994).
59. J. Darnell, Jr., I. M. Kerr, and G. R. Stark, *Science* **264,** 1415 (1994).
60. M. Muller, J. Briscoe, C. Laxton, D. Guschin, A. Ziemiecki, O. Silvennoinen, A. G. Harpur, G. Barbieri, B. A. Witthuhn, C. Schindler, and J. N. Ihle, *Nature (London)* **366,** 129 (1993).
61. D. Watling, D. Guschin, M. Muller, O. Silvennoinen, B. A. Witthuhn, F. W. Quelle, N. C. Rogers, C. Schindler, G. R. Stark, J. N. Ihle, and I. M. Kerr, *Nature (London)* **366,** 166 (1993).
62. X. Y. Fu, C. Schindler, T. Improta, R. Aebersold, and J. Darnell, Jr., *Proc. Natl. Acad. Sci. U.S.A.* **89,** 7840 (1992).
63. M. Muller, C. Laxton, J. Briscoe, C. Schindler, T. Improta, J. Darnell, Jr., G. R. Stark, and I. M. Kerr, *EMBO J.* **12,** 4221 (1993).
64. C. Schindler, X. Y. Fu, T. Improta, R. Aebersold, and J. Darnell, Jr., *Proc. Natl. Acad. Sci. U.S.A.* **89,** 7836 (1992).
65. C. Schindler, K. Shuai, V. R. Prezioso, and J. Darnell, Jr., *Science* **257,** 809 (1992).
66. K. Shuai, C. Schindler, V. R. Prezioso, and J. Darnell, Jr., *Science* **258,** 1808 (1992).
67. K. Shuai, G. R. Stark, I. M. Kerr, and J. Darnell, Jr., *Science* **261,** 1744 (1993).
68. X. Zhang, J. Blenis, H. C. Li, C. Schindler, and S. Chen-Kiang, *Science* **267,** 1990 (1995).
69. K. Shuai, C. M. Horvath, L. H. Huang, S. A. Qureshi, D. Cowburn, and J. Darnell, Jr., *Cell* **76,** 821 (1994).
70. T. Decker, D. J. Lew, J. Mirkovitch, and J. Darnell, Jr., *EMBO J.* **10,** 27 (1991).
71. Z. Zhong, Z. Wen, and J. Darnell, Jr., *Proc. Natl. Acad. Sci. U.S.A.* **91,** 4806 (1994).
72. H. Wakao, F. Gouilleux, and B. Groner, *EMBO J.* **13,** 2182 (1994).
73. J. Hou, U. Schindler, W. J. Henzel, T. C. Ho, M. Brasseur, and S. L. McKnight, *Science* **265,** 1701 (1994).
74. D. Sliva, T. J. Wood, C. Schindler, P. E. Lobie, and G. Norstedt, *J. Biol. Chem.* **269,** 26208 (1994).
75. F. Gouilleux, H. Wakao, M. Mundt, and B. Groner, *EMBO J.* **13,** 4361 (1994).
76. F. Gouilleux, D. Moritz, M. Humar, R. Moriggl, S. Berchtold, and B. Groner, *Endocrinology* **136,** 5700 (1995).
77. C. Pallard, F. Gouilleux, M. Charon, B. Groner, S. Gisselbrecht, and I. Dusanter-Fourt, *J. Biol. Chem.* **270,** 15942 (1995).
78. A. L. Mui, H. Wakao, N. Harada, A. M. O'Farrell, and A. Miyajima, *J. Leukoc. Biol.* **57,** 799 (1995).
79. C. Pallard, F. Gouilleux, L. Benit, L. Cocault, M. Souyri, D. Levy, B. Groner, S. Gisselbrecht, and I. Dusanter-Fourt, *EMBO J.* **14,** 2847 (1995).
80. F. Gouilleux, C. Pallard, I. Dusanter-Fourt, H. Wakao, L. A. Haldosen, G. Norstedt, D. Levy, and B. Groner, *EMBO J.* **14,** 2005 (1995).
81. Y. D. Wang, K. Wong, and W. I. Wood, *Mol. Endocrinol.* **9,** 303 (1995).
82. L. Goujon, G. Allevato, G. Simonin, L. Paquereau, A. Le-Cam, J. Clark, J. H. Nielsen, J. Djiane, M. C. Postel-Vinay, M. Edery, and P. A. Kelly, *Proc. Natl. Acad. Sci. U.S.A.* **91,** 957 (1994).
83. A. Sotiropoulos, S. Moutoussamy, N. Binart, P. A. Kelly, and J. Finidori, *FEBS Lett.* **369,** 169 (1995).
84. N. Billestrup, P. Bouchelouche, G. Allevato, M. Ilondo, and J. H. Nielsen, *Proc. Natl. Acad. Sci. U.S.A.* **92,** 2725 (1995).
85. A. Eilers, D. Georgellis, B. Klose, C. Schindler, A. Ziemiecki, A. G. Harpur, A. F. Wilks, and T. Decker, *Mol. Cell. Biol.* **15,** 3579 (1995).
86. Z. Wen, Z. Zhong, and J. Darnell, Jr., *Cell* **82,** 241 (1995).
87. V. S. Meier and B. Groner, *Mol. Cell. Biol.* **14,** 128 (1994).
88. B. Raught, B. Khursheed, A. Kazansky, and J. Rosen, *Mol. Cell. Biol.* **14,** 1752 (1994).

89. T. S. Schaeffer, L. K. Sanders, and D. Nathans, *Proc. Natl. Acad. Sci. U.S.A.* **92,** 9097 (1996).
90. D. C. Look, M. R. Pelletier, R. M. Tidwell, W. T. Roswit, and M. J. Holtzman, *J. Biol. Chem.* **270,** 30264 (1996).
91. P. E. Lobie, T. J. Wood, C. M. Chen, M. J. Waters, and G. Norstedt, *J. Biol. Chem.* **269,** 31735 (1994).
92. N. G. Anderson, *Biochem. Biophys. Res. Commun.* **193,** 284 (1993).
93. S. C. Souza, G. P. Frick, R. Yip, R. B. Lobo, L. R. Tai, and H. M. Goodman, *J. Biol. Chem.* **269,** 30085 (1994).
94. L. A. Winston and T. Hunter, *J. Biol. Chem.* **270,** 30837 (1996).
95. A. Le-Cam, V. Pantescu, L. Paquereau, C. Legraverend, G. Fauconnier, and G. Asins, *J. Biol. Chem.* **269,** 21532 (1994).
96. V. Rossi, J. F. Rouayrenc, L. Paquereau, M. J. Vilarem, and A. Le-Cam, *Nucleic Acids Res.* **20,** 1061 (1992).
97. P. L. Bergad, H.-M. Shih, H. C. Towle, S. J. Scwarzenberg, and S. A. Berry, *J. Biol. Chem.* **270,** 24903 (1995).
98. P. Wong, C. W. Severns, N. B. Guyer, and T. Wright, *Mol. Cell. Biol.* **14,** 914 (1994).
99. T. Kordula, J. Ripperger, K. K. Morella, J. Travis, and H. Baumann, *J. Biol. Chem.* **271,** 6752 (1996).
100. T. Kordula and J. Travis, *Biochem. J.* **309,** 63 (1995).
101. M. M. Ilondo, P. De-Meyts, and P. Bouchelouche, *Biochem. Biophys. Res. Commun.* **202,** 391 (1994).
102. S. A. Rogers and M. R. Hammerman, *Proc. Natl. Acad. Sci. U.S.A.* **86,** 6363 (1989).
103. P. Roupas, S. Y. Chou, R. J. Towns, and J. L. Kostyo, *Proc. Natl. Acad. Sci. U.S.A.* **88,** 1691 (1991).
104. P. Tollet, M. Hamberg, J. A. Gustafsson, and A. Mode, *J. Biol. Chem.* **270,** 12569 (1995).
105. J. Yuan, U. M. Wegenka, C. Lutticken, J. Buschmann, T. Decker, C. Schindler, P. C. Heinrich, and F. Horn, *Mol. Cell. Biol.* **14,** 1657 (1994).
106. D. J. Waxman, N. A. Pampori, P. A. Ram, A. K. Agrawal, and B. H. Shapiro, *Proc. Natl. Acad. Sci. U.S.A.* **88,** 6868 (1991).
107. D. J. Meyer, G. S. Campbell, B. H. Cochran, L. S. Argetsinger, A. C. Larner, D. S. Finbloom, C. Carter-Su, and J. Schwartz, *J. Biol. Chem.* **269,** 4701 (1994).

Oligonucleotides and Polynucleotides as Biologically Active Compounds

V. V. Vlassov, I. E. Vlassova,*
and L. V. Pautova

Institute of Bioorganic Chemistry
Novosibirsk 630090, Russia
**Institute of Cytology and Genetics*
Novosibirsk 630090, Russia

I. Oligonucleotide Derivatives for Targeting Biopolymers 96
   A. Oligonucleotide-Based Constructions for
      Targeting RNA and DNA 96
   B. Oligonucleotide Ligands for Targeting Proteins 104
II. Pharmacokinetics of Oligonucleotides 105
   A. Interaction with Cells 105
   B. Delivery into Cells 110
   C. *In Vivo* Studies 114
III. Biological Activities of Oligonucleotides and Polynucleotides 116
   A. Inhibition of Genetic Expression in Tissue Cultures 116
   B. Nonantisense Effects of Oligonucleotides 122
   C. Effects of dsRNA 126
   D. Oligonucleotides and the Immune System 128
IV. Concluding Remarks 132
   References 132

Nucleic acids are involved in the most important processes occurring in living cells and possess an enormous potential for specific interactions with different biopolymers. Specificity of recognition of DNA and RNA species by complementary nucleic acids is unique in the world of biopolymers. A great number of the proteins involved in processes of replication, transcription, and translation are also capable of highly specific recognition of certain nucleotide sequences and specific structural motifs in folded nucleic acids. Therefore, nucleic acids have always attracted the attention of pharmacologists dreaming of highly specific therapeutics, i.e., magic bullets targeted to specific biopolymers.

In recent years, the efforts of nucleic acid chemists have been focused on development of oligonucleotide-based biologically active compounds that can be rationally designed for targeting specific sequences in single-stranded RNA (antisense approach) and in double-stranded DNA (antigene approach). These compounds are considered as potential therapeutics capable of selectively modulating functions of specific genes and of inactivating genomes of infectious agents (1–7).

Specific interactions of oligonucleotides with different nucleic acid-related proteins suggest the design of oligonucleotide-based inhibitors for different regulatory factors and nucleic acid polymerases. Recent studies have demonstrated that oligonucleotides can interact with a broad spectrum of proteins capable of binding polyanions. The creation of libraries of oligonucleotide analogs and the development of a molecular selection approach (8–10) have opened up the possibility of generating oligonucleotide structures capable of specific binding to a variety of nonnucleic acid targets. Rapid progress in the development of protein-targeted oligonucleotide ligands with therapeutic value can thus be expected.

Several biological activities of oligonucleotides and polynucleotides are not well understood. Early work showed that injections of RNA, DNA, and nucleic acid homopolymers protect animals from viral infections and affect functions of the immune system. It was shown that double-stranded RNA triggers a few important biological processes, one of which is interferon induction. Investigation of these activities of exogenous nucleic acids may lead to discovery of new biological processes and development of new therapeutic approaches.

This article summarizes the results of investigations of the biological activities of oligonucleotides and polynucleotides that can be used for the development of therapeutics.

I. Oligonucleotide Derivatives for Targeting Biopolymers

A. Oligonucleotide-Based Constructions for Targeting RNA and DNA

Natural oligonucleotides do not meet the criteria of potential therapeutics. To achieve the goals of using oligonucleotide derivatives as biologically active compounds, approaches for synthesis of different oligonucleotide analogs and conjugates have been developed. In this section the main types of oligonucleotide derivatives and the approaches for targeting RNA and DNA are reviewed. (For detailed consideration of the chemical techniques, see Refs. 1, 2, 4, and 11.)

1. OLIGONUCLEOTIDE ANALOGS AND CONJUGATES

One of the focal points in the development of oligonucleotide derivatives for biological applications has been the replacement of the natural phosphodiester backbone with synthetic linkages, to enhance stability, to improve hybridization by removing the negative charge, and to improve the pharmacokinetics.

In the nonionic analogs of oligonucleotides, i.e., oligonucleoside methyl phosphonates and alkyl phosphotriesters (12, 13), the negatively charged phosphate oxygen of the internucleotide bond is replaced by electroneutral functions. Oligonucleoside methyl phosphonates are nuclease resistant and can enter into mammalian cells; therefore, they are considered as promising compounds for therapeutic applications. Olignucleoside methyl phosphonates synthesized by traditional methods represent a mixture of 2^n diastereomers of the oligonucleotides, where n is the number of the chiral phosphorus atoms, because the methyl phosphonate linkage is chiral and exists in either $(R)_p$ or $(S)_p$ configuration at the phosphorus. Duplexes formed between $(R)_{p-}$ isomers of the oligonucleotides and the complementary nucleic acid show an enhanced stability relative to their racemic and $(S)_{p-}$ counterparts. Therefore, attempts are made to develop methods for the synthesis of methyl phosphonate oligonucleotides with stereochemically defined internucleoside bonds.

In the phosphorothioate oligonucleotide analogs (14, 15), one of the nonbridging phosphate oxygen atoms in the internucleotide linkage is replaced by a sulfur atom. Internucleotide bonds in these analogs are chiral; however, this affects the binding properties only slightly. To avoid the chirality problem, phosphorodithioate oligonucleotide derivatives have been prepared (16). Stability of the complementary complexes formed by the analogs decreases in the following order: unmodified oligonucleotides, phosphorothioates, phosphorodithioates. These analogs have the beneficial characteristics of high stability *in vivo*, satisfactory pharmacokinetic properties, and the ability to promote cleavage of the RNA targets by RNase H (17).

Oligonucleotide analogs synthesized from α-anomeric nucleotides (α-oligonucleotides) (18) are resistant to nucleases. α-Oligonucleotides bind to the corresponding nucleotide sequences in RNA and DNA in an orientation reversed as compared to that of the natural β-anomeric oligonucleotides (19).

An important goal in the design of oligonucleotide analogs is enhancing the affinity of the oligonucleotides for their targets. This allows efficient arrest of the target functions and provides a possibility to target short complementary sequences, when the number of base pairings that can be complemented in a structured target is limited. Incorporation of 5-(1-propionyl)-2'-deoxyuridine and -cytosine in oligonucleotides yields oligomers with considerably increased binding affinity to the complementary nucleotide sequences (20, 21). The 2'-O-alkylated oligonucleotides, in particular, 2'-O-methyl and 2'-O-allyl oligonucleotides, have received attention because of the increased stability of complementary complexes of these analogs and RNA (22).

Restricting oligonucleotide conformational freedom to resemble the bound conformation has been used to stabilize their complementary complexes. It was shown that pyrimidine oligonucleotides with alternating phosphodiester and conformationally restricted riboacetal linkages display a considerable increase in affinity for a double-stranded DNA target as compared to all-phosphodiester oligonucleotide controls (23).

Electroneutral polyamide oligomers, termed peptide nucleic acids (PNAs) (24–29), in which the common DNA bases are attached by linkers to an N-(2-aminoethyl)glycine backbone, form exceptionally stable complementary complexes with RNA and DNA in low salt conditions, where natural complementary complexes are destabilized. Homopyrimidine PNAs can bind to the oligopurine strand of a dsDNA, yielding a triple-stranded PNA_2/DNA complex and displacing the homopyrimidine DNA strand. The complex can be stabilized further by replacing the cytosines in half of the bis-PNA, in the "Hoogsteen strand," with pseudoisocytosines, which allows the triple-stranded complex to form at neutral pH (29, 30). A number of other oligonucleotide analogs with nonnatural backbones have been synthesized (1, 4). However, the biological activities of these compounds remain to be investigated.

For targeting arbitrary sequences in dsDNA by triple-stranded complex formation, oligonucleotides with synthetic heterocycles capable of specific binding to different DNA base pairs have been designed (31).

Conjugation of different synthetic and natural molecules to oligonucleotides represents a powerful approach to modulating properties of the compounds for improving their stability and enhancing their uptake by the cells (32–36). Attachment of different functionalities can be achieved through the 3' or 5' termini of oligonucleotides as well as through internucleotide linkages. The C-5 or C-6 positions of pyrimidines are convenient sites for the modifications; even bulky groups introduced in these positions do not interfere with complementary complex formation.

Equipping oligonucleotides with reactive groups converts them into powerful inhibitors of target nucleic acids. A major challenge confronting chemists in developing reactive oligonucleotide derivatives is the design of reactive groups that will react efficiently with the target nucleic acid *in vivo* and yet will not affect nontarget biopolymers.

Alkylating groups, in particular derivatives of aromatic 2-chloroethylamines, are extensively used for the design of reactive oligonucleotide derivatives (2, 36, 37). Alkylating groups (38) form covalent bonds with nucleophilic centers of nucleobases, the reactivity of which decreases in the following order: N-7 atoms of guanosine > N-1 and N-3 atoms of adenine > N-3 atom of cytosine. Alkylation of heterocyclic bases, especially guanine and adenine, results in a weakening of the glycoside bonds, and the polynucleotide chain may be selectively cleaved at the alkylation sites by treatment with piperidine. An efficient alkylating group, the cyclopropapyrroloindole subunit of the antitumor antibiotic CC-1065, was conjugated to the 3' end of oligodeoxynucleotides (39). Because this group is relatively inert in physiological conditions and is activated on complementary complex formation, the conjugate is a candidate for *in vivo* applications.

Photoreactive groups provide a means to control a reaction by applying illumination. Psoralen oligonucleotide derivatives react efficiently with RNA and DNA under irradiation and have been used in the design of oligonucleotide conjugates (40, 41). Psoralens (42) contain two photoactivatable double bonds. On illumination with UV light (365 nm) they react with nucleic acids, yielding monoadducts with pyrimidine bases (thymines are the most reactive) and diadducts with pyrimidines in both strands, forming cross-links.

A disadvantage of oligonucleotide derivatives with cross-linking groups is their limited efficiency, their ability to react with biopolymers other than nucleic acids, and the possibility of *in vivo* formation of immunogenic conjugates with proteins (43). Attempts have been made to develop reactive groups capable of catalytically cleaving nucleic acids. Different groups producing active oxygen species—the EDTA·Fe group, 1,10-phenanthroline-copper, and some porphyrins—have been conjugated to oligonucleotides (2, 33–35, 44). Of these groups, when conjugated to oligonucleotides, only the copper-phenathroline and porphyrin groups can bring about efficient sequence-specific damage to the target. The DNA-cleaving antibiotic bleomycin has been conjugated to oligonucleotides. The conjugates cleave DNA efficiently at the target sequence (45, 46).

For *in vivo* applications, the groups capable of promoting hydrolytic cleavage of internucleotide bonds are the most advantageous. Some complexes of lanthanides and copper ions capable of cleaving RNA have been conjugated to oligonucleotides (47–50). Attempts have been made to design synthetic molecules mimicking active centers of nucleolytic enzymes, for development of efficient nontoxic oligonucleotide conjugates capable of cleaving the RNA targets catalytically (51–54a).

One approach to enhance the specificity of targeted chemical modification of nucleic acids is the design of binary systems of oligonucleotide conjugates. Reactive centers are formed from relatively inert precursor groups, when two oligonucleotides bearing these precursor groups bind to juxtaposed sequences of the target and bring the groups in a close contact (55).

The catalytic domains of some ribozymes (56–58) represent relatively simple structures that can be used as reactive groups for oligonucleotide constructions. Only a few ribonucleotide residues should be conserved in the structure of these domains to maintain activity. The rest of the molecule can be substituted by nucleotide analogs and some synthetic linkers that make the structure more resistant to nucleases.

Binding of some oligonucleotides with RNA triggers cleavage of the targets by cellular ribonucleases. Thus RNase H cleaves the RNA component of duplexes formed by RNA with oligodeoxynucleotides or with phosphorothioate oligodeoxynucleotides, and considerably increases the efficiency of these oligonucleotides (59).

Oligonucleotide conjugates activating 2-5A-dependent RNase L have been developed. This endoribonuclease, which mediates inhibitory effects of interferon, is activated by 5'-phosphorylated, 2',5'-linked oligoadenylates (known as 2-5A), resulting in the cleavage of single-stranded RNA. To direct the enzyme to cleave unique RNA sequences, the oligoadenylate was covalently linked to an oligonucleotide to yield a chimeric molecule triggering the enzymatic cleavage of the complementary RNA (60).

2. Targeting RNA and DNA

The affinity of oligonucleotides for different sequences of RNA depends on their involvement in the folding of the molecule. Secondary folding of nucleic acids can interfere with oligonucleotide binding to the target sequence for both kinetic and thermodynamic reasons (61–63). Folding of the target decreases the association rate because the target involved in intramolecular base pairing may not dissociate long enough for the oligonucleotide to bind to it. If the oligonucleotide and the target sequence can form an intramolecular hairpin, the equilibrium constant of duplex formation is reduced by the equilibrium constant for hairpin formation. The binding of oligonucleotides to the loops depends considerably on the position of the complementary sequence in the loop. On the 5' side, the loops have stacked residues that favor oligonucleotide binding and interfere with binding to the 3' side, which requires restructuring of the loop (58). Therefore, reasonable target sequences are represented by the 5' sides of large loops or the 5' sides and sequences in adjacent nonperfect stems of the hairpins. The affinity is increased by maximizing stacking and the total number of binding interactions in the complex of oligonucleotide and target, including the base pairs formed between the oligonucleotide and the target, and new pairs in the rearranged target.

The currently available methods allow prediction of stable hairpins in RNAs. It should be mentioned that in the cell, nucleic acids are always associated with proteins and this affects secondary structure and the availability of certain regions of RNA to oligonucleotides. Information about the native conformation of a nucleic acid can be obtained using approaches based on chemical and enzymatic probing and cross-linking with bifunctional reagents.

Apparently, the SELEX approach (8–10) (see Section I,B) can be used for identification of optimal oligonucleotides for targeting specific elements of RNA structure. A selection procedure for identification of oligonucleotides capable of tight binding to a hairpin structure was developed (63a). Libraries of oligonucleotides and RNase H can be used to cleave and thereby locate sites in RNAs that are the most accessible for oligonucleotide hybridization (63b).

For specific targeting of nucleic acids *in vivo,* some optimal length of oligonucleotide is required. On the one hand, the longer the oligonucleotide, the better the recognition of when annealing can be performed. However, under physiological conditions, long oligonucleotides will form mismatched complexes with homologous sequences in nontarget nucleic acids. For targeting specific sequences in RNA and DNA *in vivo,* oligomers 15–25 nucleotides long are used; these can form, in physiological conditions, numerous nonperfect complexes with partially complementary nucleotide sequences (64) and can produce side effects. Such side effects have been observed in experiments where specific RNAs have been targeted with antisense oligodeoxyribonucleotides in the presence of ribonuclease H. In parallel with the specific process, degradation of the nontarget cellular mRNAs occurs due to the formation of nonspecific complex formation (65, 66). This is one of the sources of reported sequence-independent effects of antisense oligonucleotides. Similar results have been obtained in experiments with a ribozyme (67). Elongation of the recognition sequence of the ribozyme did not increase the specificity of cleavage. Although looser complexes dissociated faster than the tight ones, the cleavage was too fast compared to the dissociation rate.

A few approaches have been proposed to increase the specificity and affinity of oligonucleotide derivatives. To increase the rate of association of oligonucleotides with their targets, positively charged molecules were attached to oligonucleotides (68–70).

The modification yielded enhanced association constants and accelerated hybridization. Pyrimidine triplex-forming oligonucleotides with 5-Me-dC-(N^4-spermine) residues form triple-stranded complexes with DNA at neutral pH (7.3) under conditions in which the parent nonmodified oligonucleotides failed to bind to the target (68). Incorporation in oligonucleotide-structure inosines with spermine residues attached to position C-2 greatly stabilized duplexes of oligonucleotides with complementary nucleotide sequences (69). Conjugates to oligonucleotide peptides containing four, six, or eight lysines accelerated hybridization 5-, 67-, and 48000-fold, respectively (70). Conjugation of oligonucleotides to intercalating groups (33) results in stabilization of complementary complexes of the oligonucleotides, equivalent to elongation by two to three nucleotide units.

Specificity of targeting can be increased by using oligonucleotide effectors (71, 72). An effector is an oligonucleotide complementary to a sequence adjacent to the target sequence. Due to stacking interaction between terminal bases of the effector and the oligonucleotide bound to the target sequence, the complex of the latter is stabilized. The stabilization effect is equivalent to elongation of an oligonucleotide by two or three nucleotide residues. When the effector oligonucleotide is equipped with an intercalating group, the end-to-end oligonucleotide interaction is strengthened (71, 72). By contrast to the above-mentioned sequence-nonspecific stabilization effects of intercalating and cationic groups, the stabilization effect achieved using oligonucleotide effectors is sequence specific. Effector oligonucleotides have been used for modulating specificity of targeted reactions of reactive oligonucleotide derivatives (71–73) and for modulating the efficiency of ribozyme action (74). Interaction between two oligonucleotides targeted to adjacent sequences can be stabilized further by attaching complementary terminal overhanging sequences to them. Two oligonucleotides with complementary overhangs bind to the target sequence as tight as a single long oligonucleotide of equivalent length (75, 76). The complex formed can be additionally stabilized by antibiotics capable of specific binding to sequences in the duplex formed by the overhangs (77).

Approaches for targeting single-stranded RNA by oligonucleotide constructs forming triple-stranded complexes have been developed (78–80). The constructs contain a fragment that forms Watson–Crick base pairs with the target and a fragment that engages in Hoogsteen base pairing. The fragments can be connected by a synthetic linker or by a short nucleotide sequence. The constructs form very tight complexes with RNA while maintaining good sequence specificity. An intercalating group can be placed at the oligonucleotide terminus to stabilize the assembled triple-helical structure (78). Tightly binding foldback triplex-forming oligonucleotides allow targeting of hairpins in RNA by disrupting the hairpin stem (80).

A strategy for targeting some hairpin structures in RNA containing homopurine–homopyrimidine sequences in the stem and a homopurine sequence adjacent to the stem has been developed (81). The targeting is achieved with oligomers containing a sequence capable of binding to the homopurine–homopyrimidine stem of the hairpin, and a foldback triplex-forming sequence that binds to the adjacent single-stranded homopurine sequence.

Circular oligonucleotides seem to be useful structures for targeting RNA and DNA (82). They bind more tightly and selectively to the complementary nucleic acid single strand than does the normal Watson–Crick complement. They also display a high resistance to nucleases.

Targeting of double-stranded DNA can be achieved by using the triple-stranded complex approach or through opening of the DNA structure. When a homopurine sequence is present in one strand of the double-stranded nucleic acid, and, respectively, a homopyrimidine complementary sequence is in the other strand, the third nucleotide strand can bind into the major groove of this duplex, which results in formation of the triplex structure (*83, 84*). To circumvent the requirement for homopurine–homopyrimidine sequences for triple-stranded complex formation, several strategies have been developed. A number of nucleoside analogs have been synthesized for recognition of different base pairs in dsDNA under physiological conditions (*31*). To broaden the repertoire of potential targets, oligonucleotides connected by flexible linkers that allow binding of the oligonucleotides to two adjacent sequences in two strands of DNA have been designed (*85*).

Sequences within duplex DNA can be targeted by Watson–Crick base pairing through D-loop formation. D-Loop formation is possible at any sequence that is transiently single stranded. For binding, one strand of the target sequence must be displaced. This can occur within supercoiled DNA. Alkylating derivatives of oligonucleotides react with complementary sequences in negatively supercoiled plasmids (*86*). Oligonucleotides can bind to some sequences of DNA that are in the single-stranded state in chromatin and in isolated cell nuclei (*87*). These sequences can appear in DNA regions of high negative superhelicity, where the structure can open spontaneously. The binding is facilitated when the DNA contains sequences that have the potential for alternate secondary structure. The existence of non-B-form DNA structures (such as inverted repeats), H-form DNA, AT-rich unwinding elements, matrix attachment regions, cruciform structures, and RNA polymerase open complexes increases the likelihood for transient formation of the single-stranded regions that are necessary to initiate pairing. Both PNA and phosphodiester oligonucleotides can bind to plasmid DNA by strand invasion at complementary sequences within these structures (*88*). Coupling the oligonucleotide to a positively charged peptide domain capable of associating with DNA facilitates the binding of the oligonucleotide to DNA (*88, 89*).

The development of oligonucleotide analogs that form very tight complementary complexes and can displace one of the DNA strands may allow easy targeting of any DNA sequence. Foldback triplex-forming oligonucleotides bind tightly to homopurine single-stranded sequences. They may be used for targeting this strand by displacing the other strand of the dsDNA (*90*).

PNA oligomers have shown the ability to displace one strand of duplex DNA and form PNA–PNA–DNA triplexes even at nonsupercoiled polypurine–polypyrimidine sites (*25–29*). As a result, a D-loop is formed in which the pyrimidine strand is free and the purine strand forms a very stable triple-stranded complex with two pyrimidine PNA molecules.

Triple-stranded complexes imitating intermediates formed in the process of homologous recombination can be formed in the presence of recA protein between double-stranded DNA and single-stranded oligonucleotides complementary to one of the DNA strands. Oligonucleotides conjugated to the alkylating reagent chlorambucyl cross-link to a long homologous duplex DNA in the presence of recA and ATPγS (*91*).

B. Oligonucleotide Ligands for Targeting Proteins

Oligonucleotides are short stretches of nucleic acids and therefore they can bind to different proteins capable of interacting with RNA and DNA. Specificity and strength of the interaction can vary considerably for different oligonucleotides and proteins.

Due to base pairing and other intramolecular interactions, sufficiently long oligonucleotides fold into defined three-dimensional structures with a high degree of molecular rigidity. Known simple single-stranded motifs (hairpins, pseudoknots, G-quartets) can be built from oligonucleotides about 30 nucleotides long. Synthetically prepared oligonucleotide libraries contain a bewildering number of oligonucleotides with different sequences and with different structures, representing a source of high-affinity ligands for a variety of molecular targets. Selection of oligonucleotide ligands capable of interacting with specific ligands can be performed using combinatorial approaches; the Systematic Evolution of Ligands by Exponential (SELEX) enrichment) (*8–10, 92, 93*) is one approach.

The strategy for identification of specific oligonucleotide ligands consists in preparation of a library of nucleotide sequences that can be amplified by the polymerase chain reaction, determining the binding properties of the library and selectively amplifying the more effective members of the library. The latter oligonucleotide ligands are physically partitioned prior to amplification. This cycle of selection and amplification is repeated as many times as necessary to isolate a preparation of molecules composed entirely of the active species, which can be cloned and characterized. The ligands that emerge from SELEX enrichment have been called aptamers. SELEX-derived oligonucleotides, possessing shapes that interact with target molecules, provide a great alternative to rational design of specific oligonucleotide-based ligands.

A great advantage of this approach is that it allows fast and simple selection of the tight-binding molecules from an enormous number of variants. The sequence diversity of an oligonucleotide synthetically randomized by condensing a mixture of activated monomers is n^4, where n is the number of monomer units of the oligonucleotide containing four different bases. In a single synthesis using a usual oligonucleotide synthesizer, about 10^{20} molecules can be synthesized. RNA aptamers capable of specific binding to various organic dyes, amino acids, ATP, and other different small molecules and proteins have been characterized (10).

An iterative selection procedure has been developed (94) that allows identification of oligonucleotide ligands and that can be applied to any synthetic oligomers built of a few standard monomers. According to the procedure, four libraries of n-mers are synthesized; each member has random sequences except for one position where all members of a library possess a conserved residue specific for that library. Binding of the libraries to the target is compared and the library displaying the best binding is identified. At this point, it is assumed that the specific residue at the assayed position in the winner library should be present in molecules capable of the interaction. Then four new libraries of sequences are synthesized. Oligonucleotides of all the libraries have the identical optimal residue in the position identified at the first assay. Different library-specific nucleotides are placed in another definite position. Comparison of affinities of the families to the target allows identification of the second nucleotide present in the tightly binding oligonucleotide. Repeating the synthesis of four libraries and the selection steps allows one to identify the sequence of the tightly binding oligonucleotide. Apparently, this selection approach can be used for the identification of oligonucleotides possessing any property that can be reliably tested. Thus a phosphorothioate oligonucleotide library containing all possible eight nucleotide sequences was screened for inhibition of HIV infection. Four consecutive generations were generated for maximal activity. In this case, the selection criterion was the ability to suppress virus multiplication. The maximal activity was found to be associated with oligonucleotides containing tetraguanylate sequences capable of forming guanosine quartets (95).

II. Pharmacokinetics of Oligonucleotides

A. Interactions with Cells

Many cells in tissue culture may take up oligonucleotides at pharmacologically relevant concentrations, although the compounds cannot diffuse through lipid membranes. Results of biological studies indicate that there is a mechanism allowing oligonucleotides to enter eukaryotic cells. At present, mechanisms of uptake and distribution of oligonucleotides within cells are poorly understood. Different cells show different abilities to internalize oligonucleotides; modifications of oligonucleotide structure alter uptake and behavior of oligonucleotides within cells. Investigation of natural mechanisms of oligonucleotide uptake and the effect of different modifications on interaction of oligonucleotides with the cell can result in development of efficient approaches for affecting functions of nucleic acids within cells.

Incubation of cells with radiolabeled oligonucleotides results in binding of the radioactive material to the cells and in accumulation of the material within the cells (96–98). Different cell lines show different abilities to bind oligonucleotides (98–100). Rapidly growing cells, e.g., malignant cells, take up oligonucleotides more rapidly than do normal cells. B and T cell mitogens increase cellular uptake of oligonucleotides (100a). Keratinocytes internalize oligonucleotides much more efficiently than do other cells and rapidly take them up within their nuclei, which happens in other cells only in the presence of cationic lipids. Apparently in keratinocytes there is some mechanism allowing oligonucleotides to escape or bypass the endocytotic pathway (101).

Interaction of oligonucleotides with cells is affected by the state of the cells. The uptake is influenced by the tissue culture medium (102, 103). An increase in the cell monolayer density results in a considerable suppression of oligonucleotide binding (98). Some viral infections may enhance uptake of oligonucleotides (104). In experiments with smooth blood cells, little tissue penetration and no uptake was observed when phosphorothioate oligonucleotides were introduced into the lumen of an isolated segment of normal artery. However, in balloon-injured arteries, oligonucleotides penetrated rapidly into the vessel wall and accumulated in the nuclei of the smooth muscle cells (105). Apparently, the cell membranes become nonspecifically permeable because of balloon injury.

Endocytosis seems to be a predominant mechanism of oligonucleotide uptake by eukaryotic cells. Oligonucleotides are taken up by cells in a saturable manner compatible with endocytosis. Under physiological conditions the maximal binding is achieved after 2-hr incubation. The uptake is less efficient at low temperatures and it is suppressed by known inhibitors of endocytosis: deoxyglucose, cytochalasin B, and sodium azide (97, 98). However, other mechanisms of oligonucleotide uptake may function with different efficiencies in different cells and may play an important role at low concentrations of oligonucleotides in the medium. At present, it is not clear whether the observed antisense effects in cells exposed to oligonucleotides are produced by oligonucleotides that are taken up by endocytosis and then escape from the endosomes, or by oligonucleotides that are delivered to the cytoplasm by mechanisms that bypass the destructive endosomal trap. A very rapid (within minutes) accumulation of oligonucleotides in the nuclei of fibroblasts has been detected (106, 107) and could not be explained by the endocytosis mechanism. It was speculated that, in parallel with endocytosis, potocytotic uptake of oligonucleotides takes place (106).

At high oligonucleotide concentrations, oligonucleotide uptake resembles fluid-phase endocytosis (97, 98). Under these conditions, the dependence of the limiting binding level of oligonucleotides on the extracellular oligonucleotide concentration is linear. The uptake efficiency is similar to that of polyvinylpyrrolidone, a compound known to be taken up by fluid-phase pinocytosis. The mean concentration of oligonucleotides in the cells is about $\frac{1}{10}$th of that in the medium. At low concentrations (<0.5 μM), the uptake occurs more efficiently and the mean concentration of the oligonucleotide derivatives in the cells can exceed that in the medium. This can be explained by absorption of oligonucleotides at the cell surface due to interactions with surface components.

Linear dependence of the limiting uptake level on the external oligonucleotide concentration can be explained by competition of two processes: uptake and efflux of oligonucleotides from cells; an equilibrium is reached by the second hour of incubation. Release of the bound oligonucleotides from cells was observed experimentally (97, 98, 108). About 70% of the bound labeled material can be liberated from the cells. Therefore, there are probably two pathways operating in the cellular uptake of oligonucleotides, one of them providing irreversible internalization.

No principal difference in interaction with cells was observed for oligodeoxynucleotides, phosphorothioate oligonucleotides, and chimeric oligodeoxyribonucleotides with terminal phosphorothioate or methyl phosphonate modifications (109–111). Of these oligonucleotides, phosphorothioates have the highest degree of cell binding and uptake and are retained within the cells for relatively longer periods of time (112) as compared to phosphodiester oligonucleotides, apparently because of the ability of the former to bind tightly to resident endosomal proteins.

In cells incubated with oligonucleotides, the compounds were detected both in nuclei and in cytoplasm (97, 113). In different cells, 10–40% of the material taken up was associated with cell nuclei and the rest was in the fractions containing mitochondria, lysosomes, and other vesicular structures (97). This finding is consistent with an endocytosis uptake mechanism (96, 97, 114). When incubated with primary, immortalized, or transformed human or rodent cells in culture, different fluorochrome-tagged oligomers (oligodeoxynucleotides, phosphorothioate analogs, and PNA oligonucleotides) are localized to punctate perinuclear vesicles, resembling endosomes and lysosomes (111, 115–119). Release of oligonucleotides from these vesicles occurs very slowly. A mechanism allowing oligonucleotides to escape from endosomes to the cytoplasm has not yet been elucidated.

Microinjection allows one to bypass the membrane barrier and mimics the steps following oligomer escape from the endocytotic pathway (116, 117, 120). When oligonucleotides conjugated with various fluorescent labels were injected into the cytoplasm of cells, accumulation of the compounds on nuclear structures was observed. Rapid nuclear accumulation is observed for the phosphodiester oligonucleotides, phosphorothioates, methyl phosphonate oligonucleotides, and PNA oligonucleotides in microinjected CV-1 epithelial cells and primary human fibroblasts (*116*). Distribution of oligonucleotides in the nucleus is nonuniform (*121*). Methyl phosphonate oligonucleotides colocalized with condensed regions of DNA, whereas normal and phosphorothioate oligonucleotides concentrated in regions of small nuclear ribonucleoproteins.

The reason for the nuclear concentration of oligonucleotides was proposed to be the binding of the compounds to nuclear proteins. Oligonucleotide binding to nuclear structures reduces the pool of available antisense compound. It was reported that injection of a PNA oligomer together with an excess of a nonsense PNA oligomer produced much more efficient inhibition of the target mRNA as compared to the experiment in which the specific oligomer was injected alone. It was concluded that excess nonsense oligomer saturated the nonspecific nuclear sites and allowed specific complex formation.

Fluorescence resonance energy transfer (FRET) has been used to study hybrid formation and dissociation, after microinjection into living cells of 28-mer phosphodiester oligonucleotides labeled with rhodamine and the complementary fluorescein-labeled phosphorpthioate oligonucleotides. Formation of the hybrid from sequentially injected oligonucleotides was detected by FRET transiently in the cytoplasm and later in the cell nucleus, where nearly all injected oligonucleotides accumulate. This proves that antisense oligonucleotides can hybridize to an intracellular target, in both the cytoplasm and the nucleus (*121*). Using a sensitive assay involving *in situ* reverse transcription techniques, formation of hybrid complexes between oligodeoxynucleotides and target mRNAs was observed after the cells were incubated in a medium with oligonucleotides (*122*).

Synthesis of nonionic methyl phosphonate oligonucleotide analogs was a straightforward attempt to develop nuclease-resistant compounds capable of passive diffusion through cell membranes (*123*). However the compounds can not diffuse through the membranes (*124, 125*). The predominant mechanism of cellular uptake of these compounds seems to be nonspecific adsorptive and fluid-phase endocytosis (*126*). This was confirmed by digitized videomicroscopy showing a punctate distribution of the fluorescently labeled methyl phosphonate oligonucleotides in the cells incubated with the compounds, which is consistent with the localization of the compounds within endosomal vesicles. Rhodamine-labeled methyl phosphonate oligonucleotides colocalized with fluorescein-dextran, an endosomal/lysosomal marker substance. The uptake could not be blocked by competitors, e.g., nonlabeled methyl phosphonate oligonucleotides, phosphodiester oligonucleotides, or ATP, which is in favor of the mechanism of nonspecific adsorptive endocytosis (124).

Interaction of oligonucleotides with cells of infectious agents was investigated. Phosphorothioate oligonucleotides can penetrate into cells of a live cultured metazoan parasite, the schistosome worm (127). The mechanism of oligonucleotide uptake was similar to that described for other eukaryotic cells. Oligonucleotides reached the cytoplasm and nuclei of the parasite cells. Oligonucleotide derivatives conjugated to acridine can enter trypanosomes (128); mycoplasmas take up oligonucleotides readily (98).

The features of the interaction process between oligonucleotides and eukaryotic cells suggest the existence of the cell surface nucleic acid binding receptors. This was suggested also by numerous earlier investigations of DNA binding to cells (129–134). A putative nucleic acid binding receptor with a molecular mass of 75–80 kDa has been detected at the surface of different cells (97, 135, 136). Affinity modification of the protein with oligonucleotide derivatives (97) was specific according to the competition data: single-stranded and double-stranded DNA and tRNA inhibited the modification, whereas nonspecific polyanions (heparin and chondroitin sulfates A and B) did not affect the alkylation. The calculated number of oligonucleotide-binding molecules is approximately of 10^5 per cell; the dissociation constant for oligonucleotide binding was estimated to be ~ 0.2 μM. The relative ability of organs to accumulate oligonucleotides decreases in the following order: kidney, liver, pancreas, intestine, spleen, and muscle, which is in accordance with the amounts of 75-kDa protein determined to be on the cells of the organs (87).

Recently, involvement of scavenger receptors in the internalization of oligonucleotides was suggested; these receptors have broad specificity and bind various polyanions, including polynucleotides (137, 137a). Binding of oligoguanylate to scavenger receptors on macrophages is required for oligonucleotides to augment natural killer cell activity and induce interferon (137b).

Interactions with specific receptors is suggested by the dependence of oligonucleotide uptake on oligonucleotide structure. Association of similar length oligonucleotides in the same cell line varies as a function of sequence. Cellular uptake of homopolymers decreased in the order (dG16) ≫ (dT16) > (dA16) > (dC16). The (dG16) was taken up about 40 times better than was (dC16) and was the most stable (138).

Biological functions of the nucleic acid binding receptor remain to be elucidated. The function of the proteins was suggested to be related to a salvage pathway for nucleotides derived from effete DNA appearing in the bloodstream due to death of cells (129). Malfunctioning of membrane nucleic acid binding proteins may be one of the factors important in the etiology of systemic lupus erythematosus (130).

B. Delivery into Cells

Although the reported antisense effects of oligonucleotides provide evidence that sometimes they can find a way to their targets, the bulk of oligonucleotides that enter cells by an endocytosis pathway remain trapped within endosomal or lysosomal vesicles. Inefficient escape of oligonucleotides from endosomes necessitates high concentrations of oligonucleotides in the extracellular compartment to achieve modest intracellular concentration. In some systems, potent antisense effects may be achieved using cell permeabilization techniques to introduce the compounds directly into the cytoplasm (5).

Replication-defective adenovirus p259A was used as an agent to release oligonucleotides from endocytic-like vesicles (139). It is known that adenovirus proteins cause membrane perturbation and can disrupt endosomes. Using fluorescent microscopy and the labeled oligomer, it was shown that the compound distributes diffusely over the entire cell, when used in combination with the virus. In the absence of the virus, oligonucleotide was localized in cells within endosomes, with minimal cytoplasmic and intranuclear distribution. The virus caused a considerable increase in the sequence-specific antiherpetic activity of methyl phosphonate oligonucleotides.

Eukaryotic cell walls can be made temporarily permeable to oligonucleotides by brief treatment with the pore-forming agent streptolysin O (140). KY01 myelogenous leukemia cells treated in this way accumulated over 100-fold higher intracellular levels of oligodeoxynucleotides, and, in contrast to the latter case, were observed to concentrate internalized oligonucleotides in their nuclei. Some cyclodextrin analogs—2-hydroxypropyl β-cyclodextrin, hydroxyethyl β-cyclodextrin, and a mixture of various hydroxypropyl β-cyclodextrins—increase uptake of phosphorothioate oligonucleotides by up to two- to threefold (141).

Cationic lipids and nucleic acids form complexes that bind efficiently to the anionic surface of cells and fuse with cell membranes, resulting in intracellular delivery of nucleic acids directly into the cytoplasm. A liposomal preparation (lipofectin) containing the cationic lipid N-[1-(2,3-dioleyloxy)-propyl]-N,N,N-trimethylammonium chloride (DOTMA) and dioleoyl-phosphatidylethanolamine (DOPE) increased 1000-fold the antisense effect of a phosphorothioate oligonucleotide (142, 143). In the presence of DOTMA, the association of oligonucleotides with cells increased 10-fold. The lipid markedly alters intracellular distribution of the compound. In the absence of the lipid, the fluorescently labeled oligonucleotides accumulated in discrete cytoplasmic structures consistent with endosomes or lysosomal vesicles based on punctate perinuclear cytoplasmic fluorescence. In the presence of the lipid, the oligonucleotides were accumulated rapidly in the nucleus, excluded from nucleoli, and localized in punctate cytoplasmic structures (*142*). Similar results were obtained with various cationic lipids (*141, 142*). Different cell lines differ markedly in their uptake and sensitivity to antisense oligonucleotides in the presence of cationic lipids (*145*). It should be noted that cationic compounds such as lipofectin are quite toxic to cells (*146*). Complexes of oligonucleotides with the cationic polymer polyethyleneimine are efficiently taken up by cells. In experiments with embryonic neurons, it was shown that oligonucleotides are delivered in cell nuclei. The complexes show virtually no toxic effect (*146a*).

Conjugation of various pendant moieties to oligonucleotides can be used to modify physicochemical properties of the compounds so as to facilitate binding to the cell surface or to target the compounds to specific receptors. Cell surface adsorption of oligonucleotides can be achieved simply by coupling them to lipophilic groups (*148–153*). Cholesterol-conjugated oligonucleotides bind to cells 30-fold more efficiently than do parent oligonucleotides and reach the cellular RNA and DNA. Conjugation of octyl, dodecyl, and octadecyl residues enhances cellular uptake of oligonucleotides by factors of 3, 4, and 10, respectively, as compared to the unmodified oligonucleotides in the order mentioned (*150*). A considerably greater effect of the cholesterol group, in comparison with other lipophilic groups, suggests involvement of specific cellular delivery mechanisms. Participation of a specific cholesterol-binding protein was suggested by affinity modification experiments (*154*). On the other hand, it was shown that cholesterol-linked oligonucleotides bind to plasma low-density lipoproteins (LDLs) and high-density lipoproteins (HDLs), which are responsible for the transport of hydrophobic molecules in the circulation and which are taken up by cells through receptor-mediated endocytosis (*155, 156*). Binding to these lipoproteins may allow internalization of cholesterol-conjugated oligonucleotides at least partially through the lipoprotein receptors.

Antisense cholesterol-conjugated oligonucleotides are more potent antisense agents, compared to unmodified oligonucleotides (*150, 156*). Permanent entrapment of the cholesterol residue in membranes may limit bioavailable concentrations of the cholesterol-conjugated oligonucleotides and may cause unwanted effects. Non-sequence-dependent side effects of such compounds have been reported (*156*). To overcome the problem of permanent anchoring of the compounds to cell membranes, cholesterol was conjugated to oligonucleotides via biodegradable linkers containing disulfide or ester bonds (*150, 151*). In the cells, these linkages are cleaved and free oligonucleotides are released. Conjugation of oligonucleotides to a fusogenic peptide derived from the influenza virus hemagglutinin envelope protein improved the cell binding and anti-HIV activity of the compounds by factor a of 10 (*152*).

To enhance cellular uptake, oligonucleotides were coupled to ε-amino groups of lysine residues of a known polycationic drug carrier, poly(L-lysine), which is efficiently transported into cells by nonspecific adsorptive endocytosis (*157*). In the cells, the oligonucleotides are split off the poly(lysine). The known toxicity of poly(L-lysine) can be reduced, if the poly(lysine) conjugates are used as complexes with heparin. Conjugation of specific ligands to the poly(lysine) component of the poly(lysine)-based delivery systems can be used to target them to specific cells. c-*myb* antisense oligonucleotides were complexed to poly(lysine) with conjugated folic acid and used for downregulation of c-*myb* expression. The complex resulted in greater inhibition as compared to the free oligonucleotide (*158*).

Targeting of oligonucleotide derivatives to cellular receptors has been aimed at promoting cellular uptake of the compounds. Targeting of a receptor specific for mannose 6-phosphate has been reported (*159, 160*). Fluorescein-labeled oligonucleotides conjugated to 6- phosphomannosylated proteins are internalized 20 times more efficiently by cells expressing mannose 6-phosphate-specific receptors compared to the free oligomers (*160*). Mannosylated streptavidin increases 20-fold the internalization of the biotinylated oligonucleotide into a macrophage cell line (*159*). Methyl phosphonate oligonucleotides have been conjugated to neoglycopeptide ligand, interacting with hepatic carbohydrate receptors. The conjugate was rapidly taken up by the hepatocytes (*161*).

To deliver a PNA oligonucleotide through the blood–brain barrier, which is endowed with high transferrin receptor concentrations, a biotinylated oligonucleotide was linked to a conjugate of streptavidin and an antibody to the rat transferrin receptor. Binding to the vector increased at least 28-fold the brain uptake of the oligonucleotide injected into rats (*161a*).

A straightforward approach to deliver oligonucleotides into cells consists in the use of membrane carriers, e.g., liposomes (*162*) and viral envelopes. The inner part of the membrane carrier is protected from enzymes that are present in the organism. Using liposomes with a long circulation time, controlled release of the oligonucleotides can be achieved (*163*).

A very efficient incorporation in liposomes was achieved with nonionic and cholesterol-conjugated oligonucleotides that bind to the lipid membranes. Usually, liposomes are taken up by endocytosis and release the encapsulated material in the lysosomal compartments of the cell. The pH-sensitive fusogenic liposomes promote efflux of oligonucleotides from endosomes into the cytoplasm. The membrane of these lysosomes contains an amine, which protonates at pH levels below physiological values. The vesicles are stable under physiological conditions, but in the endosomes, where the pH is reduced and protonation of the amine occurs, their membrane is destabilized, and the liposomes fuse with the endosomal membrane and deliver their contents into the cytoplasm (163, 164). Incorporation of fusogenic components of the Sendai virus envelope into liposomes promotes cell fusion and thereby facilitates cellular uptake of oligonucleotides, bypassing endocytosis and lysosomal degradation (165).

Retroviral infection can stimulate cellular uptake of pH-sensitive liposomes (166). Antisense oligonucleotides complementary to the 5' end of the mRNA coding for the env protein of Friend retrovirus were delivered into cells using pH-sensitive and regular liposomes (166). Under conditions in which the free oligonucleotide was completely inefficient, the liposome-encapsulated oligonucleotide inhibited spreading of the virus. The composition with the pH-sensitive liposomes showed higher antiviral activity compared to those with the usual carriers.

Liposomes accumulate in the reticuloendothelial system, liver, and spleen. Coupling of liposomes to various ligands that interact with cellular receptors allows them to be targeted to specific cells. Oligonucleotides complementary to the mRNA encoding protein N of the vesicular stomatitis virus were encapsulated into protein A-bearing liposomes, targeted with antibodies to the mouse major histocompatibility complex-encoded H-2K protein expressed on L cells. Due to the protein A–antibody interaction, the liposomes efficiently bound to the target cells and were internalized by endocytosis and protected the cells from the virus at $\frac{1}{100}$ th the dose as compared to the free oligonucleotide (167).

Antisense oligodeoxynucleotides targeted to the epidermal growth factor receptor were encapsulated into liposomes linked to folate through a polyethylene glycol spacer. Delivery of the oligonucleotides into cultured KB cells was 9 times higher than that achieved using nontargeted liposomes and 16 times higher than uptake of nonencapsulated oligonucleotide (168).

Reconstituted viral envelopes (RVEs) are efficient natural vehicles for the oligonucleotide delivery of various compounds into cell cytoplasm (169). A great advantage of RVEs over liposomes is their ability to deliver their contents directly into the cell cytoplasm by fusion with the plasma membrane, avoiding exposure to the lysosomal enzymes. Detailed studies have been done with the reconstituted envelopes of Sendai virus. RVEs are obtained by solubilization of viruses in detergent and reassembly of the envelope following removal of detergent. An efficient technique providing a 25–30% entrapment of oligonucleotide derivatives in these carriers has been developed (170). The RVEs loaded with the oligonucleotide derivatives bind to cells with an efficiency similar to that of the virus, which is typically about 60%. Loading of oligonucleotide derivatives targeted to the genomic RNA of tick-borne encephalitis virus into RVEs enhanced their ability to suppress multiplication of the virus in cell culture (171). Another type of natural carrier for oligonucleotides is a preparation of human erythrocyte membranes ("erythrocyte ghosts"). This can be prepared by lysis of erythrocytes by hypoosmotic shock, followed by incubation of the membranes with foreign substances in a salt solution under conditions that will spontaneously reseal the membranes (170), which now contain the foreign substance. In the presence of UV-inactivated Sendai virus, which is used as a fusion agent, the erythrocyte ghosts bind to cells efficiently and deliver the foreign substance (oligonucleotides) into the cytoplasm.

C. In Vivo Studies

Biodistribution and metabolism have been studied for oligodeoxynucleotides, phosphorothioate oligonucleotides, and the methyl phosphonate analogs (98, 172–180). Oligodeoxynucleotides are usually administered by intravenous and intraperitoneal routes. It has been found that oligonucleotides can penetrate through the mucosae when administered by intranasal, ocular, rectal, vaginal, and perioral routes. Phosphodiester oligonucleotides can even be delivered through skin, using an iontophoresis procedure (179). Methyl phosphonate oligonucleotides can diffuse through the skin (181). Independent of the administration route, oligonucleotides are rapidly distributed in organisms and enter into different tissues. Liver and kidney have the highest oligonucleotide uptake rate. The brain was found to have the lowest concentration of the compounds. The distribution patterns of oligonucleotides in organism are similar in mouse, rat, and monkey.

In cells and in the blood plasma, oligonucleotides may be degraded by nucleases. In the serum, the dominant nuclease activity is 3'-exonuclease (181, 182). A wide range of modifications has been developed to enhance the stability of oligonucleotides in organisms. Oligonucleotides with modified phosphodiester bonds are resistant to nucleases. The 3'-end-modified oligonucleotides are more resistant to degradation *in vivo*, as compared to the parent oligonucleotides. Introduction of segments with modified phosphodiester bonds at the ends of oligonucleotides and conjugation of different bulky groups to the 3' terminus will efficiently protect oligonucleotides against exonucleases (175, 184, 185). Nonmodified phosphodiester oligonucleotides are rapidly degraded in biological systems, including cell culture media and blood serum. Complete degradation occurs within 15–90 min, depending on the system. In animals, the compounds are degraded at a similar high rate. A significant amount of the oligonucleotides (in the form of intact compounds as degradation products) is excreted in urine, up to 30% at 4 hr postinjection. Degradation occurs faster in excretory organs and in reticuloendothelial systems as compared to the blood. Rapid degradation of oligodeoxynucleotides in organisms limits their use therapeutically.

Phosphorothioate oligonucleotides behave similarly to phosphodiester oligonucleotides, but they are more stable, bind to proteins with higher avidity, and circulate longer (*174, 178*). At 1.5 hr postinjection, the compounds are still present in blood, mainly in an undegraded form; in other organs, more than 50% of the intact compound is found; 50% of the phosphorothioate oligonucleotide was degraded in liver 48 hr postinjection (*174*). Up to 30% of the injected phosphorothioate oligonucleotide was excreted essentially as degradation products in mouse urine 24 hr after administration. Daily injections of a 25-mer phosphorothioate oligonucleotide (1 mg/kg/day) were sufficient to keep a steady-state concentration of the compound in rats (*174*). Compared with the results from animal studies, phosphorothioate oligonucleotides have a similar pharmacokinetic profile in humans (*178*). The pattern of tissue uptake, distribution, and elimination is similar for phosphorothioates of varying length and base composition (*173, 175, 176, 186*).

Intracerebroventricularly administered biotin- or digoxigenin-labeled antisense oligodeoxynucleotides are detectable as intact oligomers in rat brain after 4 hr (*187*). The compounds enter neurons and are detected in both the cytoplasm and nuclei of the cells. Oligonucleotides and phosphorothioate analogs are delivered with a microosmotic pump into the cerebrospinal fluid space in rats. Micromolar concentrations of intact phosphorothioate oligonucleotides can be maintained in the cerebrospinal fluid for 1 week without obvious neurologic or systemic toxicity at doses up to 15 nmol/hr for phosphodiester oligonucleotides (concentration 15 mM) and up to 1 nmol/hr (concentration 1.5 mM) for the phosphorothioate analogs. After continuous infusion, oligonucleotides are detected in many cells, especially those of presumed glial origin, indicating extensive brain penetration and marked cellular uptake of the compounds (*188*).

Intravenously injected methyl phosphonate oligonucleotides rapidly distribute in mice. The elimination half-life is 17 min. The highest concentration of the compound is achieved in the renal tissues. Oligonucleotides and degradation products are excreted in the urine within 2 hr after administration (*189*).

2'-*O*-Methyloligoribonucleotide phosphorothioates are more resistant to nucleases compared to the phosphorothioate oligonucleotide analogs *in vitro* and *in vivo*. A phosphorothioate oligonucleotide with 2'-*O*-methyloligoribonucleotide phosphorothioates at both the 3' and the 5' ends was stable in the gastrointestinal tract of a rat for up to 6 hr following oral administration (*175*). When biotinylated phosphodiester oligonucleotides were administered complexed with avidin conjugated to antibodies to the rat transferin receptor, the systemic clearance was reduced by half and the major clearance organ was the liver (*190*).

Phosphodiester, methyl phosphonate, and phosphorothioate oligonucleotides are not toxic to cells at concentrations below 100 μM (*191*) and they are well tolerated in mice and rats. In the monkey, acute hemodynamic toxicity has been observed on rapid intravenous infusion of a phosphorothioate oligonucleotide (*192, 193*); this was likely mediated by activation of C5 complement. The effect can be avoided by administering the compounds by slow infusion.

Recent studies with human HIV-1-infected subjects have demonstrated that the pharmacokinetics of phosphodiester oligonucleotides in humans is in qualitative agreement with the results of animal studies, although in humans, increased plasma disappearance and urinary secretion were observed. Phosphorothioate oligonucleotides were shown to be safe and well tolerated at a dose of 0.1 mg/kg, when administered by 2-hr intravenous infusion (*193a*).

III. Biological Activities of Oligonucleotides and Polynucleotides

A. Inhibition of Genetic Expression in Tissue Cultures

In recent years, a great number of articles have been published demonstrating the activity of oligonucleotides in a variety of systems. Clinical trials are now in progress to evaluate the therapeutic potential of antisense oligonucleotides in several human diseases, including acute myelogenous leukemia and various viral diseases (*4*).

When considering these results, one should take into account that only in a few cases has specific inhibition been rigorously demonstrated. In many studies, specificity has been inferred from the biological effects of antisense as compared to control oligonucleotides, without measuring levels of target RNA or proteins to evaluate specificity. Unintended side effects could potentially occur through a number of mechanisms. Sequence-nonspecific effects of oligonucleotides have been demonstrated in different systems. To make a conclusion about the true antisense effect, one should demonstrate selective reduction of the target mRNA and the corresponding protein in the cells exposed to the oligonucleotide. As controls, different mismatched oligonucleotides should be used; a clear demonstration of specificity is achieved when a specific effect is produced by different oligonucleotides targeted to different positions of the same RNA.

The general strategy of the antisense approach is the arrest of translation of specific mRNAs. This has been achieved with oligonucleotides complementary to sequences around the initiating codon and to the 5′-terminal cap region of the messengers (4, 33, 34). The inhibition is caused by the physical interference of the hybridized oligonucleotides with the binding of initiation factors and ribosomes. Oligonucleotides complementary to the coding regions of mRNA usually do not arrest translation because ribosomes can displace the oligonucleotides from the messenger. Translation arrest can be achieved in this case if the oligonucleotide derivative damages the messenger chemically (194–197) or triggers degradation of the mRNA by ribonuclease H (59, 65, 198, 199).

Phosphorothioate oligonucleotide analogs have been reported to cause specific antisense effects in different systems. Thus a 20-mer phosphorothioate oligodeoxynucleotide was designed to hybridize to the AUG translation initiation codon of mRNA encoding one of the murine protein kinase C isozymes. The oligonucleotide inhibited protein synthesis both *in vitro* and in mouse cells. This effect was specific for the target isozyme and was completely dependent on the oligodeoxynucleotide sequence. When administered intraperitoneally in mice, the same oligodeoxynucleotide caused a dose-dependent, oligonucleotide sequence-dependent reduction of the target mRNA in liver. The expression of the other isozymes was unaffected by this treatment. The oligonucleotide activity *in vivo* did not require the presence of cationic liposomes or any other delivery systems, although *in vitro* the oligonucleotide required cationic liposomes for inhibition of protein expression (200).

mRNAs coding for proteins taking part in cell proliferation are often targeted in order to develop approaches for controlling malignant cell growth. The gene *myc* is activated in a wide variety of neoplasms and the overexpression of this gene appears to be necessary for the rapid proliferation of these cells and may provide resistance of the cells to lysis by cytotoxic effector lymphocytes (201). Oligonucleotides complementary to the *myc* mRNA were shown to enhance susceptibility of cells to lysis by peripheral blood lymphocytes (201). The *in vivo* efficacy of continuous subcutaneous perfusion of oligodeoxynucleotides was demonstrated in experiments with athymic mice with subcutaneous xenografts of human neuroectodermal tumors. Antisense oligonucleotides directed against the *myc* RNA led to loss of the myc protein from the tumor and reduction of the tumor mass (201a).

mRNAs of mutated protooncogenes and genes causing genetic disorders represent targets that differ from normal mRNAs by a single nucleotide substitution. An example is the mRNA of the *ras* p21 protein, which takes part in the cellular signal transduction pathway. A single mutation in either the 12th or 61st codon of the cellular *ras* gene leads to a gene coding for the mutated protein that is involved in the development of the malignant cell phenotype. Antisense oligonucleotides can be used for selective arrest of translation of the mutated mRNAs. Phosphorothioate oligonucleotides (12-mers or 13-mers) centered on the mutation have a high discrimination efficiency between the mutated and the normal mRNAs. At a concentration of 10 μM the oligonucleotides inhibited proliferation of the cell line transformed with mutated Ha-*ras* gene and had no effect on the parent cell line (*196*). Methyl phosphonate oligonucleotide analogs conjugated to the photoreactive psoralen group, targeted to a mutated region of the *ras* mRNA overlapping the 12th codon, can cause 90% inhibition of synthesis of the protein (*196*).

Dodecadeoxyribonucleotides derivatized with 1,10-phenanthroline or psoralen were targeted to the point mutation (G = U) in codon 12 of the Ha-*ras* mRNA (*202*). With the conjugates, it was possible to obtain complete discrimination between the mutated oligonucleotide target [which contained a psoralen-reactive T(U) in the 12th codon] and the normal target (which contained G at the same position). When longer Ha-*ras* RNA fragments were used as targets, specific effects were achieved using longer oligonucleotides or reactive oligonucleotides in the presence of oligonucleotide effectors (see Section I,A,2.).

An attractive candidate for antisense therapy is chronic myelogenous leukemia (CML) (*203*). This hematologic stem cell disorder is characterized by the reciprocal translocation between chromosomes 9 and 22, which generates the so-called Philadelphia chromosome and the fusion gene, which is transcribed into mRNA and translated into p210 BCR-ABL protein. This gene and its product play a role in the pathogenesis of CML. Selective inhibition of leukemic cell growth by antisense oligonucleotides complementary to specific regions of BCR-ABL mRNA has been achieved in cells, using antisense oligonucleotides and methyl phosphonate oligonucleotides incorporated in liposomes (*203a, 204*). Development of lymphomas in severe combined immune-deficient (SCID) mice was inhibited when the cells were treated with the antisense oligonucleotides prior to inoculation (*204a*). Trials on patients with CML have been started (*204b*). The patients were treated with ablative chemotherapy and autologous bone marrow transplants. Before reinfusion, bone marrow mononuclear cells were incubated with a phosphorothioate oligonucleotide specific to the BCR-ABL gene. After the autograft, the proportion of Philadelphia chromosome-negative cells was observed. However, nonspecific cell growth inhibition was noted in some cases (*204c, 205*). Some control phosphodiester oligonucleotides inhibited proliferation of the leukemic cell lines. Interestingly, all the active oligonucleotides contained the TAT consensus at their 3' end, and a one-base mismatch in this sequence abolished the antiproliferative effect. This antiproliferative effect of unelucidated nature was observed in experiments with different cell lines.

Vascular smooth muscle cell (VSMC) proliferation and extracellular matrix accumulation are the principal mechanisms leading to vascular restenosis after arterial balloon dilatation. A few studies report inhibition of restenosis in a rat model of balloon angioplasty using oligonucleotides targeted to mRNAs of different factors that mediate neointima formation by activating cell cycle progression. The expression of c-*myc* is required for VSMC proliferation *in vitro* and in the vessel wall. Thus c-*myc* is a potential target for adjunctive therapy to reduce angioplasty restenosis. The effect of blocking the expression of the c-*myc* protooncogene with antisense oligonucleotides on VSMC proliferation, both *in vitro* and in a rat carotid artery injury model of angioplasty restenosis, was investigated (206, 207). Antisense c-*myc* oligonucleotides reduced average cell levels of c-*myc* mRNA and protein by 50–55% and inhibited proliferation of VSMCs when mitogenically stimulated from quiescence or when proliferating logarithmically. Corresponding sense c-*myc*, and 2-bp mismatch antisense c-*myc* oligonucleotides did not suppress c-*myc* expression or inhibit VSMC proliferation. After balloon catheter injury, peak c-*myc* mRNA expression occurred at 2 hr. Antisense c-*myc* applied in a pluronic gel to the arterial adventitia reduced peak c-*myc* expression by 75% and significantly reduced neointimal formation compared with sense c-*myc* and gel application alone.

Treatment of VSMCs with phosphorothioate antisense oligonucleotides, targeted to the mRNA of the p65 subunit of a pleiotropic trans-activator of a diverse group of genes of NF-κB, inhibited human VSMC adherence and proliferation in a concentration-dependent manner. Administration of the p65 antisense oligonucleotide significantly inhibited neointima formation in balloon angioplasty-treated rat carotid arteries (208).

The efficiency of antisense oligonucleotides in experiments with tissue cultures correlates with their ability to enter cells and their stability under experimental conditions. Design of new oligonucleotide analogs opens up more perspectives for affecting gene expression *in vivo*. Chemically modified oligonucleotides are potentially effective antisense agents. Their resistance to nucleases and enhanced binding affinities mean that they can be used at low doses, with correspondingly reduced side effects. Many of these chemically modified oligonucleotides, however, are inefficient at entering cells, and so special delivery methods are essential if their potential is to be realized.

A very efficient gene-specific antisense inhibition was achieved in mammalian cells using C-5 propynylpyrimidine (C-5 propyne) 2'-deoxyphosphorothioate oligonucleotides. These oligonucleotides have the required high binding affinity imparted by the C-5 propynyl moiety and nuclease stability properties necessary to yield potent antisense inhibition when delivered into cells. RNA inactivation was presumably achieved by an RNase H-mediated mechanism of inhibition. The oligonucleotides showed antisense effects when the target sequence was placed in a variety of sites in the target RNA.

In an attempt to create antisense oligonucleotides of very high affinity to RNA, PNA oligonucleotides and steric block oligonucleotides consisting of 2'-O-allyl ribose and C-5 propyne modifications were developed. A comparative study of a few high-affinity oligonucleotide analogs and the analogs capable of activating RNase H cleavage of RNAs was performed (209, 209a) using a CV-1 cell-based microinjection assay and antisense agents targeted to various sequences within the SV40 large-T antigen RNA. In the presence of cationic liposomes, the oligonucleotides caused a complete loss of the target protein (SV40 large-T antigen) at a concentration of 5 nM.

It was shown that although the C-5 propyne 2'-O-allyl oligonucleotide/RNA duplexes formed *in vitro* appear to be very stable, they dissociate in cells (209). Rapid dissociation occurred in both the nucleus and the cytoplasm, as evidenced by the dissociation of injected duplexes. C-5 propyne 2'-O-deoxyphosphorothioate oligonucleotides were the most potent reagents. Complexes of these oligonucleotides with RNA were stable enough to inhibit RNA translation completely, presumably by allowing rapid cleavage of the RNA by RNase H. The tested PNA oligomer was less efficient than the C-5 propynylpyrimidine deoxyphosphorothioate oligonucleotides due to the PNA slow kinetics of RNA association.

The fact, that complexes of some oligonucleotide analogs are not stable in the cells explains the low potency of some oligonucleotide analogs, including the 2'-O-allyl oligonucleotides. Unidentified dissociation factors present in the nucleus and cytoplasm of cells, and in reticulocyte lysate, were supposed to be responsible for the dissociation of the complexes.

Reverse transcription is an early event in retroviral infection and it is an attractive process for inhibition with antisense oligonucleotides. Reverse transcription can be inhibited by phosphodiester and phosphorothioate oligonucleotides by cleavage of the RNA template by retroviral RNase H and by an RNase H-independent hybrid arrest (209b).

Oligonucleotides can prevent either initiation or elongation of transcription. Oligonucleotide binding to eukaryotic promoters represses the transcription process efficiently. Binding of oligonucleotides to a nucleic acid template can arrest enzymatic synthesis of the complementary nucleic acid strand. Transcription arrest is efficient, in particular, in the case of reactive oligonucleotide derivatives forming cross-links with the target (210, 211).

The triple-helix approach was used to arrest the function of certain cellular genes. The human c-*myc* protooncogene contains a 23-bp purine–pyrimidine-rich motif within its predominant promoter P2, which is a potential target for the triplex formation. A triplex-forming oligonucleotide capable of binding to a sequence within the c-*myc* promoter was shown to decrease selectively the steady-state concentration of the *myc* mRNA (212, 213). Efficiency of the down-regulation of the mRNA was considerably improved when the cells were incubated with the oligonucleotides in the presence of 1,3-diaminopropane, which is a potent stabilizer of triplex DNA (213).

An oligonucleotide targeted to a homopyrimidine–homopurine region in the interleukin-2Rα promoter was shown to bind to the target *in vitro* and to inhibit the synthesis of the corresponding mRNA in lymphocytes. Reactive derivatives of this triplex-forming oligonucleotide with a psoralen group showed enhanced inhibitory activity (214).

The same promoter was targeted with 15-mer PNA oligonucleotides capable of binding to DNA by strand invasion. The PNA oligonucleotide bound to DNA in physiological conditions and specifically blocked interaction of the transcription factor NF-κB with the IL-2Rα NF-κB binding site. It inhibited trans-activation of the promoter when transfected into cells in culture (215). Although at moderate salt concentrations binding of PNAs to DNA by strand invasion occurs slowly, this process is considerably accelerated if the DNA is transcribed (216).

Cross-linking of the psoralen-linked triplex-forming oligonucleotides to coding sequences can interrupt transcription of the target gene (216a). Triplex-forming oligonucleotides with a psoralen group were used as site-specific mutagens (217). Targeted mutations were generated with frequencies up to 2% in a reporter gene in an SV40 vector within the monkey COS cells after treatment with the oligonucleotides and irradiation.

The antiviral activity of oligonucleotides targeted to viral nucleic acids is well documented (218, 219). Modifications improving the ability of oligonucleotides to enter cells increases the antiviral potential of the compounds. Oligonucleotides conjugated to lipophilic groups such as cholesterol demonstrate enhanced antiviral activity (220, 221). However, it has been noticed that cholesterol-conjugated control oligonucleotides having little or no base complementarity to the viral target were also effective (222). Poly(lysine)-conjugated oligonucleotides complementary to viral nucleic acids arrested multiplication of vesicular stomatitis virus and HIV-1 at concentrations much lower than active concentrations of unmodified oligonucleotides (157). Oligonucleotides conjugated to intercalating groups show enhanced antiviral activity as compared to the parent oligomers (223). It should be noted that all the mentioned modifications protect oligonucleotides against cellular nucleases and it is an important factor contributing to the enhancement of the oligonucleotide antiviral effect.

In experiments with herpes simplex virus, HIV-1, and tick-borne encephalitis virus, it was found that the antiviral properties of oligonucleotides can be greatly improved by coupling them to reactive groups, e.g., psoralen or alkylating groups (220, 224).

Phosphorothioate oligonucleotides are promising antiviral agents (225). In a few cases, efficient sequence-specific inhibition of translation (e.g., 226) and transcription (e.g., 227) of virus specific RNAs was observed with these analogs. However, in many cases, considerable antiviral activity was detected in experiments with oligonucleotides that apparently did not have perfect complementary binding sites in virus RNAs. A possibility to inhibit transcription of HIV-1 in infected cells, by oligonucleotides designed to form DNA triple helices within the virus promoter, has been reported (228). Acridine-linked oligothymidylates targeted to a homothymidine–homoadenosine region within the origin of replication of SV40 virus DNA inhibited virus multiplication in sensitive cells (229).

Different oligonucleotides complementary to genome RNA of the dengue virus were injected in cells to protect the cells from the virus. Nonmodified oligodeoxynucleotides and corresponding phosphorothioate oligonucleotides were inefficient in bringing about a significant inhibition of viral infection. Modified phosphorothioate oligonucleotides in which the C-5 atom of pyrimidines was replaced by propynyl groups caused a significant inhibition of the virus (230). Oligonucleotides targeted to virus-specific nucleic acids have some curative activity in experiments with mice infected with herpes simplex and tick-borne encephalitis viruses (13, 231). However, it is not clear if the results were due to antisense effects of oligonucleotides or if some other mechanisms, including inhibition of specific proteins or immunostimulation, provided the curative activity.

B. Nonantisense Effects of Oligonucleotides

Oligonucleotides can bind to nucleic acid-related enzymes and positive and negative regulatory protein factors interacting with RNA and DNA. Therefore, they can serve as decoys for specific proteins and inhibit or activate specific genes. Double-stranded complexes of phosphodiester and phosphorothioate oligonucleotides, representing binding sites for specific transcriptional regulatory proteins, were shown to bind proteins in cells and to interfere with transcription of specific cellular genes (232). Folded oligoribonucleotides mimicking fragments of tRNAs inhibit the cognate synthetases and can act as cell growth regulators (232a).

Inhibition of nucleic acid polymerases with polynucleotides and various analogs and derivatives of poly- and oligonucleotides is a well-known phenomenon (233, 234). Viral reverse transcriptases can be efficiently inhibited by 2'-azido analogs of polynucleotides and 2'-halogenated polynucleotide analogs. These analogs inhibit the enzymes *in vitro* and arrest multiplication of the viruses in tissue cultures. A number of studies demonstrate antiviral activity of RNA preparations against different RNA viruses (235–238). Administration of polynucleotides intranasally or by intravenous injections before infecting with influenza virus protected mice from the infection. Intravenous injections of some tRNAs and tRNA fragments suppressed multiplication of Sindbis virus in mice.

The most detailed studies have been performed with mice infected with encephalomyocarditis virus (*239, 240*). Intravenous and intraperitoneal injections of ribosomal and transfer RNA from various sources and synthetic single-stranded ribopolynucleotides protected animals from infection with encephalomyocarditis virus and Semliki forest virus. Under the conditions of the experiments, no significant interferon induction was observed, and the polynucleotides showed equal protective effects in animals infected with the interferon-sensitive strain of the virus and the wild strain, which is sensitive to interferon. The degree of protection was dose dependent and short lived, being significant for about 24 hr postinjection of the compounds, apparently because of degradation of the compounds in the organisms. The most efficient protection was conferred by single-stranded poly(C) and poly(I). The maximal protective effect was observed when the compounds were administered 4–6 hr before infection. Some protective action was noted when the compounds were administered after the infection, up to 20 hr after inoculation of the virus. These results were interpreted as inhibition of nucleic acid-related viral proteins.

Numerous observations of non-sequence-specific effects caused by oligonucleotides, in particular phosphorothioate and phosphorodithioate oligonucleotides (*241*), have stimulated investigations of oligonucleotide–protein interactions as a potential source of different biological activities. Oligonucleotides, in particular phosphorothioate oligonucleotide analogs, can bind with high avidity to different proteins of biological importance. In accordance with results of these studies, it was found that oligonucleotides, in particular phosphorothioate oligonucleotides, inhibit reverse transcriptases of retroviruses (*241a*). Short-capped ribooligonucleotides are potent inhibitors of the influenza virus RNA polymerase (*241b*). Oligonucleotides can inhibit viral proliferation by interfering with the virus life cycle by several mechanisms. Besides the effects of oligonucleotides on transcription and translation of virus-specific nucleic acids, oligonucleotides can interfere with adsorption, penetration, and uncoating of the virus; they can prevent processing and transportation of the viral nucleic acids and interfere with packaging of virions. Phosphorothioate oligonucleotides interact with nuclear proteins in non-sequence-specific fashion. Oligonucleotides with loop structures produce cytotoxic effects (*241c*).

A large contribution to the antiviral effects of oligonucleotides appears to be a blockade of adsorption and penetration of the virus into the cell. It was found that oligonucleotides interfere with binding of the HIV-1 virus to cells. The HIV-1 virus enters target cells after binding of its outer envelope glycoprotein gp120 to the CD4 receptor at the surface of T cells and macrophages. It was found that both proteins, i.e., cellular receptors CD4 and the gp120 protein, bind oligonucleotides (*242, 242a*), and the binding interferes with fixation of the virus at the surface of the cells and arrests proliferation of the virus in tissue cultures. The anti-HIV activity of different oligonucleotides correlates with their affinity for the CD4 receptor. The antiviral activity increases with increasing length of the oligonucleotides, and it is higher for phosphorothioate oligonucleotides, which bind to proteins better then natural phosphodiester oligonucleotides. Short oligodeoxyguanylates (3/4-mers) effectively prevent HIV-1-induced cytopathicity by interacting with CD4, inhibiting reverse transcriptase, and interfering with some steps after integration (*242b*).

Oligonucleotides containing oligoguanylic sequences, capable of forming tetrameric structures, bind to the gp120 protein of HIV-1 more tightly than other oligonucleotides and appear to be more efficient inhibitors of viral proliferation (*243*). A phosphorothioate oligonucleotide $T_2G_4T_2$ forms a parallel-stranded, tetrameric guanosine-quartet structure that binds tightly to the gp120 V3 loop and inhibits both cell-to-cell and virus-to-cell interactions (*244*). The compound was identified by combinatorial screening of a library of all possible octanucleotide sequences. The oligonucleotide was active against different isolates of HIV-1 and HIV-2, including all drug-resistant isolates.

It appears that tetrad-forming oligonucleotides may intervene in the progression of HIV-1 infection by at least two distinct mechanisms. The oligodeoxynucleotide containing only deoxyguanosine and thymidine, GTGGTGGGTGGGTGGGT, did not block virus entry into cells but did reduce viral-specific transcripts. It was concluded that it may function by inhibiting viral-specific transcription. The addition of a cholesterol moiety to the 3' terminus of the oligonucleotide considerably increased the antiviral activity (*245*). Conjugation of a hydrophobic dimethoxytrityl residue to guanosine-rich oligonucleotides improved their ability to block the binding of HIV-1 with the sensitive cells and to inhibit proliferation of the virus.

The binding of herpes simplex virus to the cell involves an initial attachment to cell surface heparan sulfate molecules and can be competed by addition of heparin to the medium. Phosphorothioate oligonucleotide $(SdC)_{28}$ also inhibited herpes simplex virus type 2 proliferation by blocking adsorption of the virus and by inhibiting the viral DNA polymerase (*246*). Oligonucleotides built of thymidines and guanosines very efficiently inhibited herpes simplex growth in a sequence-independent manner. Phosphorothioate oligonucleotides were more active than the phosphodiester oligonucleotides and inhibited the virus at concentrations as low as 0.02 μM (*247*).

Charged oligonucleotides are polyanions and can interact with different proteins that bind naturally occurring polyanions, e.g., the sulfated glycosaminoglycans heparin and heparan, dermatan, and chondroitin sulfate. Interactions of these polyanions with specific proteins play different physiological roles. Oligonucleotides interact with a variety of proteins in the bloodstream and in barrier fluids, tears and saliva, lactoferrin, lysozyme, different classes of immunoglobulins (248), serum albumin, protein kinase C (249), heparin-binding growth factors, and other growth factors. Oligonucleotides bind to fibronectin and to the A subunit of laminin at or near the heparin-binding site (250, 251). These interactions explain non-sequence-specific blockage of cellular adhesion by oligonucleotides. Basic fibroblast growth factor binding involves interaction with cell surface heparan sulfate. Phosphorothioate oligonucleotides bind tightly to basic fibroblast growth factor (bFGF), block interaction of the protein with cellular receptors, and inhibit bFGF-induced DNA synthesis in NIH 3T3, DU-145, and Vero cells (251). Non-sequence-specific inhibition of tumor growth by phosphorothioate oligonucleotides was explained by an antagonism between the oligonucleotides and the basic fibroblast growth factor.

Chemical modification of oligonucleotides may be an additional source of biological activities. Conjugation of cholesterol to oligonucleotides considerably improves uptake of the compounds by cells and increases their efficiency in experiments with cells (147, 150, 252). On the other hand, this modification converts oligonucleotides into sequence-nonspecific inhibitors of some proteins. Cholesterol-conjugated phosphorothioate oligonucleotides are among the most potent known inhibitors of reduced folate carrier and antagonize cytotoxicity of methotrexate (253). Cholesterol-conjugated phosphodiester oligonucleotides activate cell membrane calcium channels in HL-60 clls (253a). Self-complementary cholesterol-modified oligodeoxynucleotides cause death in certain cancer cell lines and are not toxic for other cell lines. A mechanism for this effect has not yet been elucidated (254).

SELEX against protein targets yields oligonucleotide ligands with high specificity and dissociation constants. With respect to affinities and specificities, oligonucleotide ligands identified through SELEX are as potent as antibodies (10). Oligonucleotides targeted to proteins may be used as drugs. Peptides present in blood are good therapeutic targets for oligonucleotide aptamers. The SELEX procedure was used to develop DNA ligands to thrombin. Of the identified sequences, the highest affinity was displayed by GGTTGGTGTGGTTGG. This oligonucleotide is a potent anticoagulant, perhaps of therapeutic value (255). The ligand inhibited the thrombin-catalyzed cleavage of fibrinogen to yield fibrin, and thus had anticlotting activity. The structure of this aptamer is an intramolecular G quartet (255). Oligonucleotide ligands might be used as antagonists to block functions of different growth factors important for proliferation of tumor cells and factors essential for cell differentiation and maintenance. An RNA ligand was selected that binds human nerve growth factor (*255a*).

Using the SELEX methodology, a single-stranded DNA ligand has been identified that binds to the reverse transcriptase of HIV-1 with a K_d value as low as 1 nM and inhibits the enzyme with a K_i value as low as 0.3 nM (*256*).

The nucleic acid ligands can be used for preventing abnormal targeting of self-antigens by antibodies in autoimmine patients, which leads to tissue destruction and other pathologies. This process could potentially be disrupted by small ligands that bind specifically to autoantibodies and inhibit their interaction with the target antigen. An RNA sequence was identified by the SELEX procedure that binds a mouse monoclonal antibody specific for an autoantigenic epitope of human insulin receptor and not to other mouse IgGs. The RNA can also act as a decoy, blocking the antibody from binding the insulin receptor. These results suggest that decoy RNAs may be useful in the treatment of autoimmune diseases (*257*).

C. Effects of dsRNA

An antiviral activity of polynucleotides has been detected that is due to induction of interferon, a protein that triggers a complicated antiviral cellular defense mechanism (*258, 259*). The most efficient interferon inducers are the structured viral polyribonucleotides, such as the RNA of paramyxoviruses or the double-stranded viral RNAs and complexes of synthetic polyribonucleotides, such as poly(I)·poly(C) (*260*). Triple-stranded complexes of polyribonucleotides and single-stranded polyribonucleotides capable of forming intramolecular structured regions, such as poly(I), display lower activity. Polydeoxyribonucleotides, their complexes with polyribonucleotides, and polyribonucleotides that cannot form folded structures were found to be consistently inactive (*261*). One explanation of the differences in activity is that the single-stranded polynucleotides are readily degraded by ubiquitous nucleases present in the culture media, in cells, and in organisms, whereas the double- or multiple-stranded molecules are relatively nuclease resistant.

The biological activity of dsRNAs is not restricted to interferon induction. The compounds display a number of biological activities explained by the findings as inducing a wide variety of genes involved in antiviral and growth-mediated responses, including the oncogenes c-*myc*, c-*fos*, and c-Ha-*ras* (*262–264*), and a few cytokines. They also activate directly some enzymes, such as 2′,5′-oligoadenylate (2′,5′-A) synthetase, involved in the cellular interferon-triggered antiviral defense mechanisms (*265*), and ribosome-associated dsRNA-dependent protein kinase.

On activation by dsRNA, 2′,5′-A synthetase converts ATP into a series of oligonucleotides containing 2′,5′-phosphodiester linkages (*266*). These compounds are capable of activating a latent ribonuclease, which inhibits protein synthesis by degrading RNA (277). 2-5A and 2-5A derivatives also directly inhibit HIV-1 reverse transcriptase (RT) by preventing HIV-1 RT/primer complex formation (268) and inhibit DNA topoisomerase I in HIV-1-infected cells (269).

Another interferon-induced intracellular enzyme, dsRNA-dependent/regulated protein kinase (PKR), is activated by poly(I)·poly(C) and is autophosphorylated on dsRNA activation. Once autophosphorylated, PKR phosphorylates the α subunit of eukaryotic protein synthesis initiation factor eIF-2, which prevents further polypeptide chain initiation and results in inhibition of protein synthesis (270). Unlike 2′,5′-A synthetase, PKR dose response to dsRNA is biphasic. Low concentrations of dsRNA activate it, and high concentrations inhibit it (271, 272). PKR has been reported to be a tumor suppressor gene (273). PKR activates gene transcription through regulation of transcription factors NF-κB (274) and is strongly implicated in the control of differentiation and growth (275, 276).

In addition to working through the dsRNA-dependent protein kinase, dsRNA may inhibit protein synthesis by direct interaction with eIF-2 (258). Double-stranded polyribonucleotides poly(A)·poly(U), poly(C)·poly(G), and poly(I)·poly(C) influence the behavior of cells involved in immune responses; they enhance both cell-mediated cytotoxicity and B cell response to antigens. Treatment of cells with double-stranded RNAs makes them more sensitive to the tumor necrosis factor (TNFα) and it was suggested that "labeling" of cells with RNA may be a natural mechanism that helps the factor to recognize the cells to be eliminated (277).

dsRNA activates factors DRAF1 and -2 that bind to the interferon-stimulated response element in the cells. Activation of DRAF1 and DRAF2 is independent of interferon action because it occurs in cells that are non-responsive to interferon and in cells that lack the α/β interferon locus (278). The complete mechanism by which dsRNA leads to the activation of specific cellular factors and subsequent gene induction is not yet completely clear.

dsRNA may inhibit the rate of growth of tumors (279, 280). Inhibition of cell growth may be explained by IFN induction in some cell lines (281), whereas in other cell lines sensitivity to dsRNA and IFN is often an independent phenomenon (282). It has been generally assumed that the IFN-independent antiproliferative activity of dsRNA involves activation of the dsRNA-dependent enzymes 2′,5′-A synthetase and DAI. A rapid increase in the intracellular concentration of cAMP is sufficient for dsRNA-induced growth inhibition. These results suggest that dsRNA may exert its antiproliferative effect through cAMP signal transduction.

An efficient inducer of interferon, poly(I)·poly(C) was considered potentially therapeutic; however, it was too toxic and doses predicted to be therapeutic could not be reached even when a more efficient composition of poly(I)·poly(C) with poly(L-lysine) and carboxymethyl cellulose was used (283). Later, it was found that the toxic effects are observed when the compound circulates in the organism for a long time. Because interferon induction is a fast process, modified polynucleotides were developed that induce interferon efficiently, but are less stable in the organism than is poly(I)·poly(C). The most thoroughly investigated preparation, Ampligen, is a complex of poly(I) with a copolymer of mean composition $(C_{12} U)_n$ (284, 285). In culture media and in organism, this complex is readily degraded, yielding increasing amounts of the interferon-inducing fragments that retain the biologic activity when they are at least about 50 nucleotides in size. It is as good an interferon inducer as poly(I)·poly(C), but shows substantially less toxicity. It also displays a broader specificity toward different tumor cells (286).

D. Oligonucleotides and the Immune System

Natural DNA is generally considered to be a poor immunogen; however, data generated in the past few years show that antibody response to nucleic acids is a component of normal immunity (287, 288). Although mammalian DNA elicits poor immune response, bacterial DNA with certain sequences and chemically modified DNA may exert profound effects on cells *in vivo* and *in vitro*. Immune cells appear to be especially sensitive to these activities, and there is now evidence that DNA is immunogenic as well as immunomodulatory (287–289).

Analyses of the antibodies binding nucleic acids in healthy subjects and with various disease states reveal that they are extremely variable. Two kinds of nucleic acid-binding autoantibodies have been identified. Natural autoantibodies directed against nucleic acids are present in the serum of normal, healthy individuals and individuals with autoimmune disease. The production of natural autoantibodies arises independently of any known immunization. Polyreactive antibodies of the M type, which can bind with different affinities to two or more antigens (289, 290), comprise the majority of these antibodies. Autoimmune patients, in particular patients with systemic lupus erythematosus (SLE), spontaneously produce various autoantibodies (including antibodies to nucleic acids) that differ in a number of parameters from the above-mentioned natural antibodies. They produce high-affinity autoantibodies, mainly of the IgG or IgA isotype, that are not widely cross-reactive and are encoded by somatically mutated genes. These autoantibodies are believed to have pathogenic potential (291).

The immunogenic features of nucleic acids are evidenced by successful immunizations using several double-stranded polynucleotides. Immunizations of animals have shown that poly(rA)·poly(rU) and hybrid RNA–DNA duplexes such as poly(rA)·poly(dT) are immunogenic. Monoclonal antibodies to poly(rA·poly(rU) appear to recognize a conformational epitope that is characteristic of many RNA duplexes. Monoclonal antibodies prepared against poly(rI)·poly(dC) indicate that the conformation of a duplex and not the presence of unusual bases is the primary determinant of immunogenicity (292). Chemically modified DNA or synthetic DNA bearing unusual conformations can induce antibodies in animals; left-handed Z conformation DNA is highly immunogenic (293). DNAs from different species differ in their immunologic activity. Bacterial DNA, unlike mammalian DNA, can induce significant antibody responses in mice. Normal human serum contains high titers of antibodies specific for DNA from certain bacteria. These antibodies do not cross-react with mammalian DNA or DNA from other species and they are different from lupus anti-DNA, which recognizes a conserved site and can be inhibited by essentially all DNA (288).

Nucleic acids can exert immunomodulatory effects by their influence on production of several cytokines that regulate inflammation and antigen-specific immune responses (288). That exogenous nucleic acids might influence an immune response was demonstrated first by the observation that administration of nucleic acids would restore immunocompetence to animals in which the immune response was suppressed by drugs or X-irradiation. Modulatory effects of synthetic dsRNA on both humoral and cellular immune responses have been reported (294).

It has been demonstrated that poly(I·C) enhances both natural killer (NK) activity and macrophage activity in severe combined immune-deficient mice. When poly(I·C) was administered intraperitoneally prior to murine cytomegalovirus infection of these mice, it significantly increased the life span of the mice. Intraperitoneal administration of poly(I·C) also induced both early (2 hr) and late (18 hr) types of interferon in the peritoneal fluid and blood. Attempts to investigate macrophage activation with polyribonucleotides have shown differential regulation of gene expression and the role of macrophage-directed interferon and protein kinase C (295–297) in mechanisms governing changes in the functional phenotype.

Nucleic acids are found on the surfaces of immune cells and it has been shown that this interaction has physiological consequences (301). The binding of DNA to murine spleen cells has the physiological effect of stimulating the release of significant levels of interleukin-6. By contrast, no significant increases in IL-1 activity were seen. Because IL-6 is an important mediator of B cell differentiation and antibody production, it was suggested that the secretion of this cytokine may be one factor contributing to the increased levels of immunoglobulin production observed in patients with SLE and similar autoimmune disorders (301).

The ability of bacterial DNA to induce *in vivo* proliferation of murine B cells and secretion of immunoglobulins *in vitro* and *in vivo* was investigated.

The stimulation of lymphocytes with bacterial DNA resulted in a dose-dependent response; mammalian DNA was nonimmunogenic (302). Depletion of T cells from lymphocytes did not reduce proliferation, suggesting that bacterial DNA directly triggered B cell proliferation. These studies provide evidence that DNAs are not uniform in their immunologic activities (302). Mammalian DNA differs from microbial DNA in two potentially relevant ways: it contains CpG dinucleotides at about one-fourth of the expected frequency, and most of the cytosines in CpG dinucleotides are methylated, but other cytosines are rarely methylated. In bacterial DNA, CpG dinucleotides are present at the expected frequency and are unmethylated. Unmethylated CpG in bacterial DNA (CpG motif) induces rapid polyclonal B cell proliferation and IgM secretion (303–305). The B cell stimulatory motif, consisting of an unmethylated CpG dinucleotide and flanked by two 5′-purines and two 3′-pyrimidines, was identified (303, 304). Methylation of the cytosine in this CpG motif abolished the stimulation. More than 96% of B cells were induced to enter the cell cycle when treated with oligonucleotides containing this motif. Stimulation of the cells with CpG-containing DNA was accompanied by induction of secretion of IL-6 and IgM. Unmethylated CpG motifs induce B cells to secrete IL-6, IL-12, and IgM, and induce NK and $CD4^+$ T cells to produce IFN-γ (305). Investigation of the immunomodulating properties of single-stranded DNAs has resulted in the discovery of specific folded oligonucleotides capable of affecting the immune system (305a). IFN production and augmented NK activity in murine splenocytes were induced by oligonucleotides containing a hexamer palindrome AACGTT and having a length of more than 22 nucleotides.

CpG oligonucleotides with phosphorothioate backbones stimulate B cells at concentrations more than 2 logs lower than the concentrations for stimulation by normal phosphodiester oligonucleotides (303). *In vivo* injections of nuclease-resistent phosphorothioate CpG oligonucleotides induced splenomegaly and polyclonal B cell activation manifested by increased B cell proliferation and Ly-6A and class II major histocompatibility complex (MHC) expression (304). The mechanism by which the CpG motif promotes lymphocyte activation is not understood, but it is clear that the immune system responds to bacterial DNA by activating a coordinated set of humoral and cellular responses (302). The lymphocyte activation by the CpG motif may be an antimicrobial defense mechanism that activates the immune system before development of antigen-specific reactions.

The development of antisense therapeutics targeted against specific diseases has demonstrated the potent effects that these oligomers may exert on immune cell function (306, 307). The implication of oligonucleotide-mediated induction of transcription factors *in vivo* and *in vitro* is the most striking effect. NF-κB is a mammalian transcription factor that controls a number of genes important for immunity and inflammation (308). NF-κB consists of two subunits, p50 and p65 (RelA). Both subunits are part of a larger group of transcription factors called the Rel family (308). The ability of p65 sense (relA) oligonucleotides to induce splenomegaly in vivo parallels a specific expansion of B-220+ B cells in splenocyte cultures in vitro; this was studied and it was found that sense oligonucleotides, but not antisense oligonucleotides, cause massive splenomegaly in immunocompromised mice (309). The p65 sense oligonucleotides apparently act as a potent mitogen for splenic B lymphocytes. In mice injected with phosphorothioate antisense oligonucleotides against relA, thrombocytopenia was observed. There was a toxic effect on kidney, liver, and bone marrow (310). Phosphorothioate oligonucleotides have biological effects on both white and red pulp areas of the spleen. The expansion of the white pulp areas was caused by lymphoid hyperplasia, which involved both B cell (follicular) and T cell (periarteriolar sheath) areas in relA-treated mice. Although the mechanism for the sequence-specific immune stimulation is unknown, the absence of splenomegaly in mice treated with several control phosphorothioates suggests that the effect is probably not induced by the phosphorothioate backbone.

The effect of chemical modifications of the backbone, sequence, and length of oligonucleotides on cell proliferation and antibody production by murine cells in vitro and in vivo has been investigated (311). Some of the oligonucleotides stimulate proliferation of splenocytes in a concentration-dependent manner, and IgG and IgM production. The stimulatory effect is dependent on particular sequences of oligonucleotides, modifications, and tertiary folding, and not just length. Oligonucleotides that contain fewer thioate groups induce less immune stimulatory effects both in vitro and in vivo. Oligonucleotides with four methyl phosphonate linkages at each end and oligonucleotides with 2'-O-methylribonucleotides at the ends do not induce significant cell proliferation and their ability to stimulate antibody production is lower, compared to unmodified oligonucleotides. Oligo-nucleotides with palindromic sequences showed markedly high levels of cell proliferation. Phosphorothioate oligonucleotides are much more effective stimulatory agents, compared to phosphodiester oligonucleotides and hybrid oligonucleotides with a few phosphodiester bonds. In mice, injection of the phosphorothioate oligonucleotides caused massive splenomegaly and stimulation of cell proliferation. Hybrid and phosphodiester oligonucleotides were less active.

Another example of the biological activities of phosphorothioate oligonucleotides is a rapid induction of the general transcription factor Sp1 in diverse cell types, in vitro as well as in vivo (312).

IV. Concluding Remarks

Drug development using oligonucleotide-based technologies represents a revolutionary strategy that directly targets and inhibits gene expression. The opportunities to develop oligonucleotides with novel properties have increased dramatically due to the recent successes in design of novel oligonucleotide analogs and the elaboration of molecular selection approaches.

It is still a challenge to chemists to develop chemical modifications or molecular carriers for efficient targeted delivery of oligonucleotides into cells, and to design reactive oligonucleotide derivatives capable of efficient inactivation of target nucleic acids in organisms.

Further studies of the biological effects of exogenous nucleic acids, and identification of the proteins involved in the internalization of nucleic acids and their immunomodulating effects, will open new possibilities for development of methods for antisense and gene therapy, and development of new approaches to modulate functions of the immune system.

References

1. S. T. Crooke and B. Lebleu, eds., "Antisense Research and Applications." CRC Press, Boca Raton, Florida, 1993.
2. D. G. Knorre, V. V. Vlassov, V. F. Zarytova, A. V. Lebedev, and O. S. Fedorova, "Design and Targeted Reactions of Oligonucleotide Derivatives." CRC Press, Boca Raton, Florida, 1994.
3. J. S. Cohen, ed., "Oligodeoxynucleotides. Antisense Inhibitors of Gene Expression." Macmillan, London, 1989.
4. S. T. Crooke, in "Burgers Medicinal Chemistry and Drug Discovery" (M. E. Wolff, ed.), p. 863. Wiley, New York, 1995.
5. R. W. Wagner, *Nature (London)* **372,** 333 (1994).
6. V. V. Vlassov, in "The Lock-and-Key Principle" (J.-P. Behr, ed.), p. 89. Wiley, New York, 1994.
7. A. D. Mesmaeker, R. Haner, P. Martin, and H. E. Moser, *Acc. Chem. Res.* **28,** 366 (1995).
8. C. Tuerk and L. Gold, *Science* **249,** 505 (1990).
9. A. D. Ellington and J. W. Szostak, *Nature* **346,** 818 (1990).
10. L. Gold, B. Polisky, O. Uhlenbeck, and M. Yarus, *Annu. Rev. Biochem.* **64,** 763 (1995).
11. Y. S. Sanghvi and P. D. Cook, in "Nucleosides and Nucleotides as Antitumor and Antiviral Agents" (C. K. Chu and D. C. Baker, eds.), p. 311, Plenum, New York, 1993.
12. P. S. Miller, L. Braiterman, and P. O. P. Ts'o, *Biochemistry* **16,** 1988 (1977).
13. P. S. Miller, P. O. P. Ts'o, R. I. Hogrefe, M. A. Reynolds, and L. J. Arnold, in "Antisense Research and Applications" (S. T. Crooke and B. Lebleu, eds.), p. 189. CRC Press, Boca Raton, Florida, 1993.
14. M. H. Caruthers, in "Oligonucleotides. Antisense Inhibitors of Gene Expression" (J. S. Cohen, ed.), p. 8. Macmillan, London, 1989.
15. G. Zon and T. G. Geiser, *Anti-Cancer Drug Design* **6,** 539 (1991).

16. M. H. Caruthers, G. Beaton, L. Cummins, D. Dellinger, D. Graff, Y.-X. Ma, W. S. Marshall, H. Sasmor, P. Shankland, J. V. Wu, and E. K. Yan, *Nucleosides Nucleotides* **10,** 47 (1991).
17. C. A. Stein, J. L. Tonkinson, and L. Yakubov, *Pharmacol. Ther.* **52,** 365 (1991).
18. B. Rayner, C. Malvy, J. Paoletti, B. Lebleu, C. Paoletti, and J.-L. Imbach, in "Oligonucleotides. Antisense Inhibitors of Gene Expression" (J. A. Cohen, ed.), p. 119 Macmillan, London, 1989.
19. J. S. Sun, C. Giovannangeli, J. C. Francois, R. Kurfurst, T. Montenay-Garestier, U. Asseline, T. Saison-Behmoaras, N. T. Thuong, and C. Helene, *Proc. Natl. Acad. Sci. U.S.A.* **88,** 6023 (1991).
20. B. C. Froehler, S. Wadwani, T. J. Terhorst, and S. R. Gerrard, *Tetrahedr. Lett.* **33,** 5307 (1992).
21. R. W. Wagner, M. D. Matteucci, J. G. Lewis, A. J. Gutierrez, C. Moulds, and B. C. Froehler, *Science* **260,** 1510 (1993).
22. B. P. Monia, E. A. Lesnik, C. Gonzalez, W. F. Lima, D. McGee, C. J. Guinosso, A. M. Kawasaki, P. D. Cook, and S. M. Freier, *J. Biol. Chem.* **268,** 14514 (1993).
23. R. J. Jones, S. Swaminathan, J. F. Milligan, S. Wadwani, B. C. Froehler, and M. D. Matteucci, *J. Am. Chem. Soc.* **115,** 9816 (1993).
24. M. Egholm, O. Buchardt, L. Christensen, C. Behrens, S. M. Freier, D. A. Driver, R. H. Berg, S. K. Kim, B. Norden, and P. E. Nielsen, *Nature (London)* **365,** 566 (1993).
25. P. E. Nielsen, M. Egholm, R. H. Berg, and O. Buchardt, in "Antisense Research and Applications" (S. T. Crooke and B. Lebleu, eds.), p. 363. CRC Press, Boca Raton, Florida, 1993.
26. P. E. Nielsen, M. Egholm, and O. Buchardt, *Bioconj. Chem.* **5,** 3 (1994).
27. T. A. Vickers, M. C. Griffith, K. Ramasamy, L. M. Risen, and S. M. Freier, *Nucleic Acids Res.* **23,** 3003 (1995).
28. V. Demidov, M. D. Frank-Kamenetskii, M. Egholm, O. Buchardt, and P. E. Nielsen, *Nucleic Acids Res.* **21,** 2103 (1993).
29. M. Egholm, L. Christensen, K. L. Dueholm, O. Buchardt, J. Coull, and P. E. Nielsen, *Nucleic Acids Res.* **23,** 217 (1995).
30. P. E. Nielsen, *Nature (London)* **379,** 214 (1996).
31. Y. S. Sanghivi, in "Antisense Research and Applications" (S. T. Crooke and B. Lebleu, eds.), p. 273. CRC Press, Boca Raton, Florida, 1993.
32. M. Manoharan, in "Antisense Research and Applications" (S. T. Crooke and B. Lebleu, eds.), p. 303. CRC Press, Boca Raton, Florida, 1993.
33. C. Helene and J. J. Toulme, *Biochim. Biophys. Acta* **1049,** 99 (1990).
34. J. Goodchild, *Bioconj. Chem.* **1,** 165 (1990).
35. E. Uhlmann and A. Peyman, *Chem. Rev.* **90,** 543 (1990).
36. D. G. Knorre and V. V. Vlassov, *This Series* **32,** 291 (1985).
37. D. G. Knorre and V. V. Vlassov, "Affinity Modification of Biopolymers." CRC Press, Boca Raton, Florida, 1989.
38. W. C. J. Ross, "Biological Alkylating Agents." Butterworths, London, 1962.
39. E. A. Lukhtanov, M. A. Podyminogin, I. V. Kutyavin, R. B. Meyer, and H. B. Gamper, *Nucleic Acids Res.* **24,** 683 (1996).
40. U. Pieles, B. S. Sproat, P. Neuner, and F. Cramer, *Nucleic Acids Res.* **17,** 8967 (1989).
41. J. T. Levis and P. S. Miller, *Antisense Res. Dev.* **4,** 223 (1994).
42. J. E. Hearst, *Annu. Rev. Biophys. Bioenerg.* **10,** 69 (1981).
43. E. B. Brossalina, V. V. Vlassov, and E. M. Ivanova, *Biokhimija (Russ.)* **53,** 18 (1988).
44. B. Meunier, ed., "DNA and RNA Cleavers and Chemotherapy of Cancer and Viral Diseases." Kluwer, Dordrecht, Netherlands, 1996.

45. D. S. Sergeev, V. F. Zarytova, S. V. Mamaev, T. S. Godovikova, and V. V. Vlassov, *Antisense Res. Dev.* **2**, 235 (1992).
46. D. S. Sergeev, T. S. Godovikova, and V. F. Zarytova, *Nucleic Acids Res.* **23**, 4400 (1995).
47. D. Magda, R. A. Miller, J. L. Sessler, and B. L. Iverson, *J. Am. Chem. Soc.* **116**, 7439 (1994).
48. K. Matsumura, M. Endo, and M. Komiyama, *J. Chem. Soc., Chem. Commun.*, p. 2019 (1994).
49. J. K. Bashkin, E. I. Frolova, and U. S. Sampath, *J. Am. Chem. Soc.* **116**, 5981 (1994).
50. J. Hall, D. Hüsken, U. Pieles, H. E. Moser, and R. Haner, *Chem. Biol.* **1**, 185 (1994).
51. C.-H. Tung, Z. W. Wei, M. J. Leibowitz, and S. Stein, *Proc. Natl. Acad. Sci. U.S.A.* **89**, 7114 (1992).
52. J. Smith, K. Ariga, and E. V. Anslyn, *J. Am. Chem. Soc.* **115**, 362 (1993).
53. M. A. Podyminogin, V. V. Vlassov, and R. Giege, *Nucleic Acids Res.* **21**, 5950 (1993).
54. V. V. Vlassov, G. Zuber, B. Felden, J.-P., Behr, and R. Giege, *Nucleic Acids Res.* **23**, 3161 (1995).
55. V. V. Vlassov, M. I. Dobrikov, S. A. Gaidamakov, E. K. Gaidamakova, T. I. Gainutdinov, and A. A. Koshkin, *in* "DNA and RNA Cleavers and Chemotherapy of Cancer and Viral Diseases" (B. Meunier, ed.), p. 195. Kluwer, Dordrecht, Netherlands, 1996.
56. R. A. Stull and F. C. Szoka, *Pharm. Res.* **12**, 465 (1995).
57. O. C. Uhlenbeck, *in* "Antisense Research and Applications" (S. T. Crooke and B. Lebleu, eds.), p. 83, CRC Press, Boca Raton, Florida, 1993.
58. S. M. Freier, *in* "Antisense Research and Applications" (S. T. Crooke and B. Lebleu, eds.), p. 67. CRC Press, Boca Raton, Florida, 1993.
59. J.-J. Toulme, C. Boiziau, B. Larrouy, P. Frank, S. Albert, and R. Ahmadi, *in* "DNA and RNA Cleavers and Chemotherapy of Cancer and Viral Diseases" (B. Meunier, ed.), p. 271. Kluwer, Dordrecht, Netherlands, 1996.
60. P. F. Torrence, W. Xiao, G. Li, S. Khamnei, K. Lesiak, A. Maran, R. Maitra, A. Kumar, B. Dong, B. R. G. Williams, and R. H. Silverman, *Nucleosides Nucleotides* **14**, 1073 (1995).
61. D. J. Ecker, *in* "Antisense Research and Applications" (S. T. Crooke and B. Lebleu, eds.), p. 387. CRC Press, Boca Raton, Florida, 1993.
62. M. Chastain and I. Tinoco, Jr., *in* "Antisense Research and Applications" (S. T. Crooke and B. Lebleu, eds.), p. 55. CRC Press, Boca Raton, Florida, 1993.
63. J. R. Wyatt and I. Tinoco, Jr., *in* "The RNA World," p. 465, Cold Spring Harbor Lab., Cold Spring Harbor, New York, 1993.
63a. R. K. Mishra and J.-J. Toulme, *Crit. Rev. Acad. Sci. Paris* **317**, 977 (1994).
63b. S. P. Ho, D. H. O. Britton, B. A. Stone, D. L. Behrens, L. M. Leffet, F. W. Hobbs, J. A. Miller, and G. L. Trainor, *Nucleic Acids Res.* **24**, 1901 (1996).
64. M. P. Perelroyzen and A. V. Vologodskii, *Nucleic Acids Res.* **16**, 4693 (1988).
65. R. V. Giles, D. G. Spiller, and D. M. Tidd, *Antisense Res. Dev.* **5**, 23 (1995).
66. B. Larrouy, C. Boisiau, B. Sproat, and J.-J. Toulme, *Nucleic Acids Res.* **23**, 3434 (1995).
67. D. Herschlag, *Proc. Natl. Acad. Sci. U.S.A.* **88**, 6921 (1991).
68. D. A. Barawkar, V. A. Kumar, and K. N. Ganesh, *Biochem. Biophys. Res. Commun.* **205**, 1665 (1994).
69. N. Schmid and J. P. Behr, *Tetrahedr. Lett.* **36**, 1447 (1995).
70. D. R. Corey, *J. Am. Chem. Soc.* **117**, 9373 (1995).
71. I. V. Kutyavin, M. A. Podyminogin, Yu. N. Bazhina, O. S. Fedorova, D. G. Knorre, A. S. Levina, S. V. Mamayev, and V. F. Zarytova, *FEBS Lett.* **238**, 35 (1988).
72. E. Ivanova, D. Pyshnui, I. Pyshnaya, D. Sergeev, P. Vorobjev, S. Lokhov, and V. Zarytova, *Nucleosides Nucleotides* **14**, 1065 (1995).
73. M. D. Distefano, J. A. Shin, and P. B. Dervan, *J. Am. Chem. Soc.* **113**, 5901 (1991).
74. J. Goodchild, *Nucleic Acids Res.* **20**, 4607 (1992).

75. N. Colocci, M. D. Distefano, and P. Dervan, *J. Am. Chem. Soc.* **115,** 4468 (1993).
76. E. R. Kandimalla, A. Manning, C. Lathan, R. A. Byrn, and S. Agrawal, *Nucleic Acids Res.* **23,** 3578 (1995).
77. M. D. Distefano and P. B. Dervan, *J. Am. Chem. Soc.* **114,** 11006 (1992).
78. C. Giovannangeli, N. T. Thuong, and C. Helene, *Proc. Natl. Acad. Sci. U.S.A.* **90,** 10013 (1993).
79. E. R. Kandimalla, A. N. Manning, G. Venkataraman, V. Sassisekharan, and S. Agrawal, *Nucleic Acids Res.* **23,** 4510 (1995).
80. J. C. Francois and C. Helene, *Biochemistry* **34,** 65 (1995).
81. E. Brossalina and J.-J. Toulme, *J. Am. Chem. Soc.* **115,** 796 (1993).
82. H. Yamakawa, T. Ishibashi, H. Nakashima, N. Yamamoto, K. Takai, and H. Takaku, *Nucleosides Nucleotides* **14,** 1149 (1995).
83. P. B. Dervan, *in* "Oligonucleotides. Antisense Inhibitors of Gene Expression" (J. A. Cohen, ed.), p. 197. Macmillan, London, 1989.
84. V. N. Soyfer and V. N. Potaman, "Triple-Helical Nucleic Acids." Springer-Verlag, Berlin and New York, 1995.
85. D. A. Horne and P. G. Dervan, *J. Am. Chem. Soc.* **112,** 2435 (1990).
86. V. V. Vlassov, S. A. Gaidamakov, V. F. Zarytova, D. G. Knorre, A. S. Levina, A. A. Nikonova, L. M. Podust, and O. S. Fedorova, *Gene* **72,** 313 (1988).
87. V. V. Vlassov, N. D. Kobetz, E. L. Chernolovskaya, S. G. Demidova, R. G. Borissov, and E. M. Ivanova, *Mol. Biol. Rep.* **14,** 11 (1990).
88. M. Iyer, J. C. Norton, and D. R. Corey, *J. Biol. Chem.* **270,** 14712 (1995).
89. D. R. Corey, D. Munoz-Medellin, and A. Huang, *Bioconj. Chem.* **6,** 93 (1995).
90. E. R. Kandimalla, A. N. Manning, and S. Agrawal, *J. Biomol. Struct. Dynam.* **13,** 483 (1995).
91. M. A. Podyminogin, R. B. Meyer, and H. B. Gamper, *Biochemistry* **34,** 13098 (1995).
92. J. W. Szostak and A. D. Ellington, *in* "The RNA World," p. 511. Cold Spring Harbor Lab., Cold Spring Harbor, New York, 1993.
93. L. Gold, *J. Biol. Chem.* **270,** 13581 (1995).
94. D. J. Ecker, T. A. Vickers, R. Hanecak, V. Driver, and K. Anderson, *Nucleic Acids Res.* **21,** 1853 (1993).
95. E. L. White, Jr., R. W. Buckheit, L. J. Ross, J. M. Germany, K. Andries, and R. Pauwels, *Antiviral Res.* **16,** 257 (1991).
96. J. Googchild, S. Agraval, M. P. Civeira, P. S. Sarin, D. Sun, and P. C. Zamecnik, *Proc. Natl. Acad. Sci. U.S.A.* **85,** 5507 (1988).
97. L. A. Yakubov, E. A. Deeva, V. F. Zarytova, E. M. Ivanova, A. S. Ryte, L. V. Yurchenko, and V. V. Vlassov, *Proc. Natl. Acad. Sci. U.S.A.* **86,** 6454 (1989).
98. V. V. Vlassov and L. A. Yakubov, *in* "Prospects for Antisense Nucleic Acid Therapy of Cancer and Aids" (S. T. Crooke and B. Lebleu, eds.). CRC Press, Boca Raton, Florida, 1993.
99. M. Ceruzzi and K. Draper, *Nucleosides Nucleotides* **8,** 815 (1989).
100. S. T. Crooke, *J. Drug Targeting* **3,** 185 (1995).
100a. P. L. Iversen, D. Crouse, G. Zon, and G. Perry, *Antisense Res. Dev.* **2,** 223 (1992).
101. F. O. Nestle, R. S. Mitra, C. F. Bennett, H. Chan, and B. J. Nickoloff, *J. Invest. Dermatol.* **103,** 569 (1994).
102. R. M. Crooke, M. J. Graham, M. E. Cooke, and S. T. Crooke, *J. Pharmacol. Exper. Ther.* **275,** 462 (1995).
103. S. Wu-Pong, T. L. Weiss, and C. A. Hunt, *Antisense Res. Dev.* **4,** 155 (1994).
104. W. Y. Gao, R. N. Hanes, M. A. Varquez-Padua, C. A. Stein, J. S. Cohen, and Y. C. Cheng, *Antimicrob. Agents Chemother.* **34,** 808 (1990).
105. C. L. Farrell, J. V. Bready, S. A. Kaufman, Y.-X. Qian, and T. L. Burgess, *Antisense Res. Dev.* **5,** 175 (1995).

106. P. Zamecnik, J. Aghajanian, M. Zamecnik, J. Goodchild, and G. Witman, *Proc. Natl. Acad. Sci. U.S.A.* **91,** 3156 (1994).
107. V. V. Vlassov, M. V. Nechaeva, S. I. Baiborodin, O. E. Shestova, I. V. Safronov, A. A. Koshkin, and L. A. Yakubov, *Dokl. Akad. Nauk (Russia)* **345,** 123 (1996).
108. A. M. Krieg, F. Gmelig-Meyling, M. F. Gourley, W. J. Kisch, L. A. Chrisey, and A. D. Steinberg, *Antisense Res. Dev.* **1,** 161 (1991).
109. J. Temsamani, M. Kubert, J. Tang, A. Padmapriya, and S. Agrawal, *Antisense Res. Dev.* **4,** 35 (1994).
110. P. L. Iversen, S. Zhu, A. Meyer, and G. Zon, *Antisense Res. Dev.* **2,** 211 (1992).
111. Q. Zhao, S. Matson, C. J. Herrera, E. Fisher, H. Yu, and A. M. Krieg, *Antisense Res. Dev.* **3,** 53 (1993).
112. W. Y. Gao, C. Storm, W. Egan, and Y. C. Cheng, *Mol. Pharmacol.* **43,** 45 (1993).
113. P. Hawley and I. Gibson, *Antisense Res. Dev.* **2,** 119 (1992).
114. P. C. Zamecnik, J. Goodhild, Y. Taguchi, and P. S. Sarin, *Proc. Natl. Acad. Sci. U.S.A.* **83,** 4143 (1986).
115. M. Cerruzzi, K. Draper, and J. Schwartz, *Nucleosides Nucleotides* **9,** 679 (1990).
116. D. J. Chin, G. A. Green, G.Zon, F. C. Szoka, and R. M. Straubinger, *New Biologist* **2,** 1091 (1990).
117. J. P. Leonetti, N. Mechti, G. Degols, C. Gagnor, and B. Lebleu, *Proc. Natl. Acad. Sci. U.S.A.* **88,** 2702 (1991).
118. J. L. Tonkinson and C. A. Stein, *Nucleic Acids Res.* **22,** 4268 (1994).
119. G. Marti, W. Egan, P. Noguchi, G. Zon, M. Matsukura, and S. Broder, *Antisense Res. Dev.* **2,** 27 (1992).
120. T. L. Fisher, T. Terhorst, X. Cao, and R. W. Wagner, *Nucleic Acids Res.* **21,** 3857 (1993).
121. S. Sixou, F. C. Szoka, G. A. Green, B. Giusti, G. Zon, and D. J. Chin, *Nucleic Acids Res.* **22,** 662 (1994).
122. J. C. Politz, K. L. Taneja, and R. H. Singer, *Nucleic Acids Res.* **23,** 4946 (1995).
123. E. H. Chang and P. S. Miller, in "Prospects for Antisense Nucleic Acid Therapy of Cancer and AIDS" (E. Wickstrom, ed.), p. 115. Wiley-Liss, New York, 1991.
124. Y. Shoji, S. Akhtar, A. Periasami, B. Herman, and R. L. Juliano, *Nucleic Acids Res.* **19,** 5543 (1991).
125. J. T. Levis, W. O. Butler, B. Y. Tseng, and P. O. P. Tso, *Antisense Res. Dev.* **5,** 251 (1995).
126. S. Akhtar, Y. Shoji, and R. L. Juliano, in "Gene Regulation Biology of Antisense RNA and DNA" (R. P. Erickson and J. G. Izant, eds.), p. 133. Raven Press, New York, 1992.
127. L.-F. Tao, K. A. Marx, W. Wongwit, Z. Jiang, S. Agrawal, and R. M. Coleman, *Antisense Res. Dev.* **5,** 123 (1995).
128. P. Verspieren, A. W.-C. A. Cornelissen, N. T. Thuong, C. Helene, and J.-J. Toulme, *Gene* **61,** 307 (1987).
129. R. M. Bennet, S. H. Heveneider, A. Bakke, M. Merritt, C. A. Smith, D. Mourich, and M. C. Heinrich, *J. Immunol.* **140,** 2937 (1988).
130. M. Rieber, C. Urbina, and M. S. Rieber, *Biochem. Biophys. Res. Commun.* **59,** 1441 (1989).
131. R. M. Bennet, K. A. Cornell, M. J. Merritt, A. C. Bakke, and P. H. Hsu, *Clin. Exp. Immunol.* **86,** 374 (1991).
132. R. M. Bennett, B. L. Kotzin, and M. J. Merritt, *J. Exp. Med.* **166,** 850 (1987).
133. R. M. Bennett, J. S. Peller, and M. M. Merritt, *Lancet* **1,** 186 (1986).
134. W. Emlen, A. Rifai, D. Magilavy, and M. Mannik, *Am. J. Pathol.* **133,** 54 (1988).
135. S. L. Loke, C. A. Stein, X. H. Zhang, K. Mori, M. Nakanishi, C. Subasinghe, J. S. Cohen, and L. M. Neckers, *Proc. Natl. Acad. Sci. U.S.A.* **86,** 3474 (1989).
136. D. A. Geselowitz and L. M. Neckers, *Antisense Res. Dev.* **2,** 17 (1992).
137. T. Doi, K. Higashino, Y. Kurihara, Y. Wada, T. Miyazaki, H. Nakamura, S. Uesugi, T. Ima-

nishi, Y. Kawabe, H. Itakura, Y. Yazaki, A. Matsumoto, and T. Kodama, *J. Biol. Chem.* **268**, 2126 (1993).
137a. M. S. Brown, S. K. Basu, J. R. Falck, J. K. Ho, and J. L. Goldstein, *J. Supramolec. Struct.* **13**, 67, (1980).
137b. Y. Kimura, K. Sonehara, E. Kuramoto, T. Makino, S. Yamamoto, T. Yamamoto, T. Kataoka, and T. Tokunaga, *J. Biochem.* **116**, 991 (1994).
138. A. Peyman, A. Ryte, M. Helsberg, G. Kretzschmar, M. Mag, and E. Uhlmann, *Nucleosides Nucleotides* **14**, 1077 (1995).
139. M. Kulka and L. Aurelian, *Antisense Res. Dev.* **5**, 243 (1995).
140. E. L. Barry, F. A. Gesek, and P. A. Friedman, *Biotechniques* **15**, 1016 (1993).
141. Q. Zhao, J. Temsamani, and S. Agrawal, *Antisense Res. Dev.* **5**, 185 (1995).
142. C. F. Bennet, M.-Y. Chiang, H. Chan, J. E. E. Schoemaker, and C. K. Mirabelli, *Mol. Pharmacol.* **41**, 1023 (1992).
143. C. F. Bennett, M.-Y. Chiang, H. Chan, and S. Grimm, *J. Liposome Res.* **3**, 85 (1993).
144. S. Capaccioli, G. Di Pasguale, E. Mini, T. Mazzei, and A. Quattrone, *J. Immunol.* **197**, 818 (1993).
145. K. Lappalainen, A. Urtti, I. Jaaskelainen, K. Syrjanen, and S. Syrjanen, *Antiviral Res.* **23**, 119 (1994).
146. K. Lappalainen, A. Urtti, E. Soderling, I. Jaaskelainen, K. Syrjanen, and S. Syrjanen, *Biochim. Biophys Acta* **1196**, 201 (1994).
147. A. S. Boutorin, L. V. Guskova, E. M. Ivanova, N. D. Kobetz, V. F. Zarytova, A. S. Ryte, L. V. Yurchenko, and V. V. Vlassov, *FEBS Lett.* **254**, 129 (1989).
148. R. L. Letsinger, G. Zhang, D. K. Sun, T. Ikeuchi, and P. S. Sarin, *Proc. Natl. Acad. Sci. U.S.A.* **86**, 6553 (1989).
149. A. V. Kabanov, S. V. Vinogradov, A. V. Ovcharenko, A. V. Krivonos, N. S. Metlik-Nubarov, V. I. Kiselev, and E. S. Severin, *FEBS Lett.* **259**, 327 (1990).
150. F. P. Svinarchuk, D. A. Konevetz, O. A. Pliasunuva, A. G. Pokrovsky, and V. V. Vlassov, *Biochemie* **75**, 49 (1993).
151. N. N. Polushin and J. S. Cohen, *Nucleic Acids Res.* **22**, 5492 (1994).
152. J.-P. Bongartz, A.-M. Aubertin, P. G. Milhaud, and B. Lebleu, *Nucleic Acids Res.* **22**, 4681 (1994).
153. R. G. Shea, J. C. Marsters, and N. Bischofberger, *Nucleic Acids Res.* **18**, 3777 (1990).
154. A. S. Ryte, V. N. Karamyshev, M. N. Nechaeva, Z. V. Guskova, E. M. Ivanova, V. F. Zarytova, and V. V. Vlassov, *FEBS Lett.* **299**, 124 (1992).
155. P. C. Smidt, T. L. Doan, S. Falco, and T. J.-C. van Berkel, *Nucleic Acids Res.* **19**, 4695 (1991).
156. A. M. Krieg, J. Tonkinson, S. Matson, Q. Zhao, M. Saxon, L. M. Zhang, U. Banja, L. Yakubov, and C. A. Stein, *Proc. Natl. Acad. Sci. U.S.A.* **90**, 1048 (1993).
157. G. Degols, J.-P. Leonetti, M. Benkirane, C. Devaux, and B. Lebleu, *Antisense Res. Dev.* **2**, 293 (1992).
158. G. Citro, C. Szczylik, P. Ginobbi, G. Zupi, and B. Calabretta, *Br. J. Cancer* **69**, 463 (1994).
159. E. Bonfils, C. Mendes, A.-C. Roche, M. Monsigny, and P. Midoux, *Bioconj. Chem.* **3**, 277 (1992).
160. E. Bonfils, C. Deppierreux, P. Midoux, N. T. Thuong, M. Monsignyand, and A. C. Roche, *Nucleic Acid Res.* **20**, 4621 (1992).
161. J. J. Hangeland, J. T. Levis, Y. C. Lee, and P. O. P. T'so, *Bioconj. Chem.* **6**, 695 (1995).
161a. W. M. Pardridge, R. J. Boado, and Y.-S. Kang, *Proc. Natl. Acad. Sci. U.S.A.* **92**, 5592 (1995).
162. R. L. Juliano and S. Akhtar, *Antisense Res. Dev.* **2**, 165 (1992).
163. S. Akhtar and R. L. Juliano, *J. Controlled Rel.* **22**, 47 (1992).
164. R. Straubinger, N. Duzgunes, and D. Papahadjopoulos, *FEBS Lett.* **179**, 148 (1985).
165. R. Morishita, G. H. Gibbons, Y. Kaneda, T. Ogihara, and V. J. Dzau, *Gene* **149**, 13 (1994).

166. C. Ropert, M. Lavignon, C. Dubernet, P. Couvreur, and C. Malvy, *Biochem. Biophys. Res. Commun.* **183,** 879 (1992).
167. J.-P. Leonetti, P. Machy, G. Degols, B. Lebleu, and L. Leserman, *Proc. Natl. Acad. Sci. U.S.A.* **87,** 2448 (1990).
168. S. Wang, R. J. Lee, G. Cauchon, D. G. Gorenstein, and P. S. Low, *Proc. Natl. Acad. Sci. U.S.A.* **92,** 3318 (1995).
169. R. Blumenthal and A. Loyter, *Trends Biotechnol.* **9,** 41 (1991).
170. V. V. Vlassov, E. M. Ivanova, Yu. D. Krendelev, I. V. Kutyavin, M. N. Ovander, A. S. Ryte, F. P. Svinarchuk, and L. A. Yakubov, *Biopolimery Kletka* **5,** 52 (1989).
171. V. V. Pogodina, T. V. Frolova, T. V. Abramova, V. V. Vlassov, E. M. Ivanova, I. V. Kutiavin, A. G. Pletnev, and L. A. Yakubov, *Dokl. Akad. Nauk SSSR* **301,** 1257 (1988).
172. O. M. Bazanova, V. V. Vlassov, V. F. Zarytova, E. M. Ivanova, E. A. Kuligina, L. A. Yakubov, M. N. Abdukajumov, V. N. Karamyshev, and G. Zon, *Nucleotides Nucleosides* **10,** 523 (1991).
173. H. Sands, L. J. Gorey-Feret, A. J. Cocuzza, F. W. Hobbs, D. Chidester, and G. L. Trainor, *Mol. Pharmacol.* **45,** 932 (1994).
174. S. Agrawal, J. Temsamani, W. Galbraith, and J. Tang, *Clin. Pharmacokinet.* **28,** 7 (1995).
175. S. Agrawal, X. Zhang, Z. Lu, H. Zhao, J. M. Tamburin, J. Yan, H. Cai, R. B. Diasio, I. Habus, Z. Jiang, R. P. Iyer, D. Yu, and R. Zhang, *Biochem. Pharmacol.* **50,** 571 (1995).
176. P. A. Cossum, L. Truong, S. R. Owens, P. M. Markham, J. P. Sheaand, and S. T. Crooke, *J. Pharmacol. Exp. Ther.* **269,** 89 (1994).
177. R. Zhang, R. B. Diasio, Z. Lu, T. Liu, Z. Jiang, W. M. Galbraith, and S. Agrawal, *Biochem. Pharmacol.* **49,** 929 (1995).
178. S. T. Crooke, L. R. Grillone, A. Tendolkar, A. Garret, M. J. Fratkin, J. Leeds, and W. H. Barr, *Clin. Pharmacol. Ther.* **56,** 641 (1994).
179. V. V. Vlassov, L. A. Yakubov, V. Karamyshev, L. Pautova, E. Rykova, and M. Nechaeva, in "Delivery Strategies for Antisense Oligonucleotide Therapeutics" (S. Akhtar, ed.), p. 71. CRC Press, Boca Raton, Florida, 1995.
180. J. Saijo, L. Perlaky, H. Wang, and H. Busch, *Oncol. Res.* **6,** 243 (1994).
181. H. W. Nolen III, P. Catz, and D. R. Friend, *Int. J. Pharmacokinet.* **107,** 169 (1994).
182. J. P. Shaw, K. Kent, J. Bird, J. Fishback, and B. Froehler, *Nucleic Acids Res.* **19,** 747 (1991).
183. J. Temsamani, J. Y. Tang, A. Padmapriya, M. Kubert, and S. Agrawal, *Antisense Res. Dev.* **3,** 277 (1993).
184. T. Miyao, Y. Takakura, T. Akiyama, F. Yoneda, H. Sezaki, and M. Hashida, *Antisense Res. Dev.* **5,** 115 (1995).
185. H. B. Gamper, M. W. Reed, T. Cox, J. S. Virosco, A. D. Adams, A. A. Gall, J. K. Scholler, and R. B. Meyer, *Nucleic Acids Res.* **21,** 145 (1993).
186. P. Iversen, in "Antisense Research and Applications" (S. T. Crooke and B. Lebleu, eds.), p. 461. CRC Press, Boca Raton, Florida, 1993.
187. F. Yee, H. Ericson, D. J. Reis, and C. Wahlestedt, *Cell. Mol. Neurobiol.* **14,** 475 (1994).
188. L. Whitesell, D. Geselowitz, C. Chavany, B. Fahmy, S. Walbridge, J. R. Alger, and L. M. Neckers, *Proc. Natl. Acad. Sci. U.S.A.* **90,** 4665 (1993).
189. T.-L. Chem, P. S. Miller, P. O. T'so, and M. Colvin, *Drug Metabol. Distribut.* **18,** 815 (1990).
190. Y. S. Kang, R. J. Boado, and W. M. Pardridge, *Drug Metab. Disposit.* **23,** 55 (1995).
191. R. M. Crooke, *Anti-Cancer Drug Des.* **6,** 609 (1991).
192. R. M. Crooke, *Anti-Cancer Drug Des.* **6,** 609 (1991).
193. W. M. Galbraith, W. C. Hobson, P. C. Giclas, P. J. Schechter, and S. Agrawal, *Antisense Res. Dev.* **4,** 201 (1994).
193a. K. G. Kornish, P. Iversen, L. Smith, M. Arneson, and E. Bayer, *Pharmacol. Commun.* **3,** 239 (1993).

193b. R. Zhang, J. Yan, H. Shahinian, G. Amin, Z. Lu, T. Liu, M. S. Saag, Z. Jiang, J. Temsamani, R. R. Martin, P. J. Schechter, S. Agrawal, and R. B. Diasio, *Clin. Pharmacol. Ther.* **58,** 44 (1995).
194. V. V. Vlassov, A. A. Godovikov, N. D. Kobets, A. S. Ryte, L. V. Yurchenko, and A. G. Bukrinskaya, *Adv. Enzyme Reg.* **301,** (1986).
195. V. V. Vlassov, A. A. Godovikov, V. F. Zarytova, E. M. Ivanova, and N. Yu. *Nomokonova, Molekul. Biol. (Russia)* **24,** 173 (1990).
196. I. Duroux, G. Godard, M. Boidot-Forget, G. Schwab, C. Helene, and T. Saison-Behmoaras, *Nucleic Acids Res.* **23,** 3411 (1995).
196a. J. M. Kean, A. Murakami, K. R. Blake, C. D. Cushman, and P. S. Miller, *Biochemistry* **27,** 9113 (1988).
197. H. E. Johansson, G. J. Belsham, B. S. Sproat, and M. W. Hentze, *Nucleic Acids Res.* **22,** 4591 (1994).
198. R. Y. Walder and J. A. Walder, *Proc. Natl. Acad. Sci. U.S.A.* **85,** 5011 (1988).
199. P. F. Torrence, R. K. Maitra, K. Lesiak, S. Khamnei, A. Zhou, and R. H. Silverman, *Proc. Natl. Acad. Sci. U.S.A.* **90,** 300 (1993).
200. N. M. Dean and R. McKay, *Proc. Natl. Acad. Sci. U.S.A.* **91,** 11762 (1994).
201. Y. Mizutani, B. Bonavida, M. Fukumoto, and O. Yoshida, *J. Immunother.* **17,** 78 (1995).
201a. L. Whitesell, A. Rosolen, and L. M. Neckers, *Antisense Res. Dev.* **1,** 343 (1991).
202. G. Godard, J.-C. Francois, I. Duroux, U. Asseline, M. Chassignol, N. Thuong, C. Helene, and T. Saison-Behmoaras, *Nucleic Acids Res.* **22,** 4789 (1994).
203. P. Fabritiis and B. Calabretta, *Hematologia* **80,** 295 (1995).
203a. F. X. Mahon, J. Ripoche, V. Pigeonnier, B. Jazwiec, A. Pigneux, J. F. Moreau, and J. Reiffers, *Exp. Hematol.* **23,** 1606 (1995).
204. A. M. Tari, S. D. Tucker, A. Deisseroth, and G. Lopez-Berestein, *Blood* **84,** 601 (1994).
204a. F. E. Cetter, P. Johnson, P. Hall, C. Pocock, N. Al Mahdi, J. K. Cowell, and G. Morgan, *Oncogene* **9,** 3049 (1994).
204b. P. Fabritiis, S. Amadori, M. C. Petti, M. Mancini, E. Montefusco, A. Picardi, T. Geiser, K. Campbell, B. Calabretta, and F. Mandelli, *Leukemia* **9,** 662 (1995).
204c. T. F. C. M. Smetsers, L. T. F. Locht, A. H. M. Pennings, H. M. C. Wessels, T. M. de Witte, and E. J. B. M. Mensink, *Leukemia* **9,** 118 (1995).
205. J. L. Vaerman, C. Lammineur, P. Moureau, P. Lewalle, F. Deldime, M. Blumenfeld, and P. Martiat, *Blood* **86,** 3891 (1995).
206. Y. Shi, A. Fard, A. Galeo, H. G. Hutchinson, P. Vermani, G. R. Dodge, D. J. Hall, F. Shaheen, and A. Zalewski, *Circulation* **90,** 944 (1994).
207. M. R. Bennett, S. Anglin, J. R. McEvan, R. Jagoe, A. C. Newby, and G. I. Evan, *J. Clin. Invest.* **93,** 820 (1994).
208. M. V. Autieri, T.-L. Yue, G. Z. Ferstein, and E. Ohlstein, *Biochem. Biophys. Res. Commun.* **213,** 827 (1995).
209. C. Moulds, J. G. Lewis, B. C. Froehler, D. Grant, T. Huang, J. F. Milligan, M. D. Matteucci, and R. Wagner, *Biochemistry* **34,** 5044 (1995).
209a. M. A. Bonham, S. Brown, A. L. Boyd, P. H. Brown, D. A. Bruckenstein, J. C. Hanvey, S. A. Thomson, A. Pipe, F. Hassman, J. E. Bisi, B. C. Froehler, M. D. Matteucci, R. W. Wagner, S. A. Noble, and L. E. Babiss, *Nucleic Acids Res.* **23,** 1197 (1995).
209b. C. Boiziau, S. Moreau, and J.-J. Toulme, *FEBS Lett.* **340,** 236 (1994).
210. V. V. Vlassov, V. F. Zarytova, I. V. Kutyavin, and S. V. Mamaev, *FEBS Lett.* **231,** 352 (1988).
211. B. C. F. Chu and L. E. Orgel, *Nucleic Acids Res.* **17,** 4783 (1989).
212. E. H. Postel, S. J. Flint, D. J. Kessler, and M. E. Hogan, *Proc. Natl. Acad. Sci. U.S.A.* **88,** 8227 (1991).
213. T. J. Thomas, C. A. Faaland, M. A. Gallo, and T. Thomas, *Nucleic Acids Res.* **23,** 3594 (1995).

214. M. Grigoriev, D. Praseuth, A. L. Guieysse, P. Robin, N. T. Thuong, C. Helene, and A. Harel-Bellan, *Proc. Natl. Acad. Sci. U.S.A.* **90,** 3501 (1993).
215. T. A. Vickers, M. C. Griffith, K. Ramasamy, L. M. Risen, and S. M. Freier, *Nucleic Acids Res.* **23,** 3003 (1995).
216. H. J. Larsen and P. Nielsen, *Nucleic Acids Res.* **24,** 458 (1996).
216a. V. M. Macaulay, P. J. Bates, M. J. McLean, M. G. Rowlands, T. C. Jenkins, A. Ashworth, and S. Neidle, *FEBS Lett.* **372,** 222 (1995).
217. G. Wang, D. D. Levy, M. M. Seidman, and P. M. Glazer, *Mol. Cell. Biol.* **15,** 1759 (1995).
218. J. L. Whitton, *Adv. Virus Res.* **44,** 267 (1994).
219. C. A. Stein and Y.-C. Cheng, *Science* **261,** 1004 (1993).
220. T. V. Abramova, V. M. Blinov, V. V. Vlassov, V. V. Gorn, V. F. Zarytova, E. M. Ivanova, D. A. Konevets, O. A. Plyasunova, A. G. Pokrovski, L. S. Sandakhchiev, F. P. Svinarchuk, V. P. Starostin, and S. P. Chaplygina, *Nucleotides Nucleosides* **10,** 419 (1991).
221. R. L. Letsinger, G. Zhang, D. K. Sun, T. Ikeuchi, and P. S. Sarin, *Proc. Natl. Acad. Sci. U.S.A.* **86,** 6553 (1989).
222. C. A. Stein, R. Pal, A. L. DeViko, G. Hoke, S. Mumbauer, O. Kinstler, M. G. Sarngadharan, and R. L. Letsinger, *Biochemistry* **30,** 2439 (1991).
223. A. Zerial, N. T. Thuong, and C. Helene, *Nucleic Acids Res.* **15,** 9909 (1987).
224. A. Jacob, G. Duval-Valentin, D. Ingrand, N. T. Thuong, and C. Helene, *Eur. J. Biochem.* **216,** 19 (1993).
225. A. M. Krieg, J. Tonkinson, S. Matson, Q. Zhao, M. Saxon, L.-M. Zhang, U. Bhanja, L. Yakubov, and C. A. Stein, *Proc. Natl. Acad. Sci. U.S.A.* **90,** 1048 (1993).
226. J. Lisziewicz, D. Sun, F. F. Weichold, A. R. Thierry, P. Lusso, J. Tang, R. C. Gallo, and S. Agrawal, *Proc. Natl. Acad. Sci. U.S.A.* **91,** 7942 (1994).
227. B. Bordier, M. Perala-Heape, G. Degols, B. Lebleu, S. Litvak, L. Sarih-Cottin, and C. Helene, *Proc. Natl. Acad. Sci. U.S.A.* **92,** 9383 (1995).
228. W. M. McShan, R. D. Rossen, A. H. Laughter, J. A. Trial, D. J. Kessler, J. G. Zendegui, M. E. Hogan, and F. M. Orson, *J. Biol. Chem.* **267,** 5712 (1992).
229. F. Birg, D. Praseuth, A. Zerial, N. T. Thuong, U. Asseline, T. LeDoan, and C. Helene, *Nucleic Acids Res.* **18,** 2901 (1990).
230. K. Raviprakash, K. Liu, M. Matteucci, R. Wagner, R. Riffenburgh, and M. Carl, *J. Virol.* **69,** 69 (1995).
231. V. V. Pogodina, T. V. Frolova, M. P. Frolova, T. V. Abramova, V. V. Vlassov, D. G. Knorre, A. G. Pletnev and L. A. Yakubov, *Dokl. Akad. Nauk SSSR* **308,** 237 (1989).
232. H. Wu, J. S. Holcenberg, J. Tomich, J. Chen, P. A. Jones, S.-H. Huang, and K. L. Calame, *Gene* **89,** 203 (1990).
232a. D. Hipps and P. Schimmel, *EMBO J.* **14,** 4050 (1995).
233. T. Hatta, K. Takai, and H. Takaku, *Nucleosides Nucleotides* **14,** 1145 (1995).
234. E. De Clercq, B. D. Stollar, and M. N. Thang, *J. Gen. Virol.* **40,** 203 (1978).
235. J. P. Ebel, J. H. Weil, G. Beck, and C. Bollack, *Biochem. Biophys. Res. Commun.* **30,** 148 (1968).
236. E. I. Sklyanskaya and O. P. Peterson, *Vopr. Virusol. (Russia)* **3,** 489 (1963).
237. R. Repanovici, I. Moisa, O. Mihalache, V. Iacobescu, O. Burduce, R. Iliescu, and N. Cajal, *Rev. Rhumatol. Med. Virol.* **34,** 183 (1983).
238. A. G. Stewart, C. A. Grantham, K. M. Dawson, and N. Stebbing, *Arch. Virol.* **66,** 283 (1980).
239. N. Stebbing, *J. Gen. Virol.* **44,** 255 (1979).
240. N. Stebbing and I. J. D. Lindley, *Arch. Virol.* **64,** 57 (1980).
241. A. M. Krieg and C. A. Stein, *Antisense Rev. Dev.* **5,** 241 (1995).
241a. T. Hatta, K. Takai, and H. Takaku, *Nucleosides Nucleotides* **14,** 1145 (1995).

241b. V. V. Vlassov, E. I. Frolova, T. S. Godovikova, E. M. Ivanova, A. A. Koshkin, N. B. Ledovskikh, and G. A. Nevinsky, *Nucleosides Nucleotides* **10**, 645 (1991).
241c. T. D. Y. Chung, C. Cianci, M. Hagen, B. Terry, J. T. Matthews, M. Krystal, and R. J. Colonno, *Proc. Natl. Acad. Sci. U.S.A.* **91**, 2372 (1994).
241d. G. Ehrlich, D. Patinkin, D. Ginzberg, H. Zakut, F. Eckstein, and H. Sore, *Antisense Res. Dev.* **4**, 173 (1994).
242. C. Stein, L. Neckers, B. Nair, S. Mumbauer, G. Hoke, and R. Pal, *J. AIDS* **4**, 686 (1991).
242a. C. A. Stein, A. M. Cleary, L. Yakubov, and S. Lederman, *Antisense Res. Dev.* **3**, 19 (1993).
242b. T. Fujuhashi, T. Sakata, A. Kaji, and H. Kaju, *Antisense Res. Dev.* **11**, 461 (1995).
243. J. Wyatt, T. Vickers, J. Roberson, R. Buckheit, T. Klimkait, E. DeBacts, P. Davis, B. Rayner, J. L. Imbach, and D. J. Ecker, *Proc. Natl. Acad. Sci. U.S.A.* **91**, 1356 (1994).
244. D. J. Ecker, J. R. Wyatt, T. Vickers, R. Buckheit, J. Roberson, and J.-L. Imbach, *Nucleosides Nucleotides* **14**, 1117 (1995).
245. R. F. Rando, J. Ojwang, A. Elbaggari, G. R. Reyes, R. Tinder, M. S. McGrath, and M. E. Hogan, *J. Biol. Chem.* **270**, 1754 (1995).
246. W. Y. Gao, R. N. Hanes, M. A. Varquez-Padua, C. A. Stein, J. S. Cohen, and Y. C. Cheng, *Antimicrob. Agents Chemother.* **34**, 808 (1990).
247. S. M. Fennewald, S. Mustain, J. Ojwang, and R. F. Rando, *Antiviral Res.* **26**, 37 (1995).
248. P. P. Laktionov, E. Yu. Rykova, D. V. Krepky, A. V. Bryksin, and V. V. Vlassov, *Biokhimija (Russia)* (in press) (1996).
249. S. K. Srinivasan, H. K. Tewary, and P. L. Iversen, *Antisense Res. Dev.* **5**, 131 (1995).
250. L. Benimetskaya, L. Tonkinson, M. Koziolkiewicz, B. Karwowski, P. Guga, R. Zeltser, W. Stec, and C. A. Stein, *Nucleic Acids Res.* **23**, 4239 (1995).
251. Z. Khaled, L. Benimetskaya, R. Zeltser, T. Khan, H. W. Sharma, R. Narayanan, and C. A. Stein, *Nucleic Acids Res.* **24**, 737 (1996).
251a. M. A. Guvakova, L. A. Yakubov, I., Vlodavsky, J. L. Tonkinson, and C. A. Stein, *J. Biol. Chem.* **270**, 2620 (1995).
252. J. Desjardins, J. Mata, T. Brown, D. Graham, G. Zon, and P. Iversen, *J. Drug Targ.* **2**, 477 (1995).
253. G. H. Henderson and C. A. Stein, *Nucleic Acids Res.* **23**, 3726 (1995).
253a. M. Saxon, I. Schieren, L.-M. Zang, J. L. Tonkinson, and C. A. Stein, *Antisense Res. Dev.* **2**, 243 (1992).
254. M. W. Reed, E. A. Lukhtanov, V. V. Gorn, D. D. Lucas, J. H. Zhou, S. B. Pai, Y.-C. Cheng, and R. Meyer, Jr., *J. Med. Chem.* **38**, 4587 (1995).
255. S. H. Krawczyk, N. Bischofberger, L. C. Griffin, V. S. Law, R. G. Shea, and S. Swaminathan, *Nucleosides Nucleotides* **14**, 1109 (1995).
255a. J. Binkley, P. Allen, D. M. Brown, L. Green, C. Tuerk, and L. Gold, *Nucleic Acids Res.* **23**, 3198 (1995).
256. D. J. Schneider, J. Feigon, Z. Hostomsky, and L. Gold, *Biochemistry* **34**, 9599 (1995).
257. J. A. Doudna, T. R. Cech, and B. A. Sullenger, *Proc. Natl. Acad. Sci. U.S.A.* **92**, 2355 (1955).
258. E. DeMaeyer and J. DeMaeyer-Guignard, "Interferons and Other Regulatory Cytokines," p. 1. Wiley-Interscience, New York, 1988.
259. G. C. Sen and P. Lengyel, *J. Biol. Chem.* **267**, 5017 (1992).
260. P. F. Torrence and E. De Clercq, *Methods Enzymol.* **78**, 291 (1981).
261. S. Baron, S. E. Grossberg, G. R. Klimpel, and P. A. Brunell, *in* "Antiviral Agents and Viral Diseases of Man" (G. J. Galasso, T. C. Merigan, and R. A. Buchanan, eds.). Raven Press, New York, 1984.
262. T. Taniguchi, *EMBO J.* **7**, 3397 (1988).
263. A. Ray, S. B. Tatter, L. T. May, and P. B. Sehgal, *Proc. Natl. Acad. Sci. U.S.A.* **85**, 6701 (1988).

264. A. Maran, I. D. Goldberg, and B. M. Steinberg, *Mol. Cell. Biol.* **10,** 4424 (1990).
265. P. Lenguel, *Annu. Rev. Biochem.* **51,** 251 (1982).
266. I. M. Kerr and R. E. Brown, *Proc. Natl. Acad. Sci. U.S.A.* **75,** 256 (1978).
267. M. J. Clemens and B. R. G. Williams, *Cell* **13,** 565 (1978).
268. R. W. Sobol, W. L. Fisher, N. L. Reichenbach, A. Kumar, W. A. Beard, S. H. Wilson, R. Charubala, W. Pfleiderer, and R. J. Suhadolnik, *Biochemistry* **32,** 12112 (1993).
269. H. C. Schroder, M. Kelve, H. Schacke, W. Pfleiderer, R. Charubala, R. J. Suhadolnik, and W. E. G. Muller, *Chem. Biol. Interact.* **90,** 169 (1994).
270. P. J. Ferrell, G. C. Sen, M. F. Dubois, L. Ratner, E. Slattery, and P. Lengyel, *Proc. Natl. Acad. Sci. U.S.A.* **75,** 5893 (1978).
271. A. G. J. Hovanessian, *Interferon Res.* **9,** 641 (1989).
272. S. Y. Desai, R. C. Patel, G. C. Sen, P. Malhotra, G. D. Ghadge, and B. Thimmapaya, *J. Biol. Chem.* **270,** 7, 3454 (1995).
273. R. W. Sobol, E. E. Henderson, N. Kon, J. Shao, P. Hitzges, E. Mordechai, N. L. Reichenbach, R. Charubala, H. Schirmeister, W. Pfleiderer, and R. J. Suhadolnik, *J. Biol. Chem.* **270,** 11, 5963 (1995).
274. A. Kumar, J. Haque, J. Lacoste, J. Hiscott, and B. R. G. Williams, *Proc. Natl. Acad. Sci. U.S.A.* **91,** 6288 (1994).
275. E. F. Meurs, J. Galabru, G. N. Barber, M. G. Katze, and A. G. Hovanessian, *Proc. Natl. Acad. Sci. U.S.A.* **90,** 232 (1993).
276. T. Ito, R. Jagus, and W. S. May, *Proc. Natl. Acad. Sci. U.S.A.* **91,** 7455 (1994).
277. S. H. Gromkowski, K. Mama, J. Yagi, R. Sen, and S. Rath, *Int. Immunol.* **2,** 903 (1990).
278. C. Daly and N. C. Reich, *Mol. Cell. Biol.* **13,** 3756 (1993).
279. H. B. Levy, L. W. Law, and A. S. Rabson, *Proc. Natl. Acad. Sci. U.S.A.* **62,** 357 (1969).
280. J. N. Zullo, B. J. Cochran, A. S. Huang, and C. D. Stiles, *Cell* **43,** 793 (1985).
281. M. S. Chapekar, M. C. Knode, and R. I. Giazer, *Mol. Pharmacol.* **34,** 461 (1988).
282. H. R. Hubbell, *Int. J. Cancer* **37,** 359 (1986).
283. L. Borecky, V. Lackovic, and J. Rovensky, *Texas Rep. Biol. Med.* **41,** 575 (1982).
284. W. A. Carter, D. R. Strayer, H. R. Hubbell, and I. J. Brodsky, *J. Biol. Resp. Mod.* **4,** 495 (1985).
285. W. A. Carter, H. R. Hubbell, L. J. Krueger, and D. R. Strayer, *J. Biol. Resp. Mod.* **4,** 613 (1985).
286. D. R. Strayer, P. Watson, W. A. Carter, and I. J. Brodsky, *Interferon Res.* **6,** 373 (1986).
287. D. S. Pisetsky, *Antisense Res. Dev.* **5,** 219 (1995).
288. D. S. Pisetsky, *J. Immunol.* **6,** 421 (1996).
289. M. Zouali, B. D. Stollar, and R. S. Schwartz, *Immunol. Rev.* **105,** 137 (1988).
290. S. Avrameas, *Immunol. Today* **12,** 154 (1991).
291. C. Demaison, P. Chastagner, J. Theze, and M. Zouali, *Proc. Natl. Acad. Sci. U.S.A.* **91,** 514 (1992).
292. R. P. Braun, M. L. Woodsworth, and J. S. Lee, *Mol. Immunol.* **23,** 685 (1986).
293. B. D. Stollar, *FASEB J.* **8,** 337 (1994).
294. R. E. Cone, *Pharmacol. Ther.* **8,** 321 (1980).
295. S. Ikeda, J. Neyts, and E. De Clercq, *Proc. Soc. Exp. Biol. Med.* **207,** 191 (1994).
296. D. W. H. Riches, P. M. Henson, L. K. Remigio, J. F. Catterall, and R. C. Strunk, *J. Immunol.* **141,** 180 (1988).
297. F. R. Lake, E. C. Dempsey, J. D. Spahn, and D. W. H. Ricches, *Am. J. Physiol.* **266,** 134 (1994).
298. S. H. Hefeneider, S. L. McCoy, J. I. Morton, A. C. Bakke, K. A. Cornell, L. E. Brown, and R. Bennett, *Lupus* **1,** 167 (1992).
299. Reference deleted in proof.

300. J. C. Rogers and J. W. Kerstiens, *J. Immunol.* **126,** 703 (1981).
301. S. H. Hefeneider, K. A. Cornell, L. E. Brown, A. C. Bakke, S. L. McCoy, and R. M. Bennett, *Clin. Immunol. Immunopathol.* **63,** 245 (1992).
302. J. P. Messina, C. S. Gilkeson, and D. S. Pisetsky, *J. Immunol.* **147,** 1759 (1991).
303. A. M. Krieg, A.-K. Yi, S. Matson, T. J. Waldschmidt, G. A. Bishop, R. Teasdale, G. A. Koretzky, and D. Klinman, *Nature (London)* **374,** 546 (1995).
304. A. M. Krieg, A.-K. Yi, T. J. Waldschmidt, and G. A. Bishop, *Antisense Res. Dev.* **5,** 163 (1995).
305. A.-K. Yi, J. H. Chance, J. S. Cowdery, and A. M. Krieg, *J. Immunol.* **156,** 558 (1996).
305a. T. Yamamoto, S. Yamamoto, T. Kataoka, and T. Tokunaga, *Antisense Res. Dev.* **4,** 119 (1994).
306. R. F. Branda, A. L. Moore, L. Mathews, J. McCormack, and J. Zon, *Biochem. Pharmacol.* **45,** 2037 (1993).
307. R. Narayanan, *Antisense Res. Dev.* **4,** 139 (1994).
308. C. W. Muller and S. C. Harrison, *FEBS Lett.* **369,** 113 (1995).
309. K. W. McIntyre, K. Lombard-Gillooly, J. R. Perez, C. Kunsch, U. M. Sarmiento, J. D. Larigan, K. T. Landreth, and R. Narayanan. *Antisense Res. Dev.* **3,** 309 (1993).
310. U. M. Sarmiento, J. R. Perez, J. M. Becker, and R. Naraianan, *Antisense Res. Dev.* **4,** 99 (1994).
311. Q. Zhao, J. Temsamani, P. L. Iadarola, Z. Jiang, and S. Agrawal, *Biochem. Pharmacol.* **51,** 173 (1996).
312. J. R. Perez, Y. Li, C. A. Stein, S. Majumder, A. van Oorschot, and R. Narayanan. *Proc. Natl. Acad. Sci. U.S.A.* **91,** 5957 (1994).

coexistence, controlling replication becomes necessary both for plasmid maintenance and for the well being of the host. It is therefore no surprise that, without exception, naturally occurring plasmids are found to encode control mechanisms.

This article is essentially a monograph on P1 plasmid replication with some excursions into related replicons. In particular, we have emphasized the common ground between the P1 replicon and that of the bacterium, *Escherichia coli*. Plasmids similar to P1 manage to replicate in a variety of hosts, including *E. coli*. The liberty to move around is perhaps the main benefit of being extrachromosomal. How the plasmid adapts to the host replication machinery is, therefore, basic to our understanding of the process of plasmid replication.

I. Replication Control: Definition

Cells maintain correct gene dosage by doubling their DNA once per cell cycle. The exact solution to this problem depends on the system and the state of its vegetative growth. In eukaryotes, replication is restricted to the S phase of the cell cycle (*1, 2*). Initiation of replication occurs from thousands of different loci (origins) but the same origin is rarely used more than once in the same cell cycle. In rapidly growing cells, for example during embryonic development, the number of active origins increases considerably (*3*). In other words, many potential origins can remain silent during the normal cell cycle. The choice of origins that initiate may be to some extent stochastic, but once activated a particular origin is used once per cell cycle.

In *E. coli*, the entire chromosome (about 5×10^6 bp) is replicated from one origin, but its ploidy (2^n, where $n = 0, 1, 2, 3$, or even higher integers) increases with the growth rate of the cells (*4–6*). Irrespective of the number, all origins initiate once per generation and in synchrony (within a narrow window of time compared to the generation time). For eukaryotes, synchrony may not be the norm because the origins at different parts of the chromosome appear to initiate in a programmed temporal order (*7*). These differences notwithstanding, the essence of replication control—restriction of origin usage to once per cell cycle—is the same from bacteria to humans. Also, when adventitious copies of the same origin are present, all can function (*8, 9*). Thus origin usage and not its number is the key to the control process.

The situation is quite opposite in the case of bacterial plasmids (*10–14*). The control mechanism counts origin number and not the usage frequency. Although the plasmids are found in a variety of copy numbers, each is maintained at a characteristic copy number. The control mechanisms ensure that, in individual cells, the copy number approaches but does not exceed $2N$, N being the characteristic copy number of the plasmid in a newborn cell. Because N can fluctuate in individual cells of a population (primarily due to random segregation during cell division), the average origin usage can be more, or less, than once per cell cycle, depending on whether the copy number at birth is less, or more, than N. Therefore, to reach $2N$ some plasmid copies need to be replicated more than once in the same cell cycle and hence the strict rule of once-per-cell-cycle cannot be applied for maintenance of plasmid copy number. In fact, in the normal course (when newborn cells have N plasmids), plasmids are chosen at random for replication without regard to their replication history. The control system primarily senses plasmid origin concentration and progressively reduces initiation probability to zero as the origin concentration approaches $2N$. Thus the practice of the "N to $2N$ maxim" is fundamentally different for plasmids as compared to their host.

For unit copy plasmids ($N = 1$), there is the added requirement for a positive control to ensure replication before cell division. A passive mechanism—for instance, ineffective negative control at low copy numbers, as is commonly the case with high copy plasmids before replication (15)—may not be adequate. Thus a combination of positive and negative controls and some communication with the cell cycle are logical necessities for efficient maintenance of unit copy plasmids. In what follows, we shall elaborate on the multiple modes of control that may operate on P1 and related plasmids, and how the plasmid control systems compare with those of the host bacterium, *E. coli*.

II. P1 Basic Replicon: Isolation and Structure

Plasmid P1 is a prophage. The principal claim to fame of P1 is its capacity for generalized transduction (16). This was recognized in the early days of bacterial (*E. coli*) genetics when the phage was (and still is) used routinely for moving genes between bacterial strains. Another distinguishing feature of P1 is that the prophage is maintained as a circular plasmid. In all previously studied cases, the stable prophage was found to be an integral part of the chromosome. In this condition, almost all vegetative functions, including replication functions, were repressed. The plasmid state of P1 prophage implies that some replication function must be active in the otherwise largely repressed genome. The copy number of the prophage can be as low as one per cell, similar to that of the host chromosome or the sex factor plasmid, F. In rich medium, the copy number of P1 and F can reach four or more, as does the copy number of the host chromosomal origin, *oriC* (17, 18). This is the norm for bacterial plasmids in general, where the copy number control is set with respect to the chromosomal origin rather than with respect to the cell. The media effect suggests that the concentration of initiation factors is a determinant of copy number.

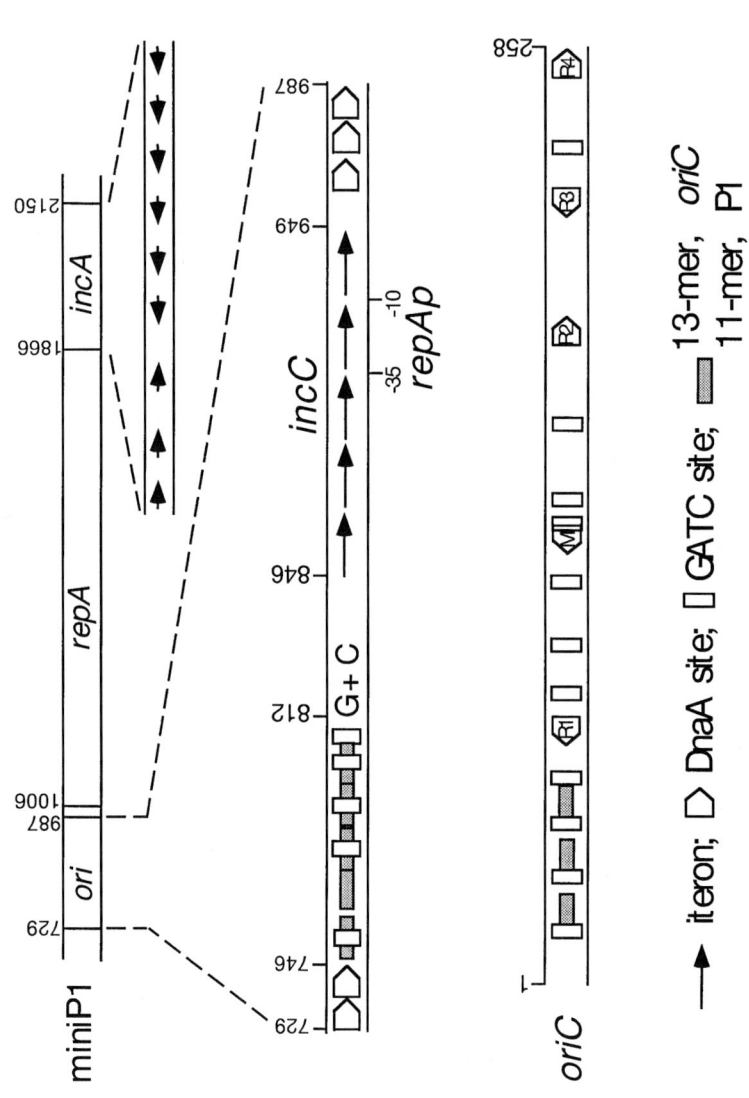

The basic replicon, the minimal region that confers replication characteristics of the parental plasmid, was identified in the case of P1 by cloning its fragments into an integration defective λ vector (19). A λ–P1 hybrid phage could be obtained that lysogenized *E. coli* as a plasmid with a copy number essentially identical to that of P1. The plasmid prophage was also stably maintained but required an additional P1 function, Par, which ensures equipartition of daughter DNAs to daughter cells (20). In the absence of partition functions, the daughter DNAs are distributed at random and, if the copy number is low, plasmid free cells can form easily. The loss rate per cell per division by random distribution (L) is given by $L = 0.5^n = 1 - V^{1/g}$, where n is the copy number at the time of cell division and V is the proportion of cells retaining the plasmid after g generations (18). The availability of λ–P1 hybrid phages was a boon in early stages of P1 plasmid research, because the 100-kb parent plasmid was difficult to manipulate. The hybrids were functionally discrete and were amenable to powerful techniques of λ genetics. More importantly, because the chimera could be grown as phages utilizing the λ replicon, and not the P1 replicon, DNA preparations became easy and there was no selection of high-copy mutant DNAs, which is otherwise a concern for particularly low-copy-number plasmids such as P1.

The P1 basic replicon has been localized to an approximately 1.4-kb piece of DNA (Fig. 1) (21–25). The region has one cis-acting element, the origin, *ori* (≈260 bp; coordinates 729–987, Fig. 1), followed by two trans-acting el-

---

FIG. 1. *Top:* The basic replicon of plasmid P1. The replicon is about 1.4 kb long and is often called mini-P1 (25). The origin (*ori*) has multiple sites for binding DnaA and RepA initiators as well as five GATC sites for Dam methylation. The term "iterons" specifically refers to RepA-binding sites. The promoter for the *repA* gene, *repAp*, also maps within the iterons. The DnaA sites map at both ends of the origin, but sites at either end suffice for origin activity (coordinates 729–949 or 746–987 can function as an origin). The left one-third of the origin is 61% A + T (between coordinates 729–812) whereas the origin region overall is 50% A + T (between coordinates 729–987). Following the GATC sites, 15 out of 18 bases are G + C. The G + C richness is functionally significant (219). The region covering the GATC sites suffices to bind the SeqA protein if the sites are hemimethylated. The GATC sites are part of an 10-mer repeat element, 5'-A/T T/C A G A/G T C C C/A T/A-3', the significance of which is not known (219). The *incA* locus contains nine iterons very similar to the ones present at the *ori* except that the spacing between the iterons is larger in *incA*. The significance of the spacing differences is not understood, especially because the origin iterons (called *incC*) can also control (25). The coordinates are from GenBank (accession number K02380). *Bottom: oriC* refers to a 258-bp segment of *E. coli* that can promote autonomous replication. The map shows the disposition of 11 GATC and 5 DnaA sites [labeled R1 to R4 and M (220)]. The left side of *oriC* is (A + T)-rich and is the region that preferentially unwinds under appropriate conditions. The region has three very similar 13-bp sequences (13-mers), each starting with a GATC. On hemimethylation the region binds SeqA with high affinity, like the 10-mers of P1. The 10-mers and the 13-mers are not homologous. The clustering of GATC sites and the hemimethylated states are sufficient for SeqA binding. The significance of the 10-mer and 13-mer sequences remains to be understood.

ements: the initiator gene, *repA* [open reading frame (ORF) = 858 bp], and the control locus, *incA* (≈285 bp; coordinates 1866–2150). Deletion of the control locus leads to an eightfold increase in plasmid copy number (*17*). The locus therefore plays only a negative role and is primarily responsible for the low copy number of the plasmid. The product of *repA* gene is a 32-kDa protein, RepA, essential for initiation. It is the only plasmid-encoded protein required for replication. *Eschericha coli* cell extracts can support replication of a supercoiled template carrying the P1*ori*, when supplemented with a purified preparation of RepA alone (*26*). The origin contains five 18-bp repeated sequences, called iterons, that bind RepA. The origin-specific binding activity and the absolute requirement for replication confer on RepA the title "initiator" (*27*). The protein is also required for control of initiation frequency. In fact, the control locus, *incA*, is nothing but a collection of nine additional iterons (*17*).

III. Iterons and Their Evolution

The *repA* gene and the 14 iterons cover 1.3 kb of the approximately 1.4-kb basic replicon (Fig. 1). Study of RepA–iteron interactions has been of major interest not only in P1 but also in several other iteron-containing plasmids. Information analysis of the iterons has been done here for 10 different plasmids and presented as sequence logos (Fig. 2). Comparison of the logos suggests how the sequences could have evolved and predicts how the protein domains that interact with the sequences could be arranged. The logo for the 14 iterons of P1 RepA, shown in the supper left-hand corner, will be used as a reference.

A sequence logo (*28*) summarizes the data in a set of aligned binding sites. At positions where the base is invariant, that base is drawn as a tall letter (e.g., position +7 in P1 RepA). Where there is more than one base, a stack of letters is made with the height of each letter being drawn proportional to the number of bases at that position. The height of the stack of letters is given in bits of information, a mathematically robust measure of sequence conservation. One bit is the choice between two equally likely possibilities. For example, when one flips a coin, the result can be reported with one bit of information. During the process of locating their binding sites, the proteins choose one base from the four possible bases. If we arrange these in a square, then we can select one of the four by answering no more than two yes–no questions: Is it on top? Is it on the left? Thus the scale on the sequence logo runs from 0 to 2 bits. When the protein accepts two bases (e.g., position +10 in P1 RepA has A or C), then it makes a 2 out of 4 choice. This is the same as a 1 in 2 choice, which is 1 bit, so the stack of letters is drawn 1 bit high. If all four bases appear at a position with equal frequency, then the protein is not making any choices and the stack is 0 bits high. Because the frequency of bases is rarely 100, 50, or 25%, a formula is used to compute the average sequence conservation in bits. The formula gives 2, 1, and 0 bits for the simple cases discussed so far but it also gives intermediate values. See the world wide web site

http://www-lmmb.ncifcrf.gov/~toms/paper/primer

for a short primer on information theory that explains how this formula works. Other useful references are Shannon's original papers (29) and the book by Pierce (30).

When sequence conservation is represented as a consensus sequence, the frequencies of bases are not evident. For example, if a position has 75% A and 25% T, it is often written as an A. This means that the consensus sequence will fail to match 25% of the sequences. Sequence logos do not require one to make any arbitrary cutoffs and so they avoid this problem. They also make no assumption as to the site size. Logos have error bars that vary inversely with the number of sequences available and show one standard deviation of the stack height. When the stack heights are near zero, and, as a consequence, close to the height of an error bar, one can no longer tell whether there is significant sequence conservation there.

Above the sequence logo is a sine wave that represents the twist of B-form DNA. The peaks are places where the major groove faces the protein. In Fig. 2, a filled circle (●) below the logo indicates positions that are protected from dimethylsulfate (DMS) methylation by the protein or that block RepA binding when methylated. Open circles (○) are bases that do not have these effects, so the minor groove of these bases faces RepA. A "+," as in R6K π binding sites, indicates enhanced methylation. When the relative orientation of the protein and DNA is not known, the sine wave is dashed. In these cases the wave is positioned by analogy with other cases or so that more information comes from the major groove. Because of the structure of DNA, more distinctions can be made between the bases by using a probe in the major groove (31). In B-form DNA the major groove can support up to 2 bits of sequence conservation, whereas the minor groove can support only 1 bit (32, 33). This simple rule allows one to orient the protein on the DNA.

P1 RepA sites have three discernible components. On the left side (positions −1 to +6) is a patch of sequence conservation containing Ts and Gs that follows the sine wave. This effect is expected because DNA is cylindrical and it is harder to evolve contacts with bases that do not face the protein directly (32). On the right side of the RepA site is a patch of Gs (34). In the middle of the RepA site is an unusual spike of sequence conservation where the base T is almost completely conserved, even though a minor groove faces the protein (33). It is unusual because, as discussed above, the structure of DNA makes it difficult to evolve conservation of more than 1 bit by contacts to the minor groove. This has been explained by proposing that the DNA is distorted away from B-form.

The upper right hand corner of Fig. 2 shows the logo for Rts1 RepA sites. It looks similar to the P1 logo, but the T spike appears to have changed into a G spike and shifted somewhat. The P1 and Rts1 RepA proteins are 60% identical (25, 35), and P1 RepA can bind to Rts1 iterons (36).

Moving down a row, on the left we have mini-F RepE binding sites. The region 0 to +3 still has a TGTG patch, but on the right side the GGG patch has been replaced with a CCC patch. Apparently the same thing has happened to the RepA sites of pCU1 on the right side of the figure. This can be explained if we imagine that these Rep proteins have two parts, one that binds a TGTG patch and the other that binds a GGG patch. Divergent evolution between the top row and the second row could represent the rotation of the GGG patch recognizer by 180°. It would then recognize a CCC patch.

This hypothesis is borne out nicely by two more examples shown on the third row. In RepHI1B, the GGG patch is conserved on the right-hand side,

---

FIG. 2. Sequence logos of iterons, sites where plasmid-encoded initiator proteins bind for initiation and control of the initiation frequency. The principles involved in generation of the logos and other details of the figure are described in Section III. The data here include plasmid names, reference sources of iteron sequences and footprinting data (where applicable), GenBank accession numbers, and coordinates and orientation (+/−) of individual iterons (the coordinates refer to the fully conserved base T at position zero of each iteron, except for λ, which is centered). **P1** (*33, 66, 221*)–K02380 (coordinates, 848+, 869+, 890+, 911+, 933+, 1868+, 1899+, 1940+, 1994+, 2025−, 2056−, 2087−, 2117−, 2148−). **Rts1** (*35, 222*)–K00053 (coordinates, 1286−, 1265−, 1244−, 1223−, 1201−, 233−, 200−, 163−, 135−, 63−, 28−) and M60191 (coordinates, 355−, 271−, 177+, 79+). The two sequences match at the *Eco*RI site at the start of both sequences. **F** (*223–225*)–J01724 (coordinates, 961+, 982+, 1015+, 1037+, 1927−, 1949−, 1982−, 2073−, 2156−). **pCU1** (*226*)–M18262 (coordinates, 406+, 428+, 450+, 472+, 494+, 1380−, 1417−, 1454−, 1492−, 1529−, 1566−, 1603−, 1640−, 1677−, 1714−, 1751−, 1788−, 1825−, 1862−). **RepHI1B** (*227*)–X68824 (coordinates, 952+, 973+, 994+, 1015+, 1038+, 2030−, 2071−, 2074+, 2192−, 2226−, 2297−, 2363−). **RK2** (*228–231*)–L27758 (coordinates, 12382−, 12405−, 12427−, 12450−, 12473−, 12610−, 12645−, 12666−, 12942−, 11864−). **λ** (*232*)–J02459 (coordinates, 39043−+, 39063−+, 39087−+, 39110−+). **R6K** (*233, 233a, 234*)–V00320 (coordinates, 208+, 230+, 252+, 274+, 296+, 318+, 340+, 477+). **RepFIB** (*235*)–M26308 (coordinates, 2066+, 2087+, 2108+, 2129+, 2152+, 3302+, 3346−, 3386−, 3513−, 3531−, 3561−, 3956−) [the sites included are BCDD′D″EFGHIJK but not A (at coordinate 1720−) because of its poor information content (*236, 237*)]. **pSC101** (*237*)–K00042 (coordinates, 586+, 607+, 639+, 680+, 698−, 716+, 734−) [we have treated each half of the inverted repeats IR-1 and IR-2 (*54*) as separate iterons as these can be a part of the origin; although the data set is small, it is clear that the conservation is poor because the IR sequences differ significantly among each other and from the three iterons].

but the left-hand TGTG patch appears to have converted to a complementary CAC patch, as if the left-hand binding domain of the recognizer molecule had been rotated by 180°. The TrfA site of RK2 appears to be a hybrid between P1 and F because it has the TGACA of F and GGG of P1. A doubly inverted Rep site (i.e., CAC-CCC or GGG-GTG) has not been observed.

A final example of this kind of possible evolutionary relationship is shown by the λ O protein sites. Here the right-hand patch of Gs exists, but the protein has dimerized and so it also recognizes a patch of Cs on the left-hand side.

We do not know how to interpret the R6K sites shown on the right side of the fourth row, though it has a TGA patch (0 to +2) like the sites of F and RK2. The degree of conservation also appears significantly higher (by five standard deviations) than required for one protein to bind there ($R_s/R_f = 1.61 \pm 0.12$) (37, 38). The sequence requirement for π binding alone could be determined by a randomization experiment (33, 39).

Finally, we include on the bottom row of Fig. 2 a logo for sites from RepFIB. This logo resembles the logo for pColV-K30 closely (not shown) (40, 41). It is not obvious if these sites and those found on pSC101 are related to the others in the figure.

In several plasmids, inverted repeat (IR) sequences constitute the operator for autogenous synthesis of initiator proteins (see Section IV) (Fig. 3). We investigated the four cases in our collection using information theory techniques (39a). In F and R6K, there is a strong IR sequence that corresponds to the left half of the iterons, whereas the right half of the site is obliterated. For F, these sites are at 1110 (+) and 1134 (−) and they correspond to positions 0 to +7 in Fig. 2. For R6K these sites are at 503 (+) and 523 (−) and they correspond to positions −5 to +4 in Fig. 2. These both cover a single major groove and one or two surrounding minor grooves. That is, if these proteins have two parts, A and B, as proposed above, in both cases the A binding site is strongly preserved in the IR sequences but the B portion is not. Information analysis shows that the A portions alone would not be bound, but that the combination of two A portions is strong enough to specifically bind DNA. This model is consistent with data that show dimer binding to IR sequences (11, 42). It seems likely that the A and B portions of the proteins can move relative to one another and that when a dimer is formed the B portions move out of the way to allow the A portions to bind the DNA. It may even be that the B portion contains the dimerization interface.

The possible overlap of DNA binding and dimerization domains has been proposed (43). These observations suggest that the two-part nature of the Rep proteins is not merely an evolutionary hangover but is an active part of their modern function. In pSC101, there are two IRs (IR-1 and IR-2) that, like F and R6K, are homologous to one-half of the iterons (positions −4 to +2, Fig.

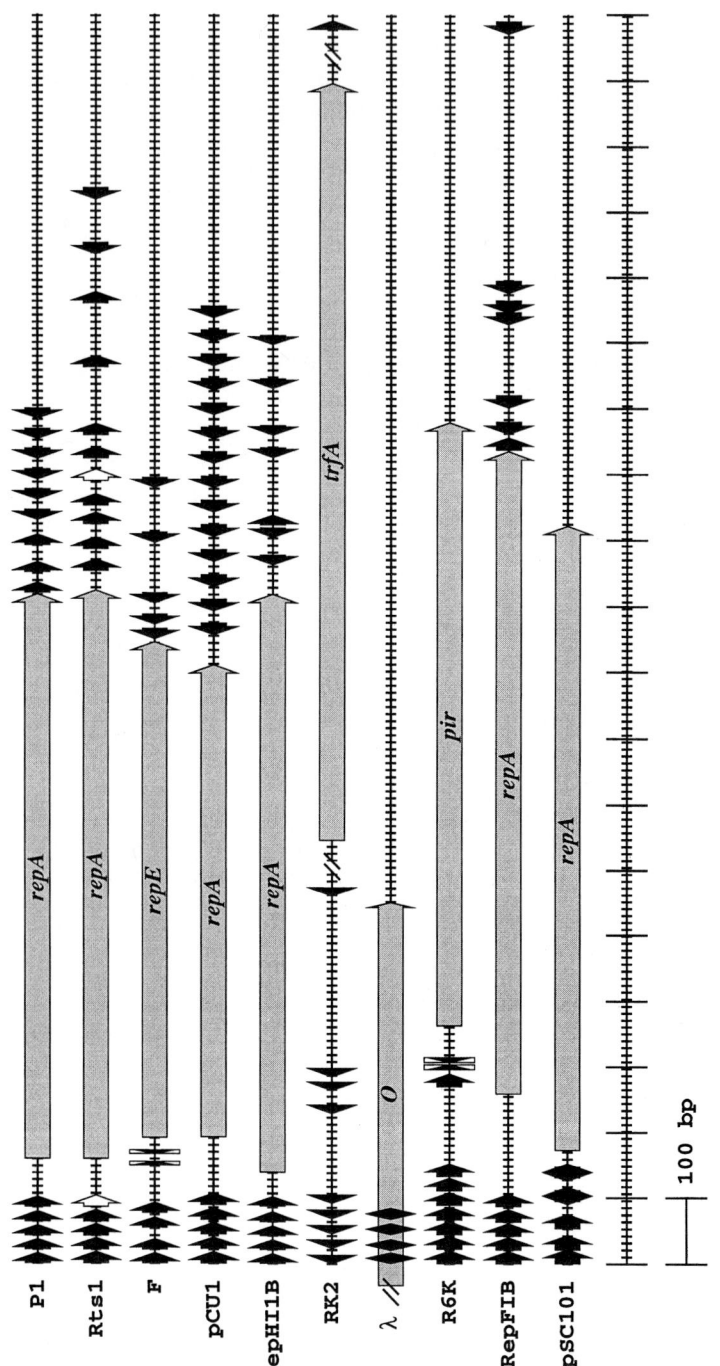

2). In this system also, dimers bind to IR sequences and monomers bind to iterons (44). Unfortunately, in pSC101 there are too few sites to do a rigorous information analysis. The IR sequence of Rts1 (coordinate 1179.5) (35) is different in that it overlaps with a strong iteron site at 1201 in our analysis. We speculate that the IR could be spurious and the iteron at 1201 could serve as the operator, like the situation in P1.

In summary, the iterons, although highly specific to their cognate initiators, appear to have in their sequence arrangements common features that allow one to make predictions about the domain organization of the initiators. We tried making a phylogenetic tree for the 10 initiator proteins using the GCG package. With various methods, the trees looked different. However, in all cases tried, P1 was closest to Rts1, and F was closest to pCU1, consistent with the expectation from the sequence logos. Other expectations were not met.

IV. Multiple Functions of Initiator Proteins

The initiator proteins of iteron-carrying plasmids are involved not only in initiation but also in regulation of their own synthesis and in the control of copy number. The structure–function analysis of the proteins has been lim-

---

FIG. 3. Plasmid maps showing relative locations of iterons (black arrows) and initiator genes (gray arrows). Iterons on the left belong to the origin. Iterons following the initiator genes (where applicable) are used exclusively for control. The length of each iteron arrow indicates the approximate length of the significant sequence conservation seen in Fig. 2. Empty arrows for Rts1 were identified in this work. The one belonging to the origin cluster at coordinate 1201 (K00053) is a well-conserved site and is included in Fig. 2. The other one at coordinate 97 (K00053) is consistently spaced with other iterons but has less conserved bases. It is likely to be a weak site. Iteron coordinates are from Fig. 2 and the reading frames for the initiator genes are from the following sources: P1 *repA* (221); Rts1 *repA* (35); F *repE* (223); pCU1 *repA* (226); RepHI1B *repA* (227); RK2 *trfA* (238) (GenBank entry Z37515); R6K *pir* (233a); RepFIB *repA* (235); pSC101 *repA* (239); λ *O* (232). At the bottom of the figure large tic marks show a 100-bp scale. Small tick marks are placed every 10.6 bp to indicate the twist of B-form DNA. These show that the iterons are frequently placed on one face of the DNA. The *trfA* gene of RK2 is 5 kb away from the origin (228), as indicated by the two slashes. To align the origin of RK2 with the other origins, the iteron at coordinate 11864 was moved "around the plasmid" and placed on the right side of the figure. The break in the sequence is indicated by two slashes. In F and R6K, the IR sequences that serve as operators for autoregulation of the initiator genes are shown as rectangles to indicate that the portion of the iteron that is conserved is the tail end of the iterons. A black triangle is drawn on top of the rectangle to indicate its orientation. The situation is similar in pSC101 except that the IRs can be a part of the origin. Each half of the IRs is therefore represented as an iteron in this analysis (they appear as diamonds because the arrows are on top of each other). Although there is an IR in Rts1, our analysis only shows evidence for a single iteron there (empty arrow at 1201 in the origin cluster).

ited (*11, 45–48*). Nonetheless, these proteins should have multiple domains to account for their multiple functions, as given in the following list.

1. The proteins can act both as monomers and dimers. The monomers are the initiators and they bind to iterons (*42–44, 49, 50*). The dimers are autorepressors and they bind to sequences (operators) with inverted repeats (Fig. 3) (*42, 44*). The operator halves are homologous to a part of the iteron sequences, indicating that the protein domains that bind to the two types of sites have regions in common (also discussed in Section III). In a small number of cases, such as P1 and ColV-K30, there are no separate operator sites and the promoters are found to be buried within the origin iterons (*24, 51*). In these cases the initiators and the repressors are likely to both be monomers. There is genetic evidence in support of this hypothesis (*52*) but, on the other hand, the autorepression and the replication initiation phenotypes of the protein have also been partially separated by mutations (*53*). As we go along it will be apparent that the interpretation of the results is not simple because of the multiple functions of the initiator. A partial defect in binding activity, detected in an autorepression assay, can be more than compensated by changes in other steps of initiation. In pSC101, an IR site also participates in origin function, indicating that both monomers and dimers serve as initiators in this plasmid (*54–56*).

2. In the case of P1 and F, the iteron binding activity of initiators requires mediation of chaperone proteins DnaJ, DnaK, and GrpE *in vivo* (*57, 58*) and *in vitro* (*59, 60*). It has been found *in vitro* that for P1 RepA, the chaperone proteins can be replaced by ClpA, the ATPase component of the ATP-dependent ClpAP protease (*61*). The *in vivo* significance of the result remains to be seen. The chaperones increase the active monomer fraction either by helping the protein to fold properly or by converting dimers to monomers, because dimers are inactive in binding to the iterons (*42, 43, 50, 62, 63*). The active form of DnaA protein in solution is also monomeric but the protein has a tendency to aggregate, more so in the presence of phospholipids or in the nucleotide-free form *in vitro* (*64, 65*). The aggregates can be dissociated by phospholipase or chaperones (DnaK or GroEL) (*64*) (J. Marszalek and J. M. Kaguni, unpublished data). The chaperones can therefore contribute positively to initiation both for DnaA and the plasmid initiators.

3. The P1 initiator has been shown to bind to two consecutive major grooves on the same face of the iteron DNA (*33, 66*). This mode of binding appears to be generally true for other plasmid initiators (Fig. 2). Because the sequences in the grooves are different, the inference is

that two separate domains of a monomer bind to the two grooves. Apparently, one of the domains is involved in binding IR sequences in plasmids that carry them (Section III). The situation is somewhat different in λ, where the initiator O has one DNA-binding domain and the protein needs to be dimeric to bind to the two grooves, as is usually the case with many transcriptional repressors (S.-J. Um and R. McMacken, unpublished data).
4. Because multiple iterons are required for initiation, the individual initiator–iteron complexes most likely organize into a higher order structure by protein–protein interactions (67–69). This scenario is also applicable to multiple DnaA binding sites in *oriC* (65, 70, 71) (Fig. 1).
5. The plasmid initiator of pSC101 can help the host initiator, DnaA, to bind to the plasmid origin (72). The binding reaction can also be mutually cooperative, as seen in the case of P1*ori* (67). Both initiators and their binding sites are required for replication of plasmids like P1 (21, 23, 73), F (73, 74), and pSC101 (75). Most likely, the two initiators contact each other directly (45).
6. The DnaA (76) and the R6K and pSC101 initiators have been shown to interact with DnaB helicase and they may therefore help the helicase to load onto the initiation complexes (77). The R6K initiator, DnaA, and DnaB also interact directly with the DnaG primase (P. Ratnakar and D. Bastia, unpublished data). Thus it appears that the loading of the helicase and the primase is facilitated by multiple protein–protein contacts that involve both DnaA and the plasmid initiators.
7. In addition to specific binding to duplex DNA, DnaA and λ O proteins can also bind to single-stranded DNA (78, 79). This activity may help stabilize origin opening, an essential requirement for helicase loading, and may prove to be widespread among plasmid initiators.
8. Several plasmid initiators can pair iterons by protein–protein interactions. The activity was first detected in the case of the R6K initiator (80–82) and subsequently for initiators of λ (83, 84), P1 (85, 86), and RK2 (87). The activity has formed the basis of the handcuffing model of plasmid copy number control (Section VI,A). DnaA has also a similar activity but its significance has not been established (88). The significance is not certain for λ O either, but it is likely that the protein–protein interaction helps the initiation process.
9. Overexpression of plasmid initiator genes can either stimulate or inhibit initiation, depending on the level of synthesis (Section VI,B). Overexpression of *dnaA* also can increase initiation from *oriC* (89, 90). This is a good justification for why these genes are usually autoregulated.

Additionally, for *oriC* replication, the initiator activity of DnaA is dependent on binding to ATP (*91, 92*). In contrast, the initiator is inhibited on binding to ADP. The acidic phopholipids can convert the ADP-bound proteins to ATP form and thereby increase its initiator activity (*93–95*). Thus, for initiation, the form of DnaA is important and not necessarily its total amount. The plasmid initiators are not known to interact with nucleotides or phospholipids. However, mutations in a lipid-related gene, *ugpA*, have been shown to destabilize mini-F plasmids, suggesting that phospholipids could be important for plasmid replication (*96*). The regulatory consequences of the various initiator activities described in this section will be discussed in more detail below, in their proper context.

V. Mode of Replication: Similarity of P1 and *oriC*

Because of the work by Schnos *et al.* (*97*) and Bramhill and Kornberg (*98, 99*), understanding initiation in the θ mode of replication has become synonymous with determining the requirements of strand opening of the origin DNA. Once 20 or so bases of single-stranded DNA can be made available for loading of DnaB helicase, the mechanisms for the remainder of the steps in initiation and elongation are generally understood. However, as discussed in Section IV, the mechanisms of how initiators direct the assembly of a multiprotein initiation complex at the origin are only beginning to be understood (P. Ratnakar and D. Bastia, unpublished data). The subsequent steps of elongation are thought to be replicon independent and to involve only host proteins (*100*). However, host factors such as DnaA, Dam, and a newly discovered protein SeqA may regulate the frequency and timing of initiation of plasmid replication, as they do for *oriC* replication. We shall discuss these aspects in detail in Section VII.

The strand-opening of *oriC* has been worked out in some detail. DnaA protein alone suffices to open the strands of *oriC*, provided that the template is adequately supercoiled and that ATP is present (*98, 101*). The unwound state has been followed by sensitivity to single-strand-specific nucleases and by reactivity to $KMnO_4$, a reagent widely used to measure DNA unstacking (*70, 98, 102*). The unwinding of *oriC* was localized to a region containing (A + T)-rich 13-mers (Fig. 1) and was shown to be a normal intermediate of the initiation reaction *in vitro* and *in vivo*. It is not known how DNA binding of DnaA leads to strand opening, but it appears that the open state may be stabilized by some single-strand binding activity of DnaA (*78*). The top strand shows less reactivity than the bottom strand, suggesting that DnaA preferentially covers the 13-mers of the top strand so that they cannot zipper back

(70, 98, 101, 103). This way it is ensured that the bottom strand stays single stranded for interaction with the DnaB–DnaC complex.

There are many similarities in the structure of P1*ori* and *oriC* (Fig. 1). Both have sites for DnaA binding that are essential for function (67, 70, 103, 104). A second commonality is the presence of multiple GATC sequences, which are sites for Dam methylation. Again, methylation seems to play a similar role in both systems (105–108) (Section VII,E). Finally, the origins have sites for binding to the SeqA protein, which functions as a negative regulator of *oriC* replication (109, 110) (Section VII,E).

As in *oriC*, the presence of DnaA alone makes supercoiled P1*ori* DNA reactive to $KMnO_4$ (Fig. 4) (67, 68). RepA alone is inactive in promoting unwinding but RepA and DnaA together make the *ori* more reactive than does DnaA alone. This effect may be mediated through the cooperative binding of the two initiators to the origin. The bias of reactivity to the two strands of the origin was extreme in the case of P1*ori in vitro*. A second striking feature was that the strand that reacted strongly with $KMnO_4$ was also the one that reacted with $KMnO_4$ in the absence of any protein, when linear DNA was used instead of supercoiled DNA. Apparently, the two strands of the origin can form very different secondary structures but require release of superhelical tension, realized either by linearization or by the binding of two initiators, RepA and DnaA. This is a remarkable finding because most secondary structure formation is favored by superhelicity. The physiological relevance of these results is being studied by *in vivo* footprinting with $KMnO_4$ (K. Park and D. K. Chattoraj, unpublished). At the moment we can only speculate that the structures of the strands are such that only one is conducive for DnaB loading. Because the helicase can move in only one direction, the strand bias of its loading could be the basis for unidirectional replication of the plasmid seen *in vitro* (26). Recently, extreme strand bias has been seen at an origin of replication of human DNA when probed with DNaseI and DMS (111). Origin opening has also been demonstrated for λ*ori* (97) and plasmid R1162 (112). The strand bias was not conspicuous in the former study and the question was not addressed in the latter.

VI. Negative Control of Replication

The N to $2N$ dictum begs for a negative control mechanism that restrains copy number to $2N$. As was originally conceived by R. H. Pritchard, an active negative control is mediated by plasmid-encoded repressors of replication whose concentration depends on plasmid copy number (5). As the repressor concentration increases with replication, the probability of subsequent initi-

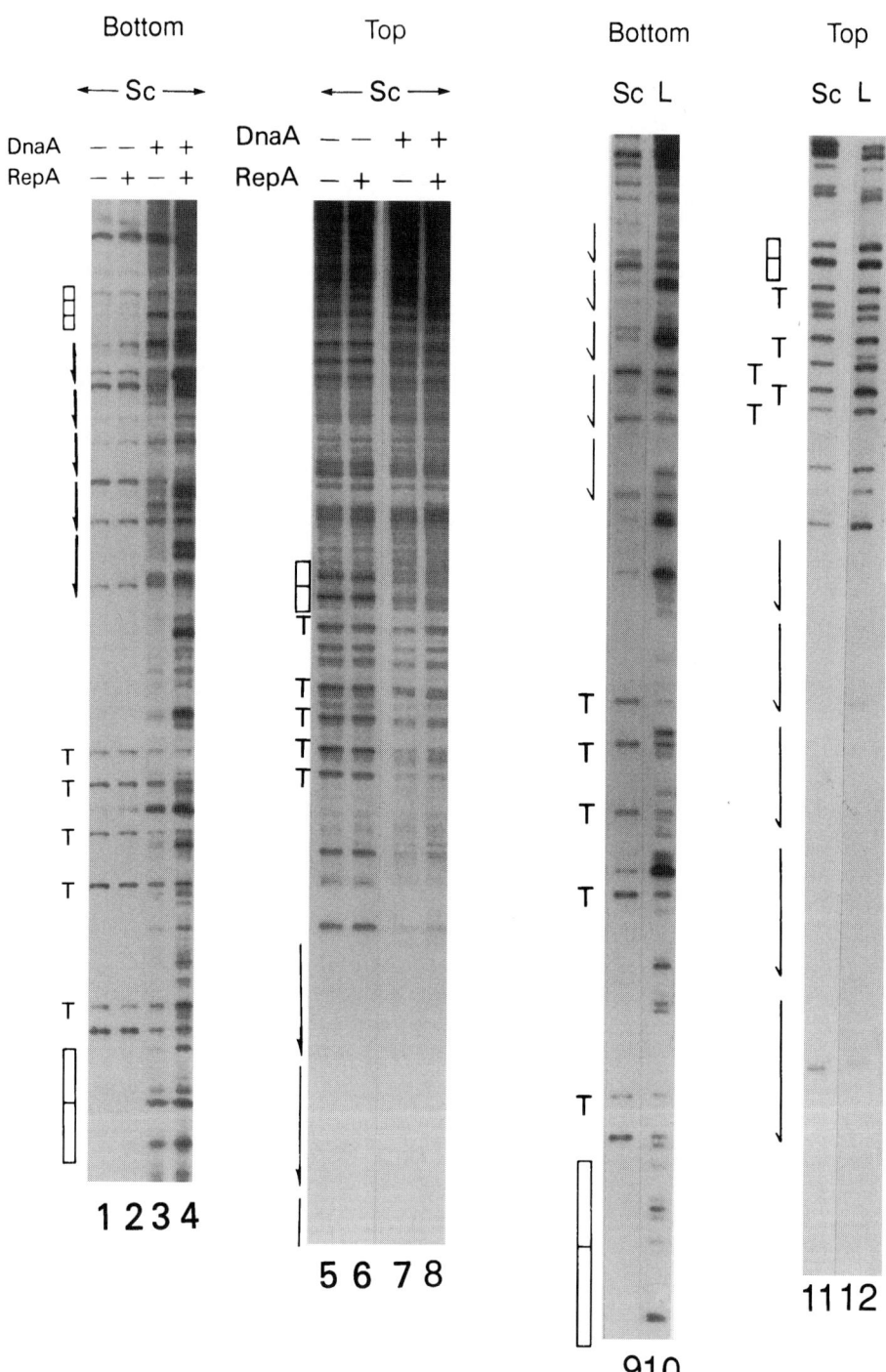

ation decreases correspondingly (negative feedback regulation). However, in order for the probability to approach zero on twofold increase in copy number, a high degree of cooperativity in the repressor action is required (5).

Repressors of replication have been found in all bacterial plasmids studied to date, thus supporting the generality of Pritchard's model (*10*). Repressors are basically of two types. In plasmids such as P1, the repressors are the iterons. Their binding to plasmid-encoded initiator proteins, an activity essential for initiation, is also the activity involved in control (*17*). As far as we know, the iterons do not make any RNA or protein product (*113, 114*). Although the iteron–initiator complexes have both positive and negative roles in initiation, they satisfy Pritchard's postulates rather well. Their concentration is obviously related to plasmid copy number and, when present approximately in equimolar to the origin in trans, they can inhibit initiation effectively *in vivo* and *in vitro* (*17, 115, 116*).

The other type of repressor is antisense RNA. There is considerable understanding of the mechanism of control by antisense RNA in plasmids such as ColE1, R1, and pT181 and the subject has also been reviewed extensively (*10, 15*).

There can also be a third class of plasmids wherein the repressor is a protein. The best studied member is derived from phage λ (*117*). However, in several plasmids, control by antisense RNA is further modulated by protein factors (*10, 15*). Iteron-carrying plasmids can be further controlled by small RNAs (*118, 119*). In this article, we shall focus on control by iterons, P1 being one of the prominent members of a large group of plasmids in which this type of control is present.

A. Models of Control by Iterons

1. INITIATOR TITRATION

The titration model posits that iterons compete for a limiting amount of initiator. As the iteron concentration increases due to replication, the control locus titrates some of the initiators, leaving less of the protein for binding to

---

FIG. 4. Strand bias of KMnO$_4$ reactivity of P1*ori in vitro*. Supercoiled (Sc) mini-P1 DNA was used *in vitro* either alone or in the presence of two initiators, RepA and DnaA. Some reactivity was seen with naked DNA but it was restricted primarily to base T of the triplet 5′-ATC-3′ (lanes 1 and 5). RepA alone did not change the background activity (lane 2) but DnaA alone did (lane 3). The reactivity was strongest in the presence of RepA and DnaA (lane 4), although RepA alone was ineffective. No such stimulation of reactivity could be seen in the top strand (lanes 4 and 8). The increased reactivity could also be seen on the naked linear (L) form of the *ori* DNA and again on the bottom strand (lanes 9 and 10) and not on the top strand (lanes 11 and 12). The KMnO$_4$-reacted bases were located by primer extension analysis. Rectangles show the positions of DnaA sites, each T shows the position of the base in GATC sites, and arrows show the positions of iterons. (Adapted from Refs. 67 and 68.)

the origin. Therefore, titration could be a mechanism of negative feedback regulation of initiation. This model was originally proposed for mini-F replication to explain the facts that deletion of the control locus increases copy number and supply of additional iterons in trans decreases copy number (*120*). Subsequently, similar results have been obtained for several other iteron-containing plasmids, including mini-P1 (*113*), Rts1 (*121*), R1162 (*122, 123*), RSF1010 (*124*), R6K (*82*), pSC101 (*75, 114*), and RK2 (*125*).

For plasmids without a separate control locus (R6K and pSC101), or when the control locus is deleted from the plasmids that carry them (mini-F, mini-P1, and RK2), the copy numbers are higher but nevertheless are controlled due to the presence of iterons at the origin. In these cases, the idea is that the initiator binding is simple (noncooperative) and the daughter origins therefore can compete for the limited amount of initiator so that no one gets a full complement of the protein required for initiation. The copy numbers are higher simply because fewer iterons are involved. To a first approximation, there is no reason to believe that addition of the control locus changes the control qualitatively. Single iterons suffice for control in cis and in trans. The copy number of P1 drops from eight to four when an extra iteron is added (*17*). To be effective in trans, the single iterons have to be present in a high-copy vector such as pUC19, both in the case of P1 and pSC101 (*113, 114*). The vector copy number can be reduced provided the affinity of the iterons for the initiator is increased, i.e., when the iterons are made to titrate more efficiently (*114*). Although for the effectiveness of control, the more iterons the better was evident from the onset (*82, 114, 120, 122, 125–128*), a recent systematic study of P1 *in vivo* and *in vitro* has essentially eliminated any major role of spacing and orientation of the iterons (Fig. 1) (*116*). It can be concluded that the total concentration of iterons, rather than their disposition, is the most relevant parameter for control by iterons. However, small effects cannot be ruled out because an isolated iteron could be intrinsically less efficient in binding and, as a result, in exerting control (*113, 114, 116, 129*).

If initiators are rate limiting, it is to be expected that supply of excess RepA would overcome the inhibitory effect of the control locus. This prediction was not always verified (*115*). Supply of RepA in excess of the physiological amount did not increase copy number even twofold when the control locus, *incA*, was present (Fig. 5). Excess initiators also did not overcome intron-mediated inhibition of P1 (*116, 126*) and RK2 (*125*) replication *in vitro*. Second, the initiator gene was found to be autoregulated in several plasmids: mini-F (*130–132*), R6K (*133, 134*), P1 (*24*), Rts1 (*135*), and pSC101 (*136, 137*). This suggested that the loss of protein due to titration should not be a concern because the free pool of the protein should be maintained by autoregulatory synthesis. For these reasons the validity of the titration model has been questioned, especially under conditions of initiator excess (*10*).

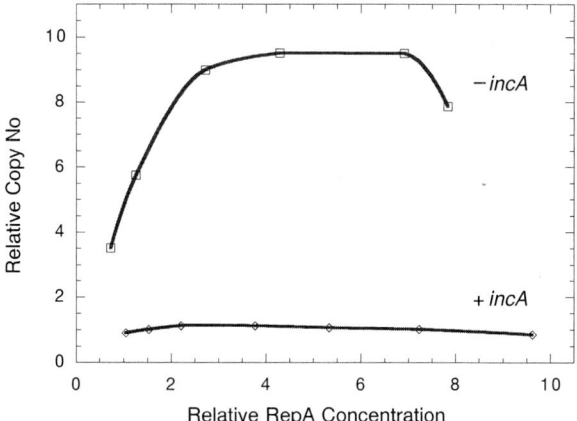

FIG. 5. Copy number of mini-P1 plasmids at different RepA concentrations. The plasmids have the P1*ori* and, additionally, the *incA* locus in cis in one of the constructs. RepA was supplied in trans from an inducible source. The copy number and RepA concentrations were expressed relative to those of a control culture carrying a wild-type mini-P1. In the absence of *incA*, it is clear that the copy number can increase when RepA concentration increases from 1, the physiological level. There is a small widow of RepA concentration (from 0 to about 3×) where the protein appears to determine the rate of replication. The copy number does not increase any further and starts to decrease after about 7× and can eventually approach 0 (data not shown). The results are qualitatively similar for *incA*-carrying plasmids, i.e., the copy number again can increase (although only to about 1.5×) and decrease depending on the concentration of RepA present. (Adapted from Ref. *115*).

However, despite autoregulation, initiator titration can still be a significant mechanism of control at physiological concentrations of the initiator in some plasmids. Several models exist that can explain how an autoregulatory protein can still be rate limiting (*86, 138–140*). In the absence of *incA*, the initiator is clearly rate limiting for P1 (Fig. 5). In and around the physiological concentration, the copy number is quite sensitive to the initiator concentration. Consistent with these results, when RepA concentration was reduced two- to threefold below the physiological level, mini-P1 plasmids could not be maintained even under selection. In these experiments a constitutive promoter was used to supply RepA (*141*). Thus an alternate prediction of the titration model—lowering of copy number due to lowering of initiator concentration below the physiological level—was satisfied. A second possible piece of evidence along these lines will be presented in Section VII,B.

Under some circumstances, the inhibitory effect of excess iterons was overcome by overproducing P1 RepA (Table I). The copy number of the plasmid R1162 was dependent on the concentration of its initiator protein, RepC, and excess initiators could overcome the inhibition due to iterons (*142*). The autoregulated initiator of pSC101 appears to be rate limiting for replication

TABLE I
Relief of incA Inhibition of Replication by Additional RepA

Additional incA	Additional RepA	Survival (42°/32°C)	No. of lysogens of	
			λ–miniP1	λ–P1*ori*
Integrative suppression of *dnaA*(ts) by (λ–mini-P1)$_{attP1}$[a]				
None	None	1	n. a.	n. a.
(λ–P1*incA*)$^2_{attλ}$	None	5×10^{-4}	n. a.	n. a.
(λ–P1*incA*)$^2_{attλ}$	7×	1	n. a.	n. a.
Lysogenization by λ–miniP1 and λ–P1*ori*[b]				
None	None	n. a.	10^8	10^3
None	7×	n. a.	10^8	10^7
(λ–P1*incA*)$^2_{attλ}$	None	n. a.	10^3	10^3
(λ–P1*incA*)$^2_{attλ}$	7×	n. a.	10^4	10^8

Note: n. a., Not applicable.
[a]The starting bacterial strain is defective for replication at 42°C and not at 32°C because of a *dnaA46*(ts) mutation. The strain survives at 42°C when a λ–mini-P1 phage is integrated into the chromosome at *attP1* near 67 min (line 1). Survival occurs because replication initiated from P1*ori* can continue into the host chromosome, and complete its replication. The temperature sensitivity returns when the functioning of the P1*ori* is blocked by *incA*, provided by a second λ vector integrated at *att*λ (line 2). The exponent 2 indicates that the block requires at least a dilysogen of λ–P1*incA*. The inhibitory effect of *incA* is overcome when sevenfold the physiological amount of RepA was supplied through a pBR322-derived vector (line 3). (Adapted from Ref. *127*.)
[b]Lysogenization of an integration-defective λ–mini-P1 phage is efficient because the prophage is stably maintained as a plasmid using the P1 replication and partition functions. When the phage has only P1*ori*, then efficient lysogenization depends on a source of RepA in trans (lines 4 and 5, last column). The background lysogenization (10^3) is most likely from random integration and from the presence of mutants in the phage stock. The *incA* inhibition of lysogeny is overcome when seven-fold excess of RepA was present. The effect was larger with λ–P1*ori* than with λ–mini-P1 (10^4 vs. 10^8, last line) because of an extra copy of *incA* in the latter and because of a stronger *incA* phenotype in cis than in trans. (Adpated from Ref. *115*.)

for several reasons: (1) The protein is preferentially cis-acting (*55, 143*); (2) a copy-up mutant has been shown to have increased binding affinity to the iterons (*144, 145*); (3) a locus, *par*, in cis increases affinity of iterons for the initiator (*114*) (*par* is also known to increase copy number); (4) finally, the initiator binding can also be cooperative (*145, 146*).

DnaA is an autoregulated protein, yet it is believed to be rate limiting for *oriC* replication and initiator titration figures prominently in a current model on the control of *oriC* replication (Section VII,F). There are a variety of posttranslational mechanisms that can modulate the initiator activity of the protein separately from its DNA binding activity (*147*) (Section VII,F). Finally, there are several indications that the DNA binding activity of RepA is subject to limitations: (1) Autoregulation limits RepA synthesis (*140*); (2) proper folding of RepA, and hence its DNA binding activity, requires mediation from chaperones (*50, 62*); (3) RepA is inactivated by dimerization and may also require chaperones for monomerization (*43*); (4) the affinity of RepA biding to iterons appears to be reduced due to an iteron-pairing activity of RepA (85) (Sections IV and VI,A,2); (5) RepA binding to iterons can be facilitated by DnaA (67). All these regulatory steps would be meaningless if the DNA binding capacity of RepA were in excess under autoregulatory conditions.

In conclusion, the titration model could very well be true but its effectiveness appears to be restricted to situations where the initiator is at or near physiological concentrations. However, the model cannot be generally applicable because in plasmid R6K, the initiator concentration can be reduced two orders of magnitude below normal physiological levels without compromising initiation (148).

2. Handcuffing of Origins

In addition to binding, plasmid initiators can also pair iterons (79–86). The activity can bring together two initiator–iteron complexes in trans as well as in cis when the sites are at a distance as in P1 (Fig. 6). It is believed that pairing can cause steric hindrance to the formation of initiation complexes (115). When the pairing is between two origins, the activity of both can be hindered ("handcuffing") (82). Following replication, as the origin concentration increases, so will the pairing probability. This mechanism will decrease initiation probability in a copy-number-dependent fashion, and, therefore, will fit the model of a negative feedback regulation. A scheme whereby initiator titration and pairing work together near physiological concentrations is detailed in the legend to Fig. 6. When the initiator concentration is no longer limiting, initiation is primarily controlled by pairing and this can explain the initiator concentration-independent mode of control seen at high initiator concentrations (Fig. 5).

A crucial prediction of the handcuffing model has been verified. Some of the high-copy mutants of R6K (149, 150), P1 (85), and RK2 (87) initiators were found to be defective in pairing *in vitro*. However, a more rigorous test of the model, correlation of pairing defect *in vitro* with increase in copy number *in vivo*, for example, remains to be verified. The model has also not been subjected to mathematical modeling. Finally, it is important to note that high-copy mutants have been found that do not show pairing defect *in vitro*, suggesting that there could be multiple modes of control (149).

B. Inhibitory Activity of the Initiator Gene

Overexpression of the initiator gene in trans leads to a decrease of plasmid copy number in plasmids R6K (148), P1 (151), Rts1 (135), and pSC101 (152), and therefore this seems to be a general property of iteron-carrying plasmids. In R6K, the current evidence is that high concentration of the initiator is inhibitory to initiation (153, 154). Furthermore, in pSC101, it has been suggested that the ratio of initiator dimers to monomers is important

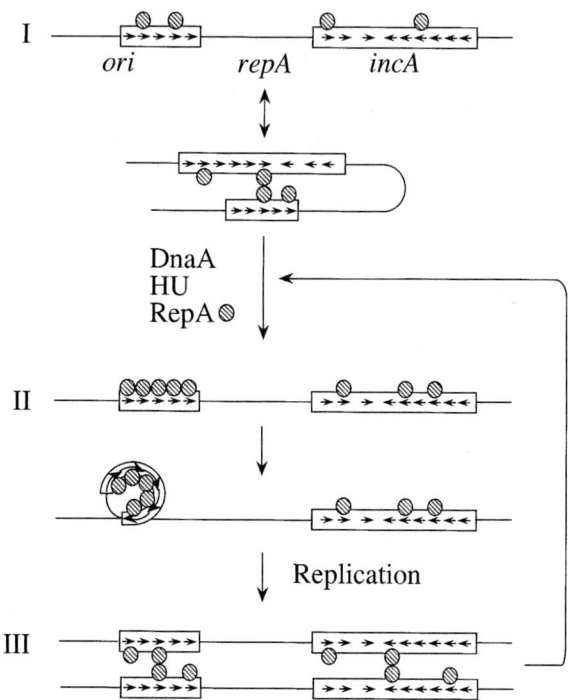

FIG. 6. Negative control of mini-P1 replication by iterons. The model shown here incorporates both initiator titration and initiator pairing and how they could be working hand in hand. (I) Assuming initiators to be limiting, a newly replicated origin is shown as partially bound with initiators. When the origin concentration is low, pairing between iterons is possible primarily in cis. This can cause steric hindrance to saturation of initiator binding. (II) Once saturation of iteron binding is achieved, possibly with the mediation of host factors such as DnaA and HU, the iterons fold into a higher order structure that allows strand opening and initiation. The folding, although still largely hypothetical, could be one way to preclude pairing and thereby commit the complexes to initiation. In this scenario, the probability of initiation is largely dependent on saturation of binding: free DnaA and RepA would favor binding, and initiator pairing would cause steric hindrance to binding. (III) Following replication, because of the presence of multiple iterons, the initiators are distributed randomly to daughter origins so that essentially none is saturated (initiator titration). The daughters are also expected *in vivo* to be sequestered temporarily to the membrane (not shown). On release from the membrane, pairing in trans is possible (handcuffing). As the copy number increases so does the opportunity to pair in trans. The paired complexes give rise to initiation complexes, as with steps I and II, and the cycle continues. The model accounts for several facts: (1) RepA can bind and bend iterons (68). In the presence of all five iterons this can lead to change of one superhelical turn, consistent with a folded structure of RepA–iteron complexes. (2) There are two modes of control—RepA concentration dependent and RepA concentration independent (see Fig. 5). The latter mode is mediated by iteron pairing in the model. (3) Pairing can reduce RepA binding to iterons (85). (4) RepA and DnaA possibly cooperate in the binding of each other (67). The demonstration of a (DnaA mediated?) transition from a pairing-sensitive to a pairing-insensitive initiation complex remains

for the inhibitory activity, implying that the activity is coming from the dimers (152, 155). In P1, the situation is more complex. The inhibitory activity remains intact when the initiator gene is transcribed but not translated, indicating that the inhibitor may be an RNA or other unidentified products of the *repA* gene (151). Consistent with this view is the finding that excess purified RepA does not inhibit replication *in vitro* (26, 116). The integrity of the initiator structural gene is not essential for the inhibitory activity in any of the initiators tested (135, 151, 153). It should be stressed that the inhibitory activity could be seen only with a few fold overexpression (Fig. 5). It is possible that some inhibitory activity is present even under physiological conditions. At about 40-fold overproduction, replication from P1*ori* is shut off completely. Mutant initiators that allow replication under these conditions are found and they show a high-copy phenotype without overproduction (S. Pal and D. K. Chattoraj, unpublished results). The inhibitory activity thus could be physiologically important.

The presence of the inhibitory activity has been a major complication in determining the copy number over a wide range of initiator concentrations (Fig. 5). This is especially important to rule out the titration model convincingly. It is quite possible that the stimulatory and inhibitory activities parallel as RepA concentration is increased a few fold, giving the false impression that RepA is not rate limiting (Fig. 5).

In summary, we feel that there could be at least two ways to restrain overreplication of iteron-carrying plasmids, the effectiveness of each depending on initiator and origin concentrations. It is important to note that the mechanisms involved are not necessarily mutually exclusive. As mentioned earlier, the possibility of multiple modes of control is also suggested from the characterization of high-copy mutants. Although only a few have been analyzed, not all are defective in handcuffing (149).

As stated in Section IV, the initiator activity of DnaA is inhibited when it complexes with ADP (91, 92). An activity has been found in cell extracts that

---

the major challenge facing the handcuffing mode of control. Some supporting evidence may have been found already. The *incC* iterons were found to be less effective in exerting control when the remainder of P1*ori* sequences were added (a mutant *ori* was used for technical reasons, but the mutation was well outside of *incC*) (116). These results can be taken to mean that initiation complex formation was blocking the capacity of the *incC* iterons to exert control. The relationship of pairing probability with saturation of iteron binding is not known, but is likely to be directly proportional. The relationship, however, must break down at saturation of binding in order for initiation to happen. This transition is formally a cooperativity step that was postulated to be required for control (5). It should be noted that in the present discussion transition from the unfolded to the folded state is an example of positive cooperativity. Irrespective of the sign, it is important that there is a cooperativity step in the control circuit.

renders DnaA inactive for *oriC* replication, presumably by enhancing hydrolysis of ATP bound to DnaA (*156*). DnaA is also inactivated when DnaA–ADP molecules are rejuvenated by phospholipids in the absence of *oriC* DNA. Thus initiator inactivation and its recycling are important parameters in the control of *oriC* replication.

A novel form of initiator inactivation has been discovered recently in plasmid pT181 (*157, 158*). In this system, the initiators are not recycled at the end of one cycle of replication, but are inactivated by covalent binding to a newly synthesized piece of origin DNA. As expected, the initiator is rate limiting for replication in this plasmid (*159*). The mechanism of inhibition in P1-type plasmids is expected to be different from both *oriC* and pT181 systems. RepA is not known to bind ATP and the mode of replication that is responsible for inactivation in the case of pT181 is not applicable to iteron-type plasmids. *In vitro*, RepA is inactivated on dimerization (Section VI,A,1) (*43*). Because the local concentration of bound RepA is high at the origin, a passage of the replication fork may inactivate the released protein by dimerization. There could be additional mechanisms operating when inactivation is seen in the absence of *repA* translation (*151*).

VII. Positive Control and Coordination with Cell Cycle

In this section we shall review what guarantees exist for initiation to occur before cell division. It is known that the control locus *incA* reduces the copy number of P1 by an order of magnitude (*17*). In other words, positive initiation factors cannot be rate limiting; it is the overriding ability of the negative control that prevents the copy number from reaching its maximal value. In view of such strong negative control, the question then arises as to how the plasmid is so efficient in initiation. We are interested to know whether there are any systems of positive control that function to deal with the situation when the copy number is N or less than N. A case in point is plasmid R1, a system wherein the initiator concentration is unquestionably rate limiting (*15*). The initiator can be transcribed from two promoters, but the stronger one is normally kept repressed by a plasmid-encoded protein, CopB. However, if the copy number drops to $<N$, not enough CopB is made to effect repression, and as a result, the initiator concentration increases dramatically (*15, 160*). In the plasmid context, this is a true example of positive control—presence of a plasmid-encoded system dedicated to correct the low-copy situation. This is conceptually different from passive mechanisms whereby the low-copy situation is rescued by an increase in initiation potential due to an increased ratio of initiators to origin, and decreased trans reactions such as

handcuffing. In P1, if there are elements of positive control, they are yet to be found. However, there are many possible candidates.

A. Autoregulation of Initiator Synthesis

Initiator genes of iteron-regulated plasmids are often autoregulated (Section VI,A,1). Autoregulation is perhaps the most economic way to regulate the synthesis of a protein. The control system maintains the free concentration of a protein independent of the growth rate of the cell or the copy number of the gene in question. In terms of initiation, autoregulation contributes positively, although in terms of synthesis it is a negative control. Initiators are essential. Their concentration can be rate-limiting for initiation, and their overproduction even severalfold above the physiological (autoregulatory) level can be inhibitory for initiation (Fig. 5). Therefore, maintenance of an optimal initiator concentration appears to be a necessity. Because the inhibitory activity does not dominate until there is about a severalfold overproduction of the protein (Fig. 5), the autoregulatory level can be set slightly in excess of what would be required to ensure initiation.

The concentration of DnaA is also maintained by autoregulation (89, 90, 161). Increase of the protein from an inducible promoter increases origin concentration, implying that the protein is rate limiting for initiation (89, 90). However, increased initiation does not result in an increase of overall DNA concentration, most likely because the protein retards the elongation rate by binding to its multitude of sites strewn around the chromosome (90).

The rate of transcription of an autoregulated gene is expected to vary with growth rate and on replication. As the cell grows in size, so should the rate of transcription to maintain the free concentration of the protein, reaching a maximum just prior to replication (162). On replication, there should be a twofold drop of transcription per gene due to a twofold increase in gene dosage. The sensitivity of the transcription rate to copy number and cell growth rate implies that the autoregulatory circuit can play roles beyond maintaining the concentration of the free pool of the protein. For example, if initiation is sensitive to transcriptional activation, as is often the case (163–167), the maximal transcription rate prior to gene duplication can itself contribute to initiation. Additionally, because an autoregulatory circuit responds to changes in cell size, it follows that the initiator synthesis would also be under the cell cycle regulation.

In the case of broad host range plasmids, autoregulation may dampen fluctuation of initiator synthesis that would otherwise happen due to intrinsic differences in promoter strength in different hosts. In *E. coli*, the autoregulated activity of the P1*repA* promoter is about 100-fold lower than its constitutive activity (in the absence of RepA) (24). In a different host, the constitutive activity can be two orders of magnitude lower and still be enough to maintain initiator concentration. However, the fluctuation of initiator concentration would be much greater if no autoregulation operated to account for cell growth and copy number changes. In R6K, the autoregulatory level is inhibitory to the activity of the $ori\gamma$ but stimulatory to $ori\beta$ and $ori\alpha$. It has been proposed that the $ori\gamma$ can help in the establishment of the plasmid but when the autoregulatory level is set, replication can switch to $ori\beta$ and $ori\alpha$ (11).

In the discussion above we have only considered repression to be a function of the free concentration of the initiator. In real life the situation can be more complicated due to preferential cis action of the initiator (55, 143), the effect of initiator pairing on DNA binding (85), the mediation of chaperones (58) and other host factors such as DnaA on initiator binding (67, 72), membrane sequestration (147, 162), and cooperative interactions (51). We will return to some of these points.

B. Initiator Activation by Molecular Chaperones

Among the iteron-carrying plasmids, replication of mini-P1, mini-F, and Rts1 has so far been shown to be facilitated by chaperone proteins (63). The activity of the chaperones is seen in the absence of stress (i.e., at low temperatures) and under conditions where *oriC* replication is apparently unaffected. The chaperones can significantly increase iteron binding of the initiators by increasing the active fraction of the protein and thereby also help replication significantly (43, 50). DnaK can also disaggregate DnaA *in vitro* and thereby activate the protein (64). (Note that the mechanism of this activation is different from that mediated by acidic phospholipids, as discussed at the end of Section IV.) The extent of aggregation of DnaA may not be significant *in vivo*, explaining why the absence of DnaK does not cause major problems for *oriC* replication. *In vitro oriC* replication with purified components and with wild-type DnaA shows no requirements for chaperones (64). Use of activated RepA also obviates the need for chaperones of P1 replication *in vitro* (168). The requirement for chaperones is, therefore, indirect. Nonetheless, by increasing the active fraction of initiators, the chaperones could contribute positively to initiation, assuming the initiators to be rate limiting.

Evidence to this effect is independently suggested by replication behavior of mini-P1 and mini-F plasmids in the absence of chaperones. For these plasmids, the requirement of chaperones is not absolute but facilitatory (169, 170). The plasmids are maintained, though unstably, in *dnaJ* mutant cells, and as expected, their copy numbers are lower. Nonetheless, some active initiators must form in *dnaJ* mutants but they are not enough for stable maintenance. In other words, even from an autoregulatory circuit enough active initiators could not be made. Control experiments in the same genetic background show that the wild-type copy number can be achieved when a 40-fold excess of (total) initiator polypeptides is provided in trans from a constitutive promoter (52). Obviously, transcription could not reach this level from the autoregulatory promoter. These results suggest that (active) initiator concentration can be limiting from an autoregulatory source. This argument is based on the assumption that the initiator and the autorepressor forms are identical in mini-P1. There is genetic and physical evidence in support of this notion (52).

Mutant initiators that can bind to iterons independent of the chaperones have been found (52). As expected, they show a higher affinity for iterons *in vitro* in the absence of chaperones but, somewhat unexpectedly, often show a high-copy phenotype (85). In the case of the mini-F initiator, chaperone-independent binding and high-copy phenotypes were also found to be correlated (170). The P1 mutants have been shown to be pairing defective *in vitro*, and this readily explains their high-copy phenotype by the handcuffing model (85). In the wild-type initiator, the binding defect in the absence of chaperones could then be associated with the pairing activity of the initiators. The result may be important because it shows that the two activities of the initiator proteins, DNA binding and DNA pairing, are intimately (and inversely) related. In fact, the high-copy phenotype is also consistent with the titration model. Because the mutants show increased affinity for iterons, this can be interpreted to mean that the high-copy phenotype could be simply due to better binding (85). A high-copy mutant of pSC101 is also found to be increased in DNA binding affinity (145). In both P1 and pSC101, the effect of the mutations is on the on-rate of binding and not on off-rates. Similar off-rates might mean that the DNA-binding domains have not changed in the mutants. A second activity of the proteins (pairing in the present context) must be involved in the overall binding reaction and this activity directly or indirectly influences the copy number. Understanding the basis of this connection of on-rate of DNA binding to copy number is therefore crucial not only from the point of view of replication control but also in terms of revealing the role of chaperones in initiator activation. Unlike the situation in P1 and pSC101, the pairing defect does not increase iteron binding in R6K (150). The two initiator activities may not be connected in all cases.

C. Cooperativity of Initiator Binding

The presence of multiple binding sites in iteron-containing plasmids raises the possibility of cooperative iteron–initiator interactions. As already mentioned, an all-or-none control by a twofold change in gene dosage or cell volume is hard to imagine without invoking some form of cooperativity (4, 5). The iterons of the origin are most likely exploited in the formation of a higher order nucleoprotein complex required for strand opening (68, 69). In R6K, out of seven origin iterons, precise deletion of one of the iterons does not affect origin function (*171*). But some point mutations in the iterons were not tolerated. These results indicate that although six iterons suffice, they need to be adjacent. A possible reason could be the formation of higher order complexes. It is possible that negative control systems function by hindering the steps leading to strand opening, and one way to overcome the disruptive forces of titration/handcuffing is to make the initiator–iteron interactions positively cooperative. Once all iterons are occupied, the complexes can assume a form that is committed to strand opening and can no longer be disrupted (Fig. 6). The degree of cooperativity would depend on the strength of negative control forces and cannot be too high or runaway replication would occur.

Demonstration of cooperative binding of P1 RepA to iterons has not been successful (J. Dibbens and D. K. Chattoraj, unpublished), although favorable results were obtained in other iteron-carrying plasmids *in vitro*: R6K (*69, 172*), pSC101 (*145, 146, 173*), and RK2 (*129*). In pColV-K30, a replicon very similar to P1, evidence consistent with cooperative initiator binding was obtained *in vivo* (*51*). It has been found that cooperative interactions can be lost easily if the buffer conditions are not optimal, and such may be the case for P1 studies to date (*174*). Even if P1 RepA binding to iterons is not cooperative, in the presence of DnaA, binding of the two proteins was found to be mutually cooperative (*67*). Cooperativity thus remains a potential mechanism by which initiation can be controlled positively.

D. DnaA

As is the case for *oriC*, DnaA is essential for replication of several of the iteron-containing plasmids. These include pSC101 (*75*), mini-P1 (*73*), mini-F (*73, 74*), R6K (*175, 176*), Rts1 (*177*), and RK2 (*178*). Here we present arguments in favor of DnaA as the prime mover of plasmid replication in the cell cycle.

1. From *in vitro* studies, it is amply clear that the protein is involved in strand opening of the P1*ori* DNA (*67*). Preliminary *in vivo* evidence has also been obtained in support of these results (K. Park and D. K. Chattoraj, unpublished). The P1 RepA greatly stimulates the reaction. It remains to be seen how general this role of DnaA will be for other plasmids.
2. DnaA can interact with DnaB and, therefore, may shoulder the responsibility of loading this helicase onto the origin (*76*).
3. DnaA is also involved in loading of the primase, DnaG. Initially, primase loading was shown to require special sequences called a primosome assembly site (PAS). The iteron-carrying plasmids normally do

not contain PAS sequences in their origin, although exceptions can be found (179). It was quite comforting when an alternate route of primase loading was discovered through DnaA (180). The primosome assembly at PAS sequences involves seven proteins: PriA, PriB, PriC, DnaT, DnaB, DnaC, and DnaG. The first four proteins and the PAS sequences are not required by the DnaA route (181, 182) and this could be advantageous, especially for broad host range plasmids. Evidence for a direct interaction between DnaA and DnaG has been obtained (P. Ratnakar and D. Bastia, unpublished data).

4. *In vitro*, DnaA facilitates iteron-binding activity of plasmid-encoded initiators of plasmids P1 and pSC101 (67, 72). The binding reactions can also be mutually cooperative. Most likely, the two proteins interact with each other directly. This possibility is also suggested in the case of plasmids R1 (183) and Rts1 (177), which require the DnaA protein but can function without the DnaA-binding sites. The role of the sites may simply be to increase the efficiency of DnaA action, for example, by increasing local concentration of the protein. Direct interactions of the initiators is also suggested in pSC101 replication (45). Some DnaA mutants have been found that specifically affect the plasmid but not the host replication. The authors speculated that these mutants could be defective in the domain that interacts with the plasmid initiator.

5. Overproduction of DnaA can overcome inhibition of replication due to excess RepA in plasmids R6K (175) and pSC101 (152). In the latter plasmid, excess DnaA could also compensate for replication deficiency due to suboptimal concentrations of RepA or absence of a replication enhancer locus *par* (55). Taken together, these results indicate that DnaA not only participates in initiation directly but also has the potential to override situations nonpermissive for initiation.

If DnaA can guarantee replication of *oriC*, it ought to be even more true for plasmids because the requirements for DnaA seem to be more relaxed in the case of plasmids. First, DnaA-requiring plasmids such as mini-P1 and mini-F can integratively suppress *dnaA46*(ts) strains (Table I) (127). In other words, at temperatures nonpermissive for host replication, the residual activity of the mutant protein suffices for initiation from plasmid origins (184). We note that DnaA46 protein is defective in binding to ATP or ADP and it is either the presence of the ATP form or the absence of the ADP form of wild-type DnaA that is required for unwinding of *oriC* (98, 185). Because DnaA is also believed to unwind the P1*ori*, it is possible that the interactions with the plasmid initiator could be relaxing adenine nucleotide requirements. Replication of pSC101 has been shown to require less DnaA protein than that of *oriC* (45). Second, the ADP form of DnaA can support P1 plasmid replication but not *oriC* replication (22, 92). Interactions of DnaA with the plasmid initiator could again be the reason for the activity of DnaA–ADP molecules specifically in plasmid replication. By these criteria, DnaA should not be rate limiting for plasmid replication unless under extreme conditions. P1 and F replication can be inhibited when cells are grown in poor medium and when they carry p*oriC* (186). These conditions most likely reduce the availability of DnaA for plasmids.

Overcoming both replication deficiency and replication inhibition of pSC101 by excess DnaA is at present the only evidence suggesting that the protein can be a positive regulator of plasmid replication, as it is for *oriC* replication.

E. Methylation and Sequestration

The functions of Dam and SeqA are interdependent and our understanding of them has been a fascinating development in the field of control of *oriC* replication (4, 105, 110, 187). These proteins most likely perform similar functions in P1 replication (106, 109).

Dam methylates N^6 of adenine bases only at GATC sites (188). There is a cluster of such sites both in P1*ori* and *oriC* (Fig. 1). Methylation serves two functions: (1) It can increase initiation efficiency, a positive function. Compared to methylated DNA, replication of unmethylated *ori* DNAs is reduced severalfold both *in vivo* and *in vitro* (106). That Dam can contribute positively is also indicated by the fact that its absence aggravates temperature sensitivity of *dnaA*(ts) strains and causes lethality when SeqA is overproduced (4). Methylation could stimulate replication by enhanced bending and unwinding of the DNA (189–191) and by recycling initiators (105). It has been proposed that DnaA is released from newly replicated origins and made available selectively to unreplicated origins only in Dam^+ cells (Section VII,F). (2) Methylation can also delay initiation, apparently a negative function. When methylated DNA becomes hemimethylated after one round of initiation, it is sequestered in the cell membrane. Specific sequestration of hemimethylated DNA in the membrane has been documented *in vitro* (192). In a Dam^+ host, the sequestration lasts about 30–40% of the cell cycle (irrespective of the growth rate) and contributes significantly to interinitiation time and to synchrony of initiation, as will be discussed. The sequestration is relieved when the hemimethylated *ori* DNAs become methylated again (193). This inhibitory role of hemimethylation can reduce transformation efficiency of methylated DNA in Dam^- cells by several orders of magnitude (106, 188). In a Dam^- host, sequestration is essentially a dead end, because there is no Dam activity to salvage the sequestered DNA (188). Replication of methylated DNA, therefore, does not proceed more than one round. In contrast to methylated DNA, unmethylated DNA transforms Dam$^-$ cells efficiently except that the copy number of the plasmid in transformed cells is severalfold lower because of missing positive functions of Dam.

The *seqA* gene was identified by mutations that allow efficient transformation of methylated *oriC* plasmids in a Dam$^-$ host (*4, 194*). SeqA protein was, therefore, thought to be involved in sequestering hemimethylated DNA to membranes, preventing replication temporarily in Dam$^+$ cells and permanently in Dam$^-$ cells. The basic premise has been borne out by *in vitro* experiments. The protein bound specifically to hemimethylated DNA (*109, 110*). However, some binding to methylated DNA was also seen. Hemimethylated *ori* DNAs replicate *in vitro* but not when bound to membranes (*106, 195, 196*). The membrane binding was also shown to be absolutely dependent on SeqA (*110*). SeqA binding to hemimethylated DNA, however, was not specific to P1*ori* or *oriC* DNA. Other DNAs with a clustering of GATC sites were also efficient substrates. Sequestration, on the other hand, was specific to *oriC* (and the promoter region of the *dnaA* gene) (*193*). Therefore, signals other than hemimethylation and SeqA are required for *ori*-specific sequestration. That other factors may be involved was also suggested by the fact that in the Δ*seqA* strain *ori*-specific sequestration was not totally eliminated (*4*).

Membrane binding *in vitro* and sequestration are not identical: the former is dependent on SeqA but not on DnaA (*110*). Sequestration, on the other hand, may involve both proteins, as suggested by strong genetic interactions between them (*4*). It has also been suggested that SeqA and Dam may function independently. Overproduction of SeqA is lethal to Dam$^-$ cells (*4*). This result may mean that SeqA can inhibit initiation in the absence of hemimethylation. Similarly, Δ*seqA* strain growth, which is aberrant most likely due to overinitiation, improves significantly in Dam$^-$ cells. This indicates that Dam was responsible for overinitiation (through some of its positive roles) in the complete absence of SeqA.

Although the process of sequestration is not entirely clear and almost surely involves other factors, the demonstration of sequestration is a conceptual breakthrough in our understanding of how reinitiation of *oriC* replication can be prevented in the same cell cycle. It should be noted that sequestration by itself is not enough to determine the interinitiation time, because it covers only about one-third of the cell cycle (*4*). However, the time period of sequestration could be long enough for all *oriC* DNAs to complete initiation and to remain protected from reinitiation when the initiation potential is still high (*147*).

The role of sequestration for *oriC* and P1 replication could be very similar except that the control by sequestration does not appear to have a copy-number-sensitive negative feedback mechanism. This may have warranted plasmids to include additional control mechanisms through iterons.

F. Synchrony

Chromosomal replication may initiate when the concentrations of some critical initiating factors reach a threshold (*161*). In wild-type cells this happens at a specific time of the cell cycle. At that time, replication initiates once from all chromosomal origins in synchrony (*6*). A hallmark of synchrony is that the origin number per cell seldom deviates from 2^n (n = 0, 1, 2, 3, or more). The timing of initiation is maintained even when multiple copies of *oriC* plasmids are present (*9*). Thus, although the formation of initiators is believed to be rate limiting, all origins replicate irrespective of the copy number, and yet when the initiation potential is high, an origin is seldom used more than once. These seemingly contradictory facts have been explained (*105*). The availability of active DnaA molecules to *oriC* is believed to be the factor controlling initiation because overproduction of DnaA from an inducible promoter triggers initiation prematurely (*89*). Synchrony is explained by an initiation cascade. The model exploits the finding that the affinities of DnaA for its four sites (R1 to R4) on *oriC* are different, R3 being the weakest of all (*197*). It is assumed that R1, R2, and R4 sites are filled in all origins before any of the R3 sites are filled. Initiation is triggered as soon as a R3 site is filled in one of the randomly chosen origins. The evidence for cell-cycle-specific filling of R3 has been obtained (*198*). It is assumed that bound DnaA proteins are released from newly replicated origins, which, being hemimethylated, get sequestered to the membrane. The released DnaA proteins would thus become selectively available to unreplicated origins. Replication and sequestration of these origins make even more DnaA available and this leads to a cascade of initiation that completes initiation of all origins in a short interval of time. The time to complete all initiations is apparently shorter than the time of sequestration, explaining why origins do not reinitiate even when the initiation potential is high. When sequestration ends, the number of origins has already doubled, and the DnaA/origin ratio becomes too low for initiation to occur.

The synchrony phenotype is lost in Dam$^-$ (*147*) or SeqA$^-$ mutants (*4*). This is also explained by the initiation cascade model. In the absence of sequestration, the newly replicated origins can compete with the unreplicated origins for the released DnaA proteins. Because new origins are created on replication, the DnaA/origin ratio decreases following each initiation event that thwarts the initiation cascade. However, in the absence of sequestration, there is also no barrier for once-replicated origins to be used again when the DnaA level catches up, for a new initiation event and origins are chosen at random. By the same token, some origins are also not used at all. Thus it is easily explained why in Dam$^-$ or SeqA$^-$ cells origin numbers differ from 2^n.

Synchrony is also lost in DnaA (*199*) and FIS mutants (*199, 200*). This is not surprising considering that these proteins are key players in the initiation cascade model. FIS interferes with DnaA binding to R3 (201) and inhibits opening of oriC (202). Characterization of several DnaA mutants has revealed that the initiation efficiency and initiation synchrony need not be correlated (4, 203). Thus, there could be other players in synchrony, such as the FIS protein. As has been pointed out earlier, synchrony does not appear to be a requirement for initiation but an outcome of initiator titration, sequestration, and other control processes yet to be identified (147).

Escherichia coli is viable and does reasonably well when synchrony is lost in any of the mutants discussed above. This could be due to tight control of DnaA availability due to autoregulation; modulation of DNA binding by IHF, HU, and FIS; titration by a large number of DnaA-binding sites on the chromosome; posttranslational control of the formation of the active form of DnaA by phospholipids and chaperones; cell-cycle-specific transcription of promoters in the vicinity of oriC (204–207); and, finally, feedback regulation—cells born with fewer origins enjoy a higher DnaA/origin ratio and vice versa (105). There can be additional controls resembling plasmid handcuffing, contributing to the interinitiation time. Thus, in view of the multiple regulatory elements, the minor perturbation of the replication program in the absence of sequestration is easily rationalized. Similarly, DnaA-dependent plasmids such as F and pSC101 have only a single GATC site in the origin and replicate normally in Dam⁻ strains (108) (C. Miller and S. Cohen, unpublished data).

Synchronous replication is not expected for plasmids because of the requirement for correcting fluctuation in copy number. However, because of the DnaA connection, some cell cycle entrainment cannot be ruled out. Synchronous replication may also result due to plasmid partition, as will be discussed in the following section. P1 and F plasmids have been tested for synchrony with contradictory results from different laboratories (186, 208–212). The resolution of this controversy may reveal the importance of host factors, especially DnaA, in the positive control of plasmid replication.

G. Plasmid Partition

A partition (Par) locus by definition functions by distributing daughter DNAs to daughter cells at cell division. Because Par works even when the copy number of plasmids is one to two per cell, the act of partition must separate handcuffed daughters as well (126). This reduces the barrier to initiation from handcuffing in a new cell. Indeed, Par-defective mini-P1 plasmids are distributed in a "worse-than-random" fashion at cell division, most likely because the plasmid DNAs form a network by handcuffing at multiple iteron sites (18, 213). The networking in effect reduces partitionable units and thereby gives the false impression of a worse-than-random distribution.

These results indicate that Par functions, although not required for replication, can contribute to replication initiation by reducing plasmid networking. Because Par is believed to function at cell division, it also has the potential to connect replication to the cell cycle (126). These attractive ideas remain to be substantiated.

Several plasmids are stabilized in *topA* strains in the absence of Par (214). It is believed that in *topA* strains, increased negative superhelicity of the plasmids somehow makes them less amenable to networking. This can convert a worse-than-random distribution to a random distribution. In other words, the plasmids still distribute by random diffusion and not actively by a new Par system turned on in a *topA* strain. However, the average copy number of the plasmids does not appear to increase very much in *topA* strains, as would be expected if reduced networking were the only mechanism (18, 214). Therefore, additional mechanisms cannot be ruled out. We shall argue below that decreased handcuffing may be compensated by increased initiator titration because increased superhelicity can increase the affinity of initiator–iteron interactions and incompatibility phenotype of the iterons (114).

From the preceding discussion it is clear that iteron–initiator complexes, other than their roles in initiation and control of copy number, can also affect plasmid distribution at cell division. This role is most evident in the case of plasmid pSC101. The *par* locus of pSC101 maps outside of the basic replicon and, although involved primarily in plasmid distribution, may only function by modulating initiator–iteron interactions at the origin. The locus binds to DNA gyrase (215) and this results in increased negative superhelicity of the plasmid (214). The effect of changed superhelicity could be localized to the origin (216). One established outcome is increased initiator affinity for the iterons in cis (114). This would help initiation if the initiator is rate limiting for replication. Additionally, the nature of initiation complexes in the presence of *par* is altered in such a way that the plasmids are distributed at random and not in worse-than-random fashion during cell division. Again the assumption is that *par* in cis is increasing diffusion due to reduced networking through handcuffing. In other words, *par* can help by helping the initiator, first, to bind to the origin and, second, to commit the complexes toward initiation by reducing handcuffing (Fig. 6). The latter effect helps partition but apparently affects copy number nominally because of the possibility of increased initiator titration. The complex phenotypes of the *par* locus of pSC101 perhaps can therefore be understood once both initiator titration and handcuffing are taken into consideration. The *par*-mediated increased iteron binding and reduced handcuffing, if true, would again suggest that the two activities of the initiator could be inversely related (Section VII,B). In any event, it is becoming clear that the role of pSC101*par* in plasmid replication is direct, and in plasmid distribution is indirect. In contrast, the function of

Par in the case of low-copy-number plasmids such as P1 and F is to distribute daughter DNAs actively at cell division, which may contribute to replication indirectly.

Phenotypically, the *cmp* locus of *Staphylococcus aureus* plasmid pT181 is very similar to pSC101*par*. The *cmp* locus is extraneous to the origin but is believed to contribute to initiation by promoting initiator–origin interactions (217). The locus, however, has not been implicated in plasmid distribution. It binds a host factor but the protein is not a gyrase (M. Gennaro, personal communication). The phenotypes of *par* and *cmp* are most evident under competitive situations where the initiators are apparently limiting. These plasmid-encoded elements, such as the CopB of R1, may therefore be called positive regulators that help guarantee initiation.

VIII. Summary Remarks and the Future

The proposal from work in *oriC* that the precise control of replication initiation may involve a balance between positive and negative elements (4) is extended in this article to plasmid replicons such as P1. In conventional thinking, replication control in plasmids invariably means negative control. The finding of a genetically distinct control locus (Fig. 3) (15) in many plasmids established an obligatory role of negative control in maintenance of plasmid copy number. However, the negative control need not be of one kind. We favor the idea that both initiator titration and handcuffing work hand in hand in regulating replication of iteron-carrying plasmids such as P1 and pSC101 under physiological conditions whereby the initiator concentration can be rate limiting (Fig. 6). The initiator activities that underlie the two control processes, iteron binding and iteron pairing, appear to be interdependent in P1 and pSC101, but not in R6K (Sections VII,B and VIII,G). Overproduction of plasmid initiators can repress replication by yet another (unknown) mechanism (Fig. 5) (Section VI,B).

In view of the efficient negative controls, we have emphasized the requirement for some active processes to ensure replication in the cell cycle and have identified some possible candidates (Section VII). As for *oriC*, DnaA seems to be an important initiation factor in plasmids such as P1, F, and pSC101. A cooperative interaction between DnaA and the plasmid initiators, with the help of architectural elements such as IHF, HU, and FIS, appears to be the key by which plasmids invite the host initiator to their origin (67, 72, 198, 218). The cooperative interaction can reduce the demand on the host initiator for plasmid replication and thus competition with host replication is avoided. Moreover, the interactions can allow efficient interfacing with the rest of the host initiation apparatus through DnaA. Other than these direct roles, overproduction of DnaA can also allow initiation under conditions that would otherwise be nonpermissive for initiation (152, 175). Because of the involvement of DnaA and factors such as Dam, SeqA, and architectural proteins, initiation mechanisms of plasmids such as P1 are likely to be similar to that of *oriC*. We have emphasized this closeness in this article, so that thinking about control in one may benefit from the lessons learned in the other.

We have also argued that the control in the two systems cannot be entirely similar. Although *oriC* control is insensitive to copy number, maintenance of copy number within a characteristic twofold range is the concern of the plasmid control systems. They must have some degrees of freedom from the host control so that fluctuation in copy number can be corrected. This is where the plasmid-encoded iterons come into the picture.

Although the iterons perform several positive functions such as specifying the origin, inviting DnaA and other host factors such as DnaB and DnaG, it is through the iterons that a copy-number-sensitive negative feedback loop is provided. Iterons may also provide the opportunity for positively cooperative interactions with initiator proteins, and this can be the basis for switching replication on and off. The short answer for "why iterons?" is, we believe, control.

Comparison of iteron sequences from different plasmids shows significant homology, yet each binds to its cognate initiator with considerable precision (Fig. 2). One of the major remaining unknowns in these systems is the identification of the DNA-binding domains of the initiators. Comparison of the structures of the proteins, and the iterons, may provide insight into the evolutionary process of this protein–DNA interaction system.

The study of iterons has provided fascinating insights into their role in initiation and its control. How the same primary initiator–iteron interaction faithfully cycles through two opposite reactions remains to be understood. In-depth analysis of the reactions in one system may help in understanding the basis of broad host range.

For the control of *oriC* replication, the primary interest is how the timing of initiation is determined in the cell cycle. The question is of direct interest in the plasmid field as well, because initiator titration has been proposed to be the primary delay mechanism in both cases. In other words, the rate-limiting step appears to be the saturation of initiator binding to the origin. But how the titration is made effective in view of autoregulatory synthesis of the initiators needs to be better understood. The mechanism seems to have many components: it involves membrane, modulation of DNA structure, cell-cycle-specific transcription, and initiator activation and inhibition involving chaperones and nucleotide cofactors, to name a few. Considering the challenges, it is encouraging that there are similarities in mechanisms of control in plasmids and in chromosomal replication in the world of *E. coli*.

Acknowledgments

The article is dedicated to the memory of Nat Sternberg, whose groundbreaking construction of λ–P1 hybrids and characterization of the P1 plasmid replicon have made room for us to fill in the details. We are grateful to Stuart Austin, Deepak Bastia, Stan Cohen, Elliott Crooke, Justin Dibbens, Ramon Diaz, Roy Magnuson, Roger McMacken, Chris Miller, Peter Papp, and Michael Yarmolinsky for critical reading and many helpful suggestions, and to Don Helinski, Charles Helmstetter, and Walter Messer for making available their unpublished chapters for the *E. coli* book. We thank Nate Herman for collecting the F RepE binding sites.

References

1. J. M. Roberts and H. Weintraub, *Cell* **52,** 397 (1988).
2. K. A. Heichman and J. M. Roberts, *Cell* **79,** 557 (1994).
3. A. B. Blumenthal, H. J. Kriegstein, and D. S. Hogness, *CSH Symp. Quant. Biol.* **38,** 205 (1974).
4. M. Lu, J. L. Campbell, E. Boye, and N. Kleckner, *Cell* **77,** 413 (1994).
5. K. von Meyenburg and F. G. Hansen, in "*Escherichia coli* and *Salmonella typhimurium. Cellular and Molecular Biology*" (F. C. Neidhardt *et al.*, eds.), p. 1555. American Society of Microbiology, Washington, D.C., 1987.
6. K. Skarstad, E. Boye, and H. B. Steen, *EMBO J.* **5,** 1711 (1986).
7. B. J. Brewer *et al.*, *CSH Symp. Quant. Biol.* **58,** 425 (1993).
8. M. R. Jensen, A. Lobner-Olesen, and K. V. Rasmussen, *J. Mol. Biol.* **215,** 257 (1990).
9. C. E. Helmstetter and A. C. Leonard, *J. Bacteriol.* **169,** 3489 (1987).
10. D. R. Helinski, A. E. Toukdarian, and R. P. Novick, in "*Escherichia coli* and *Salmonella typhimurium.*" (F. C. Neidhardt *et al.*, eds.), p. 2295. American Society of Microbiology, Washington, D.C., 1996.
11. M. Filutowicz *et al.*, *Prog. Nucleic. Acid Res. Mol. Biol.* **48,** 239 (1994).
12. B. L. Kittel and D. R. Helinski, in "*Bacterial Conjugation*" (D. B. Clewell, ed.), p. 223. Plenum, New York, 1993.
13. K. Nordstrom, *Cell* **63,** 1121 (1990).
14. R. P. Novick, *Microbiol. Rev.* **51,** 381 (1987).
15. E. G. Wagner and R. W. Simons, *Annu. Rev. Microbiol.* **48,** 713 (1994).
16. M. B. Yarmolinsky and N. B. Sternberg, in "*The Bacteriophages*" (R. Calendar, ed.), p. 291, Vol. 1. Plenum, New York, 1988.
17. S. K. Pal, R. J. Mason, and D. K. Chattoraj, *J. Mol. Biol.* **192,** 275 (1986).
18. S. J. Austin and B. G. Eichorn, *J. Bacteriol.* **174,** 5190 (1992).
19. S. Austin, F. Hart, A. Abeles, and N. Sternberg, *J. Bacteriol.* **152,** 63 (1982).
20. S. Austin and A. Abeles, *J. Mol. Biol.* **169,** 353 (1983).
21. A. L. Abeles, L. D. Reaves, and S. J. Austin, *J. Bacteriol.* **172,** 4386 (1990).
22. S. Wickner, J. Hoskins, D. Chattoraj, and K. McKenney, *J. Biol. Chem.* **265,** 11622 (1990).
23. S. J. Austin, R. J. Mural, D. K. Chattoraj, and A. L. Abeles, *J. Mol. Biol.* **183,** 195 (1985).
24. D. K. Chattoraj, K. M. Snyder, and A. L. Abeles, *Proc. Natl. Acad. Sci. U.S.A.* **82,** 2588 (1985).
25. A. L. Abeles, K. M. Snyder, and D. K. Chattoraj, *J. Mol. Biol.* **173,** 307 (1984).
26. S. H. Wickner and D. K. Chattoraj, *Proc. Natl. Acad. Sci. U.S.A.* **84,** 3668 (1987).
27. F. Jacob, *CSH Symp. Quant. Biol.* **58,** 383 (1993).
28. T. D. Schneider and R. M. Stephens, *Nucleic Acids Res.* **18,** 6097 (1990).

29. N. J. A. Sloane and A. D. Wyner, "Claude Elwood Shannon: Collected Papers." IEEE Press, Piscataway, New Jersey, 1993.
30. J. R. Pierce, "An Introduction to Information Theory: Symbols, Signals and Noise." Dover Publ., New York, 1980.
31. N. C. Seeman, J. M. Rosenberg, and A. Rich, *Proc. Natl. Acad. Sci. U.S.A.* **73,** 804 (1976).
32. T. D. Schneider, *Methods Enzymol.* **274,** 445 (1996).
33. P. P. Papp, D. K. Chattoraj, and T. D. Schneider, *J. Mol. Biol.* **233,** 219 (1993).
34. M. Filutowicz et al., in "Plasmids in Bacteria" (D. R. Helinski, S. N. Cohen, D. B. Clewell, D. A. Jackson, and A. Hollaender, eds.), p. 125. Plenum, New York, 1985.
35. Y. Kamio et al., *J. Bacteriol.* **158,** 307 (1984).
36. A. Tabuchi, M. Ohnishi, T. Hayashi, and Y. Terawaki, *J. Bacteriol.* **177,** 4028 (1995).
37. R. M. Stephens and T. D. Schneider, *J. Mol. Biol.* **228,** 1124 (1992).
38. T. D. Schneider, G. D. Stormo, L. Gold, and A. Ehrenfeucht, *J. Mol. Biol.* **188,** 415 (1986).
39. T. D. Schneider et al., *Nucleic Acids Res.* **17,** 659 (1989).
39a. T. D. Schneider, In preparation.
40. J. F. Perez-Casal, A. E. Gammie, and J. H. Crosa, *J. Bacteriol.* **171,** 2195 (1989).
41. M. D. Gibbs, A. J. Spiers, and P. L. Bergquist, *Plasmid* **29,** 165 (1993).
42. M. Ishiai, C. Wada, Y. Kawasaki, and T. Yura, *Proc. Natl. Acad. Sci. U.S.A.* **91,** 3839 (1994).
43. S. Wickner, J. Hoskins, and K. McKenney, *Proc. Natl. Acad. Sci. U.S.A.* **88,** 7903 (1991).
44. D. Manen, L. C. Upegui-Gonzalez, and L. Caro, *Proc. Natl. Acad. Sci. U.S.A.* **89,** 8923 (1992).
45. M. D. Sutton and J. M. Kaguni, *J. Bacteriol.* **177,** 6657 (1995).
46. A. Roth and W. Messer, *EMBO J.* **14,** 2106 (1995).
47. F. Matsunaga et al., *J. Bacteriol.* **177,** 1994 (1995).
48. J. L. Cereghino, D. R. Helinski, and A. E. Toukdarian, *Plasmid* **31,** 89 (1994).
49. A. Toukdarian, D. R. Helinski, and S. Perri, *J. Biol. Chem.* **271,** 7072 (1996).
50. S. DasGupta, G. Mukhopadhyay, P. P. Papp, M. S. Lewis, and D. K. Chattoraj, *J. Mol. Biol.* **232,** 23 (1993).
51. A. E. Gammie and J. H. Crosa, *Mol. Microbiol.* **5,** 3015 (1991).
52. S. Sozhamannan and D. K. Chattoraj, *J. Bacteriol.* **175,** 3546 (1993).
53. A. E. Gammie, M. E. Tolmasky, and J. H. Crosa, *J. Bacteriol.* **175,** 3563 (1993).
54. S. Ohkubo and K. Yamaguchi, *J. Bacteriol.* **177,** 558 (1995).
55. C. A. Miller, H. Ingmer, and S. N. Cohen, *J. Bacteriol.* **177,** 4865 (1995).
56. D. Manen, G. Xia, and L. Caro, *Mol. Microbiol.* **11,** 875 (1994).
57. Y. Kawasaki, C. Wada, and T. Yura, *Mol. Gen. Genet.* **220,** 277 (1990).
58. K. Tilly, S. Sozhamannan, and M. Yarmolinsky, *New Biol.* **2,** 812 (1990).
59. S. H. Wickner, *Proc. Natl. Acad. Sci. U.S.A.* **87,** 2690 (1990).
60. Y. Kawasaki, C. Wada and T. Yura, *J. Biol. Chem.* **267,** 11520 (1992).
61. S. Wickner et al., *Proc. Natl. Acad. Sci. U.S.A.* **91,** 12218 (1994).
62. D. K. Chattoraj, R. Ghirlando, K. Park, J. D. Dibbens, and M. S. Lewis, *Genes Cells* **1,** 189 (1996).
63. D. K. Chattoraj, *Genet. Eng. (N.Y.)* **17,** 81 (1995).
64. D. S. Hwang, E. Crooke, and A. Kornberg, *J. Biol. Chem.* **265,** 19244 (1990).
65. E. Crooke, R. Thresher, D. S. Hwang, J. Griffith, and A. Kornberg, *J. Mol. Biol.* **233,** 16 (1993).
66. P. P. Papp and D. K. Chattoraj, *Nucleic Acids Res.* **22,** 152 (1994).
67. G. Mukhopadhyay, K. M. Carr, J. M. Kaguni, and D. K. Chattoraj, *EMBO J.* **12,** 4547 (1993).
68. G. Mukhopadhyay and D. K. Chattoraj, *J. Mol. Biol.* **231,** 19 (1993).
69. S. Mukherjee, I. Patel, and D. Bastia, *Cell* **43,** 189 (1985).

70. B. Woelker and W. Messer, *Nucleic Acids Res.* **21,** 5025 (1993).
71. B. E. Funnell, T. A. Baker, and A. Kornberg, *J. Biol. Chem.* **262,** 10327 (1987).
72. T. T. Stenzel, T. MacAllister, and D. Bastia, *Genes Dev.* **5,** 1453 (1991).
73. E. B. Hansen and M. B. Yarmolinsky, *Proc. Natl. Acad. Sci. U.S.A.* **83,** 4423 (1986).
74. B. C. Kline, T. Kogoma, J. E. Tam, and M. S. Shields, *J. Bacteriol.* **168,** 440 (1986).
75. D. Manen and L. Caro, *Mol. Microbiol.* **5,** 233 (1991).
76. J. Marszalek and J. M. Kaguni, *J. Biol. Chem.* **269,** 4883 (1994).
77. P. V. A. L. Ratnakar, B. K. Mahanty, M. Lobert, and D. Bastia, *Proc. Natl. Acad. Sci. U.S.A.* **93,** 5522 (1996).
78. B. Y. Yung and A. Kornberg, *J. Biol. Chem.* **264,** 6146 (1989).
79. J. H. LeBowitz, M. Zylicz, C. Georgopoulos, and R. McMacken, *Proc. Natl. Acad. Sci. U.S.A.* **82,** 3988 (1985).
80. S. Mukherjee, H. Erickson, and D. Bastia, *Proc. Natl. Acad. Sci. U.S.A.* **85,** 6287 (1988).
81. S. Mukherjee, H. Erickson, and D. Bastia, *Cell* **52,** 375 (1988).
82. M. J. McEachern, M. A. Bott, P. A. Tooker, and D. R. Helinski, *Proc. Natl. Acad. Sci. U.S.A.* **86,** 7942 (1989).
83. M. Schnos and R. B. Inman, *Virology* **183,** 753 (1991).
84. M. Schnos, K. Zahn, F. R. Blattner, and R. B. Inman, *Virology* **168,** 370 (1989).
85. G. Mukhopadhyay, S. Sozhamannan, and D. K. Chattoraj, *EMBO J.* **13,** 2089 (1994).
86. D. K. Chattoraj, R. J. Mason, and S. H. Wickner, *Cell* **52,** 551 (1988).
87. A. Blasina, B. Kittel, A. E. Toukdarian, and D. R. Helinski, *Proc. Natl. Acad. Sci. U.S.A.* **93,** 3559 (1996).
88. R. S. Fuller, B. E. Funnell, and A. Kornberg, *Cell* **38,** 889 (1984).
89. A. Lobner-Olesen, K. Skarstad, F. G. Hansen, K. von Meyenburg, and E. Boye, *Cell* **57,** 881 (1989).
90. T. Atlung and F. G. Hansen, *J. Bacteriol.* **175,** 6537 (1993).
91. K. Sekimizu *et al.*, *Cell* **50,** 259 (1987).
92. E. Crooke, C. E. Castuma, and A. Kornberg, *J. Biol. Chem.* **267,** 16779 (1992).
93. J. Garner and E. Crooke, *EMBO J.* **15,** 3477 (1996).
94. W. Xia and W. Dowhan, *Proc. Natl. Acad. Sci. U.S.A.* **92,** 783 (1995).
95. C. E. Castuma, E. Crooke, and A. Kornberg, *J. Biol. Chem.* **268,** 24665 (1993).
96. B. Ezaki, H. Mori, T. Ogura, and S. Hiraga, *Mol. Gen. Genet.* **223,** 361 (1990).
97. M. Schnos, K. Zahn, R. B. Inman, and F. R. Blattner, *Cell* **52,** 385 (1988).
98. D. Bramhill and A. Kornberg, *Cell* **52,** 743 (1988).
99. D. Bramhill and A. Kornberg, *Cell* **54,** 915 (1988).
100. T. A. Baker and S. H. Wickner, *Annu. Rev. Genet.* **26,** 447 (1992).
101. H. Gille and W. Messer, *EMBO J.* **10,** 1579 (1991).
102. S. Sasse-Dwight and J. D. Gralla, *Methods Enzymol.* **208,** 146 (1991).
103. D. S. Hwang and A. Kornberg, *J. Biol. Chem.* **267,** 23083 (1992).
104. D. S. Hwang *et al.*, *J. Biol. Chem.* **267,** 2209 (1992).
105. A. Lobner-Olesen, F. G. Hansen, K. V. Rasmussen, B. Martin, and P. L. Kuempel, *EMBO J.* **13,** 1856 (1994).
106. A. Abeles, T. Brendler, and S. Austin, *J. Bacteriol.* **175,** 7801 (1993).
107. A. L. Abeles and S. J. Austin, *Gene* **74,** 185 (1988).
108. A. L. Abeles and S. J. Austin, *EMBO J.* **6,** 3185 (1987).
109. T. Brendler, A. Abeles, and S. Austin, *EMBO J.* **14,** 4083 (1995).
110. S. Slater *et al.*, *Cell* **82,** 927 (1995).
111. D. S. Dimitrova *et al.*, *Proc. Natl. Acad. Sci. U.S.A.* **93,** 1498 (1996).
112. Y. J. Kim and R. J. Meyer, *J. Bacteriol.* **173,** 5539 (1991).
113. P. P. Papp, G. Mukhopadhyay, and D. K. Chattoraj, *J. Biol. Chem.* **269,** 23563 (1994).

114. C. A. Miller and S. N. Cohen, *Mol. Microbiol.* **9,** 695 (1993).
115. S. K. Pal and D. K. Chattoraj, *J. Bacteriol.* **170,** 3554 (1988).
116. A. L. Abeles, L. D. Reaves, B. Youngren-Grimes, and S. J. Austin, *Mol. Microbiol.* **18,** 903 (1995).
117. D. D. Womble and R. H. Rownd, *J. Mol. Biol.* **191,** 367 (1986).
118. K. Kim and R. J. Meyer, *Nucleic Acids Res.* **14,** 8027 (1986).
119. I. Patel and D. Bastia, *Cell* **47,** 785 (1986).
120. H. Tsutsui, A. Fujiyama, T. Murotsu, and K. Matsubara, *J. Bacteriol.* **155,** 337 (1983).
121. Y. Kamio and Y. Terawaki, *J. Bacteriol.* **155,** 1185 (1983).
122. L. S. Lin and R. J. Meyer, *Plasmid* **15,** 35 (1986).
123. L. S. Lin, Y. J. Kim, and R. J. Meyer, *Mol. Gen. Genet.* **208,** 390 (1987).
124. C. Persson and K. Nordström, *Mol. Gen. Genet.* **203,** 189 (1986).
125. B. L. Kittell and D. R. Helinski, *Proc. Natl. Acad. Sci. U.S.A.* **88,** 1389 (1991).
126. A. L. Abeles and S. J. Austin, *Proc. Natl. Acad. Sci. U.S.A.* **88,** 9011 (1991).
127. D. Chattoraj, K. Cordes, and A. Abeles, *Proc. Natl. Acad. Sci. U.S.A.* **81,** 6456 (1984).
128. A. Tolun and D. R. Helinski, *Cell* **24,** 687 (1981).
129. S. Perri, D. R. Helinski, and A. Toukdarian, *J. Biol. Chem.* **266,** 12536 (1991).
130. L. A. Rokeach, L. Sogaard-Andersen, and S. Molin, *J. Bacteriol.* **164,** 1262 (1985).
131. L. Masson and D. S. Ray, *Nucleic Acids Res.* **16,** 413 (1988).
132. F. Bex *et al.*, *J. Mol. Biol.* **189,** 293 (1986).
133. W. Kelley and D. Bastia, *Proc. Natl. Acad. Sci. U.S.A.* **82,** 2574 (1985).
134. M. Filutowicz, G. Davis, A. Greener, and D. R. Helinski, *Nucleic Acids Res.* **13,** 103 (1985).
135. Y. Terawaki *et al.*, *J. Bacteriol.* **172,** 786 (1990).
136. C. Vocke and D. Bastia, *Proc. Natl. Acad. Sci. U.S.A.* **82,** 2252 (1985).
137. K. Yamaguchi and Y. Masamune, *Mol. Gen. Genet.* **200,** 362 (1985).
138. D. D. Womble and R. H. Rownd, *J. Mol. Biol.* **195,** 99 (1987).
139. J. D. Trawick and B. C. Kline, *Plasmid* **13,** 59 (1985).
140. D. K. Chattoraj, A. L. Abeles, and M. B. Yarmolinsky, *Basic Life Sci.* **30,** 355 (1985).
141. J. A. Swack, S. K. Pal, R. J. Mason, A. L. Abeles, and D. K. Chattoraj, *J. Bacteriol.* **169,** 3737 (1987).
142. K. Kim and R. J. Meyer, *J. Mol. Biol.* **185,** 755 (1985).
143. P. Linder, G. Churchward, X. Guixian, Y. Yi-Yi, and L. Caro, *J. Mol. Biol.* **181,** 383 (1985).
144. G.-X. Xia *et al.*, *Mol. Microbiol.* **5,** 631 (1991).
145. G. Xia, D. Manen, Y. Yu, and L. Caro, *J. Bacteriol.* **175,** 4165 (1993).
146. C. Vocke and D. Bastia, *Cell* **35,** 495 (1983).
147. E. Boye and A. Lobner-Olesen, *Cell* **62,** 981 (1990).
148. M. Filutowicz, M. J. McEachern, and D. R. Helinski, *Proc. Natl. Acad. Sci. U.S.A.* **83,** 9645 (1986).
149. A. Miron, I. Patel, and D. Bastia, *Proc. Natl. Acad. Sci. U.S.A.* **91,** 6438 (1994).
150. A. Miron, S. Mukherjee, and D. Bastia, *EMBO J.* **11,** 1205 (1992).
151. K. Muraiso, G. Mukhopadhyay, and D. K. Chattoraj, *J. Bacteriol.* **172,** 4441 (1990).
152. H. Ingmer and S. N. Cohen, *J. Bacteriol.* **175,** 7834 (1993).
153. I. Levchenko, D. York, and M. Filutowicz, *Gene* **145,** 65 (1994).
154. A. Greener, M. S. Filutowicz, M. J. McEachern, and D. R. Helinski, *Mol. Gen. Genet.* **224,** 24 (1990).
155. H. Ingmer, E. L. Fong, and S. N. Cohen, *J. Mol. Biol.* **250,** 309 (1995).
156. T. Katayama and E. Crooke, *J. Biol. Chem.* **270,** 9265 (1995).
157. A. Rasooly and R. P. Novick, *Science* **262,** 1048 (1993).
158. A. Rasooly, P. Z. Wang, and R. P. Novick, *EMBO J.* **13,** 5245 (1994).
159. J. Bargonetti, P. Z. Wang, and R. P. Novick, *EMBO J.* **12,** 3659 (1993).

160. D. D. Womble and R. H. Rownd, *J. Mol. Biol.* **192**, 529 (1986).
161. J. Herrick, M. Kohiyama, T. Atlung, and F. G. Hansen, *Mol. Microbiol.* **19**, 659 (1996).
162. V. Norris, *Mol. Microbiol.* **15**, 985 (1994).
163. B. Learn, A. W. Karzai, and R. McMacken, *CSH Symp. Quant. Biol.* **58**, 389 (1993).
164. K. Mensa-Wilmot, K. Carroll, and R. McMacken, *EMBO J.* **8**, 2393 (1989).
165. K. Skarstad, T. A. Baker, and A. Kornberg, *EMBO J.* **9**, 2341 (1990).
166. T. A. Baker and A. Kornberg, *Cell* **55**, 113 (1988).
167. S. L. Beaucage, C. A. Miller, and S. N. Cohen, *EMBO J.* **10**, 2583 (1991).
168. S. Wickner, D. Skowyra, J. Hoskins, and K. McKenney, *Proc. Natl. Acad. Sci. U.S.A.* **89**, 10345 (1992).
169. K. Tilly and M. Yarmolinsky, *J. Bacteriol.* **171**, 6025 (1989).
170. M. Ishiai, C. Wada, Y. Kawasaki, and T. Yura, *J. Bacteriol.* **174**, 5597 (1992).
171. M. J. McEachern, M. Filutowicz, and D. R. Helinski, *Proc. Natl. Acad. Sci. U.S.A.* **82**, 1480 (1985).
172. M. Filutowicz, D. York, and I. Levchenko, *Nucleic Acids Res.* **22**, 4211 (1994).
173. S. Sugiura, M. Tanaka, Y. Masamune, and K. Yamaguchi, *J. Biochem. (Tokyo)* **107**, 369 (1990).
174. M. Urh, D. York, and M. Filutowicz, *Gene* **164**, 1 (1995).
175. F. Wu, I. Levchenko, and M. Filutowicz, *J. Bacteriol.* **176**, 6795 (1994).
176. T. W. MacAllister, W. L. Kelley, A. Miron, T. T. Stenzel, and D. Bastia, *J. Biol. Chem.* **266**, 16056 (1991).
177. Y. Itoh and Y. Terawaki, *Plasmid* **21**, 242 (1989).
178. P. J. Gaylo, N. Turjman, and D. Bastia, *J. Bacteriol.* **169**, 4703 (1987).
179. K. Tanaka *et al.*, *J. Bacteriol.* **176**, 3606 (1994).
180. W. Seufert and W. Messer, *Cell* **48**, 73 (1987).
181. T. A. Baker, K. Sekimizu, B. E. Funnell, and A. Kornberg, *Cell* **45**, 53 (1986).
182. T. A. Baker, B. E. Funnell, and A. Kornberg, *J. Biol. Chem.* **262**, 6877 (1987).
183. S. Ortega-Jimenez, R. Giraldo-Suarez, M. E. Fernandez-Tresguerres, A. Berzal-Herranz, and R. Diaz-Orejas, *Nucleic Acids Res.* **20**, 2547 (1992).
184. F. G. Hansen, *Mol. Microbiol.* **15**, 133 (1995).
185. D. S. Hwang and J. M. Kaguni, *J. Biol. Chem.* **263**, 10625 (1988).
186. J. E. Grimwade, A. C. Leonard, and C. E. Helmstetter, *J. Cell. Biochem.* **17E**, 293 (1993).
187. E. Crooke, *Cell* **82**, 877 (1995).
188. D. W. Russell and N. D. Zinder, *Cell* **50**, 1071 (1987).
189. M. Collins and R. M. Myers, *J. Mol. Biol.* **198**, 737 (1987).
190. H. Yamaki, E. Ohtsubo, K. Nagai, and Y. Maeda, *Nucleic Acids Res.* **16**, 5067 (1988).
191. T. Kimura, T. Asai, M. Imai, and M. Takanami, *Mol. Gen. Genet.* **219**, 69 (1989).
192. G. B. Ogden, M. J. Pratt, and M. Schaechter, *Cell* **54**, 127 (1988).
193. J. L. Campbell and N. Kleckner, *Cell* **62**, 967 (1990).
194. U. von Freiesleben, K. V. Rasmussen, and M. Schaechter, *Mol. Microbiol.* **14**, 763 (1994).
195. A. Landoulsi, A. Malki, R. Kern, M. Kohiyama, and P. Hughes, *Cell* **63**, 1053 (1990).
196. E. Boye, *J. Bacteriol.* **173**, 4537 (1991).
197. C. E. Samitt, F. G. Hansen, J. F. Miller, and M. Schaechter, *EMBO J.* **8**, 989 (1989).
198. M. R. Cassler, J. E. Grimwade, and A. C. Leonard, *EMBO J.* **14**, 5833 (1995).
199. K. Skarstad, K. von Meyenburg, F. G. Hansen, and E. Boye, *J. Bacteriol.* **170**, 852 (1988).
200. E. Boye, A. Lyngstadaas, A. Lobner-Olesen, K. Karstad, and S. Wold, *in* "DNA Replication and the Cell Cycle" (E. Fanning, R. Knippersand, and E.-L. Winnacker, eds.), p. 15. Springer-Verlag, Berlin and New York, 1993.
201. H. Gille, J. B. Egan, A. Roth, and W. Messer, *Nucleic Acids Res.* **19**, 4167 (1991).
202. H. Hiasa and K. J. Marians, *J. Biol. Chem.* **269**, 24999 (1994).

203. K. Skarstad and E. Boye, *Biochim. Biophys. Acta* **1217,** 111 (1994).
204. J. A. Bogan and C. E. Helmstetter, *J. Bacteriol.* **178,** 3201 (1996).
205. T. Ogawa and T. Okazaki, *J. Bacteriol.* **176,** 1609 (1994).
206. P. W. Theisen, J. E. Grimwade, A. C. Leonard, J. A. Bogan, and C. E. Helmstetter, *Mol. Microbiol.* **10,** 575 (1993).
207. T. Asai, C. P. Chen, T. Nagata, M. Takanami, and M. Imai, *Mol. Gen. Genet.* **231,** 169 (1992).
208. A. Eliasson, R. Bernander, and K. Nordström, *Mol. Microbiol.* **20,** 1023 (1996).
209. A. C. Leonard and C. E. Helmstetter, *J. Bacteriol.* **170,** 1380 (1988).
210. L. J. Koppes, *J. Bacteriol.* **174,** 2121 (1992).
211. J. D. Keasling, B. O. Palsson, and S. Cooper, *J. Bacteriol.* **173,** 2673 (1991).
212. J. D. Keasling *et al.*, *J. Bacteriol.* **174,** 4457 (1992).
213. W. T. Tucker, C. A. Miller, and S. N. Cohen, *Cell* **38,** 191 (1984).
214. C. A. Miller, S. L. Beaucage, and S. N. Cohen, *Cell* **62,** 127 (1990).
215. E. Wahle and A. Kornberg, *EMBO J.* **7,** 1889 (1988).
216. D. L. Conley and S. N. Cohen, *Nucleic Acids Res.* **23,** 701 (1995).
217. M. L. Gennaro, *Proc. Natl. Acad. Sci. U.S.A.* **90,** 5529 (1993).
218. T. Ogura, H. Niki, Y. Kano, F. Imamoto, and S. Hiraga, *Mol. Gen. Genet.* **220,** 197 (1990).
219. T. Brendler, A. Abeles, and S. Austin, *J. Bacteriol.* **173,** 3935 (1991).
220. S. Schaper and W. Messer, *J. Biol. Chem.* **270,** 17622 (1995).
221. A. L. Abeles, *J. Biol. Chem.* **261,** 3548 (1986).
222. H. Nozue, K. Tsuchiya, and Y. Kamio, *Plasmid* **19,** 46 (1988).
223. T. Tokino, T. Murotsu, and K. Matsubara, *Proc. Natl. Acad. Sci. U.S.A.* **83,** 4109 (1986).
224. L. Masson and D. S. Ray, *Nucleic Acids Res.* **14,** 5693 (1986).
225. T. Murotsu, K. Matsubara, H. Sugisaki, and M. Takanami, *Gene* **15,** 257 (1981).
226. P. P. Papp and V. N. Iyer, *J. Mol. Biol.* **246,** 595 (1995).
227. P. Gabant, A. O. Chahdi, and M. Couturier, *Plasmid* **31,** 111 (1994).
228. W. Pansegrau *et al.*, *J. Mol. Biol.* **239,** 623 (1994).
229. M. H. Larsen and D. H. Figurski, *J. Bacteriol.* **176,** 5022 (1994).
230. S. Perri and D. R. Helinski, *J. Biol. Chem.* **268,** 3662 (1993).
231. D. M. Stalker, C. M. Thomas, and D. R. Helinski, *Mol. Gen. Genet.* **181,** 8 (1981).
232. K. Zahn and F. R. Blattner, *EMBO J.* **4,** 3605 (1985).
233. D. M. Stalker, R. Kolter, and D. R. Helinski, *Proc. Natl. Acad. Sci. U.S.A.* **76,** 1150 (1979).
233a. D. M. Stalker, R. Kolter, and D. R. Helinski, *J. Mol. Biol.* **161,** 33 (1982).
234. J. Germino and D. Bastia, *Cell* **34,** 125 (1983).
235. D. Saul *et al.*, *J. Bacteriol.* **171,** 2697 (1989).
236. A. J. Spiers and P. L. Bergquist, *J. Bacteriol.* **174,** 7533 (1992).
237. A. J. Spiers, N. Bhana, and P. L. Bergquist, *J. Bacteriol.* **175,** 4016 (1993).
238. C. A. Smith and C. M. Thomas, *J. Mol. Biol.* **175,** 251 (1984).
239. C. Vocke and D. Bastia, *Proc. Natl. Acad. Sci. U.S.A.* **80,** 6557 (1983).

Changes in Gene Structure and Regulation of E-Cadherin during Epithelial Development, Differentiation, and Disease

JANUSZ A. JANKOWSKI,*,‡
FIONA K. BEDFORD,†
AND YOUNG S. KIM‡

*Gastrointestinal Gene Group
Department of Medicine
University of Birmingham
Birmingham B15 2TH, England
†Department of Physiology
University of California
San Francisco, California 94121
‡Colorectal Cancer Program
University of California
San Francisco, California 94121

I. Cadherin Structure	188
II. Diversity of Action and Interaction	194
A. Interactions between E-Cadherin and Other Cell Adhesion Molecules	194
B. Downstream Signal Transduction of E-Cadherin via Cytoskeleton (Modulation of Response)	194
C. α-Catenin: Master Modulator of Adhesion	198
III. Cellular Properties of E-Cadherins Conserved during Development, Differentiation, and Disease	198
A. Adherens Junctions	198
B. Cell-Specific Recognition	199
C. Embryogenesis and Organogenesis	199
D. Tissue and Glandular Differentiation	201
E. Cell Phenotype/Polarity	201
F. Cytokinesis–Invasion	202
G. Epithelial/Leukocyte Interactions	203
H. Cell Adhesion and Apoptosis	203
IV. Decreased or Defective E-Cadherin Expression in Colorectal Disease	204
A. Defects in Cell Adhesion during Colonic Inflammation—Spatial Changes of E-Cadherin *in Vivo*	204
B. Structural Changes of E-Cadherin during Tumorigenesis	204
C. Role of Other Cadherins in Generation of a "Cancer Phenotype"	206
V. Conclusions	208
References	209

The cell has four main routes of biological adaptation to its surrounding environment: proliferation, differentiation, senescence, and migration (*1*). These processes are temporally preprogrammed in the genome but they are also under the influence of several signal transduction pathways initiated by contact with adjacent cells. One of the major differences between multicellular organisms and prokaryotes is that the former require complex and multiple cell adhesion mechanisms. The extended family of cell adhesion molecules is growing rapidly and cadherins are one of the important types found in higher eukaryotes. Cadherins are expressed ubiquitously in all epithelial tissues and in many mesenchymal tissues. There are three main subsets of the cadherin family: classical cadherins; protocadherins, including desmocollins and desmogleins; and molecules with cadherin-type domains. Cadherins are expressed in a differential spatiotemporal pattern throughout development, differentiation, and tumorigenesis. During tumorigenesis there is great biological heterogeneity that can be accounted for due to the increasing knowledge of cadherin function.

I. Cadherin Structure

There are three main types of cell–cell junctions: occluding (tight), anchoring [desmosomal and zonula adherens (ZA)], and communication (GAP junctions or chemical synapses) (Fig. 1). Tight junctions seal off the basolateral aspects of the cell from fluids at the apical border, and this is usually achieved by a unique set of molecules, including ZO-1 and -2. Two subtypes of anchoring junctions are adhesion belts (cell–cell interaction) and basal focal contacts (cell–matrix interactions). The adhesion belts ("belt desmosome") insert just under the tight junctions and are cadherin rich.

Cadherins are a large family of Ca^{2+}-dependent cell–cell adhesion molecules whose functions are essential for the induction and maintenance of intercellular connections (*2*). They are characterized in the extracellular domain by four conserved repeated amino acid sequences (cadherin repeats) that have a conserved α-helix–loop–β-barrel structure (*3*). Several subclasses of cadherins have been described, including the classical cadherins; protocadherins, including desmosomal cadherins; and those with cadherin-like domains (*4, 5*) (Table I). E- and P-cadherins share homology with related molecules at two partially conserved domains, the HAV sequence of the extracellular domain and the cytoplasmic catenin-binding site. Homology is also highly conserved between human E-cadherin and different species—78% with mouse E-cadherin and 55% with chicken E-cadherin. E-Cadherin cDNA homology across species is 60–80% at the DNA level and 70–80% at the amino acid level. Furthermore, homology between E-cadherins and P-a

CHICKEN N-CADHERIN	DQKKIEDIIFPWQQYKDSSHLKRQKR	DWVIPPINLPENSRG
MOUSE P-CADHERIN	LTRGTVQGGKDAMHSPPTRILRRRKR	EWVMPPIFVPENGKG
MOUSE E-CADHERIN	DPASESNPELLMFPSVYPG LRRQKR	DWVIPPISCPENEKG
HUMAN E-CADHERIN	HQASVSIQAELLTFPNSSPGLRRQKR	DWVIPPISCPENEKG

b

CHICKEN N-CADHERIN	QIASFHLRAHAVDVNGNQVEN
MOUSE P-CADHERIN	KIVKYELYGHAVSENGASVEE
MOUSE E-CADHERIN	AIAKYILYSHAVSSNGEAVED
HUMAN E-CADHERIN	RIATYTLFSHAVSSNGNAVED

c

CHICKEN N-CADHERIN	RPIHAEPQYPVRSAAP	HPGDAGDFANEGLKAADNDPTAPPYDSL
MOUSE P-CADHERIN	VPTFIPTPMYRPRPA	NPDEIGNFIIENLKAANTDPTAPPYDSL
MOUSE E-CADHERIN	APTLMSVPQYRPRPA	NPDEIGNFIDENLKAADSDPTAPPYDSL
HUMAN E-CADHERIN	APTLMSVPRYLPRPA	NPDEIGNFIDENLKAADTDPTAPPYDSL

FIG. 1. Homology map of cadherins across species in three conserved domains. (a) Precursor peptide and cleavage site of mature protein. (b) EC1 region containing HAV recognition sequence. (c) Cytoplasmic domain containing catenin-binding site. (E-cadherin sequence from Ref. 183).

TABLE I
MEMBERS OF THE CALCIUM-DEPENDENT HUMAN CADHERIN SUPERFAMILY
OF ADHESION MOLECULES

Class of adhesion molecule	Description	Ref.
Classical	*Conserved HAV, four cadherin repeats, and cytoplasmic domains*	
E-Cadherin	Epithelium (120 kDa)	183
P-Cadherin	Placenta and epidermis (118 kDa)	184
N-Cadherin	Heart and neurons (130 kDa)	185
R-Cadherin	Retinal cadherin, pancreas	186, 187
T-Cadherin	Neuronal tissue	186, 188
K-Cadherin	Kidney epithelium	116
Protocadherins	*Partially conserved extracellular and cytoplasmic domain or increased number of cadherin repeats*	
M-Cadherin	Embryonic and adult muscle	189
Desmocollins and desmogleins (DSG1-3)	Desmosomes	174, 190
Cadherin-5 (VE)	Endothelium	191
Cadherin-11 (OB-1)	Mesenchyme	192
LI-Cadherin	Liver and intestine	96
Unrelated molecules	*Containing cadherin sequences but not adhesive*	
C-Ret protooncogene	Tyrosine kinase-containing cadherin repeats	—
hpt-1	Colon (peptide transport)	—

cadherin is 61% at the DNA level (76% at the protein level); for N-cadherin homology is 61% at the DNA level (67% at the protein level) (Fig. 1). Protocadherins such as PC42, PC43, M-cadherin, and neuronal and desmosomal cadherins, however, exhibit a completely different cytoplasmic domain (*6, 7*) and may also lack some of the cysteine residues in the extracellular domain. More distantly related molecules termed "cadherin-like" molecules may even lack the transmembrane and the cytoplasmic domains completely and have substantial alterations to the HAV sequence (*8*).

E-Cadherin is the major mediator of cell–cell adhesion in epithelial cells and is required for the formation of intermediate and adherens junctions (*1*). Evidence for this comes in part from studies indicating that when E-cadherin is functionally expressed, the inactivation of other cell–cell adhesion molecules has little effect. Conversely, when E-cadherin is functionally down-regulated, cell dispersal and increased epithelial permeability frequently occur even with the preservation of all other adhesive forces (*9–11*).

The classical cadherin genes vary in size: E-cadherin is 100 kb (chromosome 16q22) (*2, 12*), N-cadherin is 200 kb (chromosome 18) (*13*), and P-cadherin is 45 kb (chromosome 16q22) (*14;* M. Bussemakers, personal correspondence); the intron–exon boundaries, however, are highly conserved throughout these genes (*14*). The E-cadherin gene and the other cadherins are composed of 16 exons, the last being the largest (*15*). Polymorphisms are common in exons 3, 4, 11, and 13 and less common in 10 and 14 (*11, 16*) (Fig. 2).

The promoter of E-cadherin has been characterized and is a GC palindromic sequence that is homologous to the regulatory elements of keratin genes. The promoter lies 10 bp 5′ to the transcription initiation (CAAT) site (*17*) and 200 bp upstream of the E-cadherin gene. Four different cis-acting elements have been identified for the E-cadherin gene: two copies of the CANNTG consensus DNA binding motif for binding helix–loop–helix transcription factors, two AP2-binding motifs recognized by Jun–Jun leucine zippers, an SP1 zinc finger binding motif, and an HMG binding motif (*18*). The minimal promoter contains the E-pal region. The positive regulatory element in the E-cadherin promoter is the CCAAT-box and it is specifically bound by transcription factors in E-cadherin-expressing but not nonexpressing cells. All-*trans*-retinoic acid can increase E-cadherin expression *in vitro* (MCF-7 and HCT-8 cells) at even modest concentrations of 1 µM (*19*). Recent evidence has suggested that c-erbB2 may be a transcriptional regulator of E-cadherin *in vitro* (*20*), but evidence is still lacking *in vivo* (*21*). Exogenous growth factors may also regulate cadherin expression, for example, insulin growth factor, sex steroids, transforming growth factors, and cytokines. Recent evidence suggests that either androgen or estrogen may up-regulate E-cadherin expression in tissues with hormone responsiveness (*22*). Interferon-γ and γ-linolenic acid also both increase E-cadherin expression (*23, 24*). Other positive regulators of E-cadherin include the adenovirus E1A 12S gene, which can immortalize epithelial cells in part by increased cadherin expression (*25*). Conversely, the silencing of the E-cadherin promoter during epithelial mesenchymal transition and tumor progression may be due to a loss of transcription factor binding *in vivo* and to chromatin rearrangement in the regulatory region (*26, 27*). In addition, E-cadherin may also be silenced by CpG methylation (*28*). TGF-β has recently been shown to result in epithelial mesenchymal transdifferentiation and the generation of a spindle tumoral phenotype in transformed mouse epidermal cells in part by this latter mechanism (*29*).

The structure of E-cadherin cDNA has five separate regions: (1) a noncoding region followed by a signal sequence containing a 5′ CAP to prevent exonucleolytic degradation [30 amino acids (aa)], (2) a precursor segment that contains the start codon AUG and is histidine rich (90 aa), (3) the coding sequence of mature peptide (610 aa), (4) an untranslated trailer region

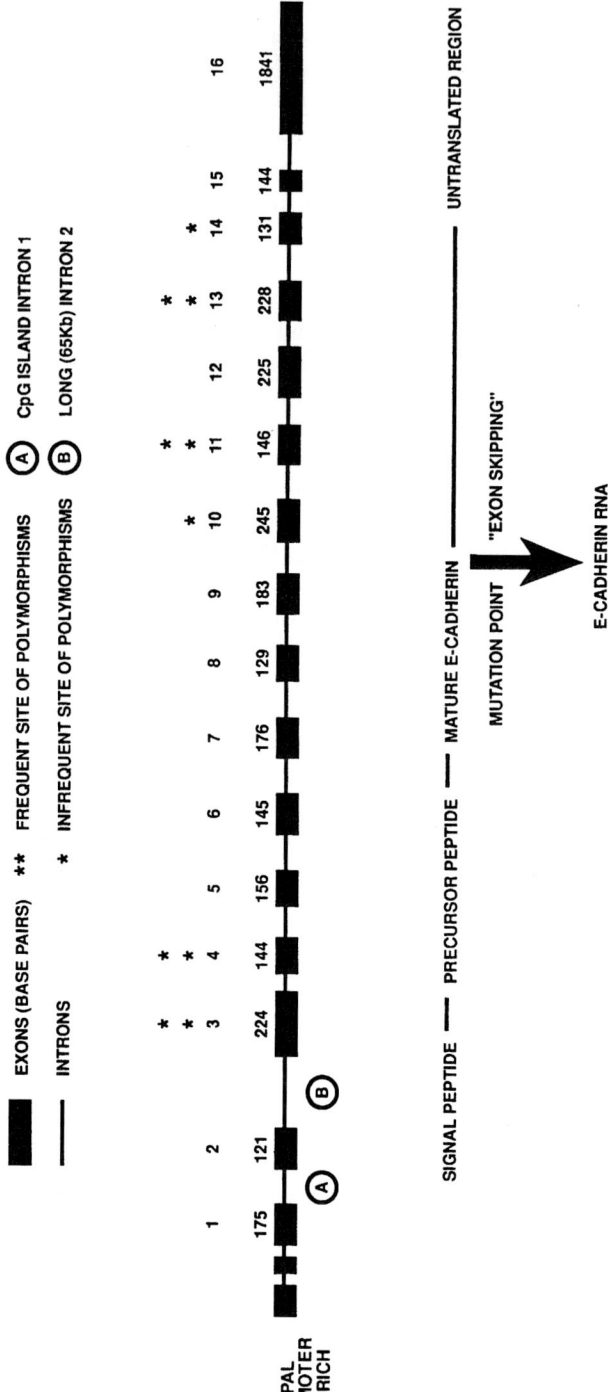

Fig. 2. E-cadherin gene (chromosome 16q22.1). The sizes of each of the 16 exons of the E-cadherin gene and sites of frequent polymorphisms are shown.

that contains the poly(A) signal AAUAAA, and (5) a poly(A) tail that prevents degradation at the 3' end.

The presequence for each cadherin is unique, whereas the rest of the protein is well conserved, compared with the homology shared between the mature peptides. Specific proteases have been suggested to cleave the precursor peptide from the mature cadherin molecule *in vitro* at the Arg-any amino acid-Lys-Arg site (RXKR) to activate the molecule (*12, 14*). Uncleaved E-cadherin or cleavage at the wrong site leads to dysfunctional protein membranes; all of these proteins have an indistinguishable molecular mass of 120 kDa. This may be due to inhibitory effects of small precursor peptides released by the action of other proteases, such as trypsin and factor Xa protease, which cuts at the LRIEGR/DWV site instead of at LRRQKR/DWV (*12, 14*).

The mature protein contains an N-terminal domain (33 aa and four homologous glycosylated extracellular domains (EC) repeats (each approximately 80 aa); the most distal repeat domain (EC 1) contains the HAV site and determines the homophilic cellular binding specificity (*2, 14*). The crystal structure of N-cadherin has revealed that the N-terminal domain of cadherins assembles and disassembles in a zipper fashion (*30*), and it is likely that almost all cadherins with extracellular domains have a common mechanism of molecular interaction (*31*). Inhibition of glycosylation by tunicamycin does not appear to affect function, although it may protect E-cadherin from proteolysis. However, desialylation of the sialic acid in the E-cadherin external domain of MCF-7/AZ cells by neuraminidase does decrease their aggregation as determined by the fast aggregation assay (*32*). The EC 1–4 domains have DXD and DXNDN residues representing putative Ca^{2+} binding sites flanking the EC boundaries. The premembrane domain is cysteine rich (60 aa), the transmembrane region is leucine rich (30 aa), and the cytoplasmic domain is glycine and glutamic acid rich (160 aa).

The two most important functional domains are the outer two EC domains and the internal catenin-binding domain. Evidence for this comes from many sources. Decapeptides such as LFGHAVSSNG inhibit the adhesive function. Interestingly, E-cadherin without the extracellular domain may still weakly remodel (polarize) the membrane, whereas E-cadherin without the cytoplasmic domain may still encode weak cell–cell adhesion (*33*). Each domain will, however, act as a dominant negative for the other domain; for example, a lost or dysfunctional external domain will act as a dominant negative for adhesion (*34, 35*), whereas an absent or dysfunctional internal cadherin domain may act as a dominant negative for membrane polarization (*36*). In other cadherins, such as vascular endothelial (VE) cadherin, the external domain alone is sufficient for cell recognition and weak adhesion, whereas the internal domain is necessary to provide strength and cohesion to the adhesive junction (*37*).

II. Diversity of Action and Interaction

A. Interactions between E-Cadherin and Other Cell Adhesion Molecules

In squamous epithelium, integrins and P-cadherin expression occur in the basal cell layer. In the parabasal layers E-cadherin is expressed as P-cadherin and the integrins are down-regulated (38, 39). Although the exact mechanism of this regulation is not fully elucidated, it has been reported that cotransfection of rat 3Y1 cells with E-cadherin results in the expression of only one type of cadherin and the suppression of endogenous P-cadherin (40). Furthermore, expression of E-cadherin catenin complexes decreases the expression of fibronectin and laminin receptors in *Xenopus* XTC cells (41). It appears that the converse action is also possible because the expression of ECM proteins can down-regulate the expression of E-cadherin mesoderm during gastrulation in the mouse embryo (42).

It is recognized that E-cadherin and its avian homolog LCAM may not always have homotypic adhesion, because they can also bind other membrane receptors, including the $\alpha E \beta 7$ integrin in man and B-cadherin in chicken, thereby eliciting a diversity of transduction signals in neighboring cells (43, 44).

The gene products of the adenomatosis polyposis coli gene and the E-cadherin gene both utilize α and β catenins (45–47). Although *in vitro* they show competition for the catenins (48, 49), it seems less likely *in vivo* because APCs and E-cadherin form independent heteromeric complexes with α-catenin (50) (Fig. 3).

Protein zero [P(0)] is an immunoglobulin gene superfamily glycoprotein that mediates the self-adhesion of neuronal cells. This molecule, when transfected into HeLa cells, stimulates the adherens and desmosomal junctions by up-regulation of cadherin-based adhesion. It therefore seems that other adhesion membrane molecules can act downstream on the cadherin catenin complexes (51).

E-Cadherin and F-actin may also have complex interactions with the proteoglycan syndecan-1 because normal murine mammary cells, when transfected with the antisense syndecan-1 cDNA, have an elongated fusiform invasive phenotype deficient in extracellular matrix (ECM) and adherens junction proteins (52). Other molecules present in the adherens junctions are α-actinin, talin, vinculin, and focal adhesion kinase (pp125FAK) (53).

B. Downstream Signal Transduction of E-Cadherin via Cytoskeleton (Modulation of Response)

E-Cadherin associates with three catenins: 102-kDa α-catenin (chromosome 5q21–22) (54–56), 92-kDa β-catenin (chromosome 3p22) (57, 58), and

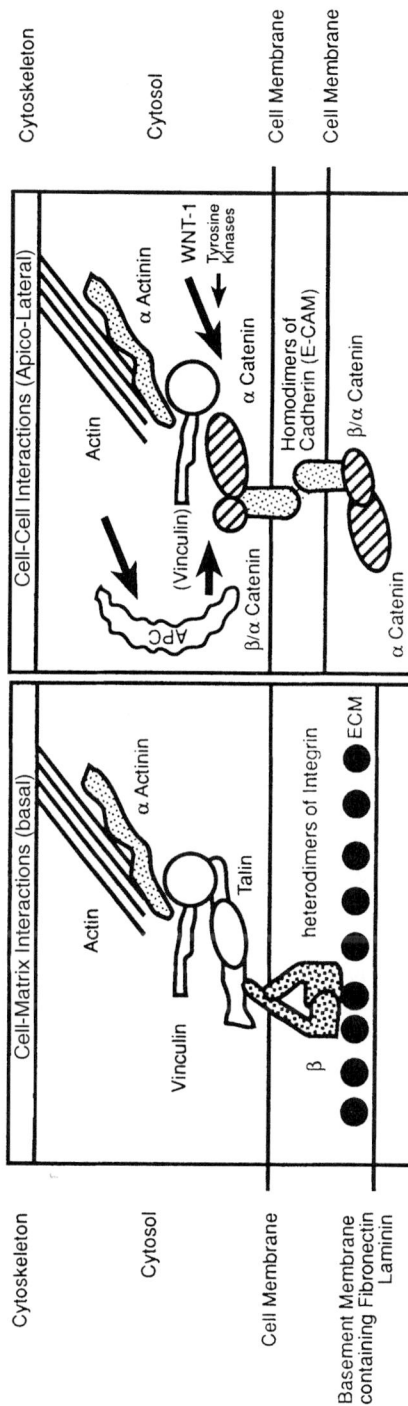

FIG. 3. Cell–cell and cell–matrix interactions in epithelial cells. Both integrins and cadherins interact with the actin cytoskeleton by different pathways. The cadherins interact with catenins then vinculin and p120cas before connecting with the actin cytoskeleton. Molecules such as Wnt-1, APC, and tyrosine kinases may interfere with the cadherin-generated signal.

83-kDa γ-catenin/plakoglobin (59). In neurons, plakoglobin replaces the role of β-catenin. Catenin binding sites are present in the 70-aa carboxy domain of the cytoplasmic tail, and when these are deleted, E-cadherin cannot function correctly, even when it is located on the cell membrane (60). β-Catenin may transduce a cascade of cellular signals whereas α-catenin may be important for cytoskeletal binding (61) (Fig. 3). α-Catenin has an identical homology with the cytoskeletal component vinculin (62) and is located adjacent to its pseudogene on chromosome 5q21–22. The α-catenin gene can have loss of heterozygosity (LOH) or can be deleted, resulting in inadequate function (63). β-Catenin (3p21–22) has 70% homology with plakoglobin a major component of human desmosomes and may signal E-cadherin effects (47).

The downstream signaling pathways can be regulated by several independent but related mechanisms: catenin affinity, catenin bioavailability, catenin phosphorylation, formulation of multisubunit catenin complexes, and other as yet undefined mechanisms. First, γ-catenin (plakoglobin) seems to bind E-cadherin in keratinocytes with highest affinity, suggesting both temporal, functional, and tissue-specific modulation of keratinocytes by cadherins (40). β-Catenin appears to bind E-cadherin most tightly immediately after the formation of E-cadherin and subsequently α-catenin binds to this complex, simultaneously linking the catenin/cadherin complex to the cytoskeleton (TX-100-soluble fractions). Cadherins such as P-cadherin may bind the cytoskeleton and the catenins more weakly than E-cadherin because the former can be more easily extracted by detergents, suggestive of less permanent interactions (64). In addition, P- and E-cadherins associate in separate complexes with catenins α and β (65).

Second, different catenin/cadherin complexes may each have specific and independent functions and can be formed from each, other allowing rapid adaptation of signal transduction (66). The actions of these complexes are in turn increased by many other molecules, including actin, fodrin, vinculin, and the ezrin, radixin and moesin (ERM) proteins, but are decreased by the Wnt-1 protooncogene product (67), via its membrane receptor (68). There are also regional differences in cells; evidence suggests that only the APC molecules present in the lateral cytoplasm of mouse intestinal cells bind catenins, whereas APC molecules in the apical cytoplasm and microvilli have functions independent of catenins (69). It is unclear whether the apical APC molecules interact with microtubules (70); however, mutated APCs do not bind actin (71).

Third, intrinsic tyrosine kinase activity may also inhibit cadherin-induced modulation via phosphorylation of E-cadherin, β-catenin, and plakoglobin (72, 73). This can be induced experimentally by pervanadate, a tyrosine phosphatase inhibitor that causes rapid opening of tight junctions (74, 75). All the adhesive actions can be restored by utilizing a tyrosine kinase inhibitor, herbimycin A. Implicated tyrosine kinases include the receptors for acidic fibroblast growth factor and hepatocyte growth factor (HGF) (8, 76–78), c-erbB2 (20), acidic fibroblast growth factor (79), and src and yes kinases. HGF acts on c-met (7q31), which also has tyrosine kinase activity and may activate downstream targets such as c-src. This phenomenon appears to be tissue specific; for example, HGF appears to loosen adherens junctions in mucin-secreting cells as well as Madin–Darky canine kidney (MDCK) cells, but not in keratinocytes that express both E- and P-cadherins (77, 80).

Fourth, E-cadherin-mediated calcium-dependent aggregation appears to be independent of epithelial phenotype. In this context short-term adhesion requires E-cadherin and α- and β-catenin, whereas epithelial morphology and polarization require all the parts of the complex including plaque proteins and γ-catenin (40).

A fifth group of as yet undefined molecules, including the repulsins and cytotactin (81), disaggregate cells, but the mechanisms are not adequately elucidated. Furthermore, cell adhesion regulatory molecules (CARs) (chromosome 16q) may also mediate integrin and E-cadherin coordination (82, 83)

p120cas can bind directly via domains 1–10 to E-cadherin but not to the adenomatous polyposis coli protein or α-catenin. p120cas is absent in the HCT116 colorectal cancer cell line but is expressed in certain well-differentiated cell lines such as the PC3 prostate cell line (84). p120 and p100 proteins can also be regulated by growth factors, thereby modulating the downstream interactions of the cadherin/catenin complexes (85, 86). Transformation of epithelial cells by ras leads to elevated phosphotyrosine in many proteins, including β-catenin and p120cas (72). p120cas displaces the β-catenin from cadherin complexes, allowing the majority of B-catenin to partition into the detergent-soluble fraction, suggesting it is not tightly bound to the actin cytoskeleton (72). Recent evidence also indicates that c-erbB2 gene product associates directly with β-catenin and plakoglobin through its cytoplasmic domain and may act as a signaling pathway (87).

Protein kinase C is up-regulated by cell–cell adhesion, but the exact nature of this downstream signaling pathway is unclear. Application of protein kinase inhibitors staurosporine and bisindolylmaleimide can have major morphoregulatory effects in neural tissue, increasing epitheliomesenchymal transformation in part by loss of N-cadherin junctions (88). Paradoxically, the protein kinase C activator tetra-O-decanoylphorbol-16-acetate (TPA) is reported to decrease both the total amount of E-cadherin and the ratio of E-cadherin bound to the actin cytoskeleton in HT29 M6 cells (89). Interestingly, EGF has been shown to temporarily release contact inhibition mediated by cadherins in thyroid epithelial cells, but when combined with TPA the effect is long lasting (90).

Intact microfilaments and intermediate filaments are both important in the generation of desmosomal junctions, but the role of intermediate filaments in the adherens junctions are not yet fully elucidated (91). Recent evidence has indicated that a catenin may bind actin F directly, thereby organizing and tethering actin filaments at the zones of E-cadherin-mediated cell–cell contact (92). Furthermore, the N-cadherin/α-catenin complex colocalizes with α-actinin but not vinculin, but in epithelial cells both molecules associate with the E-cadherin/α-catenin complex (93).

C. α-Catenin: Master Modulator of Adhesion

The importance of catenins in modulating the adhesive response has been shown by several separate experiments. First, deletion of the catenin-binding domain on cadherins results in adhesion-defective cells (36, 56, 94). Second, overexpression of the same catenin-binding domain creates a dominant-negative mutant that competes with endogenous cadherin for catenins (35). Third, alteration or down-regulation of catenins can negate E-cadherin functions such as changing the localization of E-cadherin, ZO-1, desmoplakin, integrins, and laminins from a random distribution in cell surface to polarized sites (80). The PC9 lung cancer line, however, has an α-catenin mutation but normal cadherin expression and yet it is only when the α-catenin is transfected that adhesive functions are restored (94). Fourth, E-cadherin is decreased or absent in 50% of colorectal tumors where α-catenin is absent in 75% of colorectal tumors. Interestingly α-catenin levels can increase dramatically when E-cadherin is transfected, resulting in adhesion and suggesting increased α-catenin stability or production (95). Fifth, the role of β-catenin in cell adhesion can be bypassed by creating α-catenin cadherin fusion molecules (56).

III. Cellular Properties of E-Cadherins Conserved during Development, Differentiation, and Disease

A. Adherens Junctions

The classical cadherins have widespread tissue distribution (12, 14), in contrast to the liver–intestine (LI) cadherins (96) (see Table I). The classical cadherins are present in the junctional complexes (both adherens and desmosomal junctions) and organize the actin cytoskeleton at the apical junctional complex. LI cadherins, on the other hand, exert adhesive properties in an even distribution on the basolateral membranes, probably at the desmosomal junctions (96).

B. Cell-Specific Recognition

Cadherins act as cell–cell specific recognition receptors in order to keep the epithelium together and prevent mechanical damage or migration. Four patterns of cadherin recognition are recognizable: cell condensation, cell sorting, cell stratification, and spheroid formation (97).

1. Cell clumping (cell condensation) occurs when cells express the same amount of an identical cadherin and organize into cellular aggregates or monolayers (Fig. 4A).
2. Cell sorting is the phenomenon when different populations of cells each express different cadherins, which results in their growth as segregated clumps with no intervening adhesion.
3. Cell stratification occurs when cells express varying amounts of two different types of classical cadherin. Cells *in vitro* with predominantly one type of either cadherin will aggregate together, whereas those with varying amounts of both cadherins will bridge the gap between the other cellular islands. *In vivo*, cells aggregate into distinct layers, such as occurs in squamous or renal epithelia. In the skin, P-cadherin is expressed in the basal layer and may inhibit the expression of E-cadherin until the cells divide and move into more superficial layers. In addition, interference of P- and E-cadherin function by utilization of blocking antibodies has been shown to disrupt epithelial stratification in squamous epithelia and organogenesis in mucin-secreting tissues (98, 99).
4. Cell lines in suspension expressing different amounts of the same cadherin organize into multicellular spheroids, with denser cadherin expression on the centrally sited cells and weaker cadherin expression on the peripherally sited cells. Therefore, both qualitative and quantitative differences in cellular cadherin expression can contribute to sorting (14, 100) (Fig. 4B).

C. Embryogenesis and Organogenesis

All newly formed epithelial cells express high levels of E-cadherin (101). The different mechanisms of cell–cell interaction enable the highly complex tissue patterning that occurs during embryogenesis and organogenesis (102). Homologous recombination of a mutated E-cadherin in murine embryonic stem cells, if homozygous results in failure of preimplantation, whereas, if heterozygous, the animals appear normal and are fertile (103). It seems that, in the oocyte, intracellular transport, proteolytic processing, and secretion are

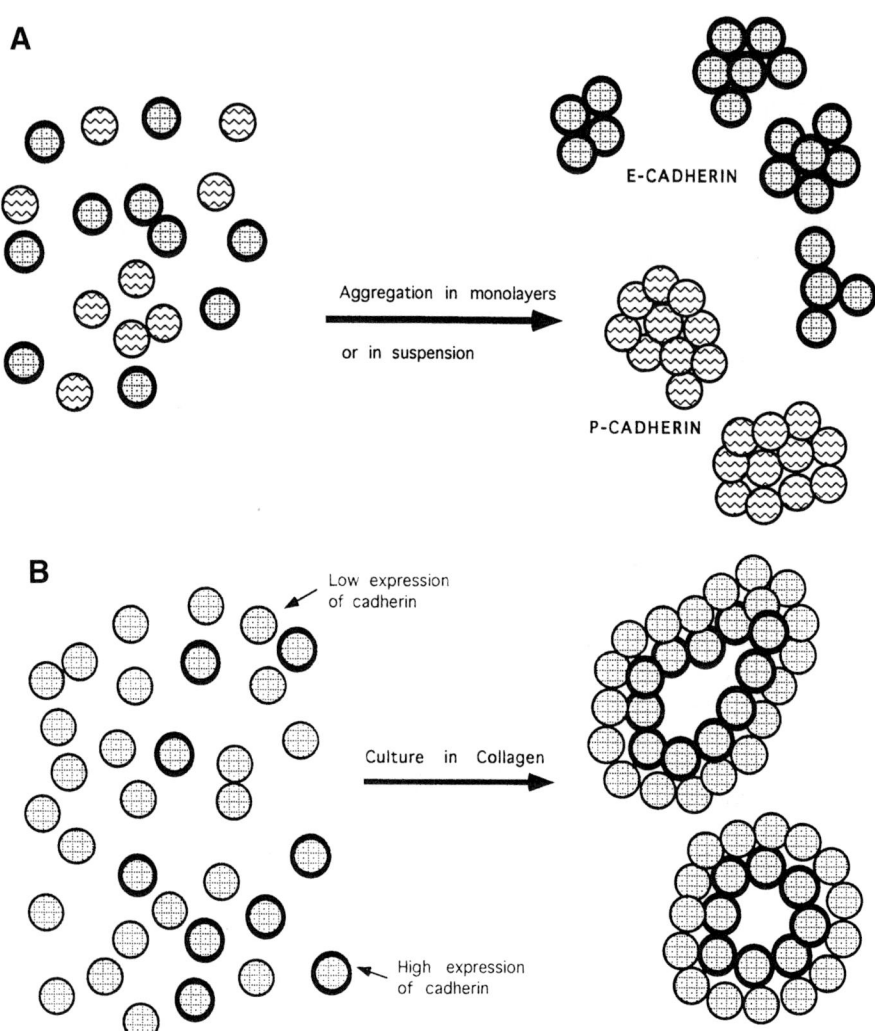

FIG. 4. Mechanisms of cell–cell recognition by cadherins. (A) Cell sorting in either plastic or cell suspension culture—the cells expressing similar cadherins aggregate into cellular clumps. (B) Spheroid formation in cell suspension culture—cells expressing differential levels of the same cadherin will form multilayered hollow spheroids with a gradient of cadherin expression from inside out.

the most important factors that affect functional expression, rather than E-cadherin production (104). In humans, E-cadherin is present in the morula at 1–3 days (8–32 cells), then neural and retinal cadherins are expressed in the blastocyst (>64 cells). During organogenesis, tissue-specific cadherins, such as P and LI cadherins, are expressed; specifically during renal development in humans P-cadherin specifies cell sorting, tissue spreading, and specific spatial patterning through differential expression levels of cadherins between the medulla and the cortex (105). Furthermore, podocytes in the developing kidney loose some of the cadherins and express only P-cadherin (106). Subsequently more tissue-specific renal cadherins are expressed, resulting in a mosaic of cadherin distribution. During epithelialization of early tissues, mesenchyme can be transformed by E-cadherin (107), and it is of interest that when E-cadherin is blocked by functional antibodies to the EC1 domain (108), the epithelial cells revert to a spindle-shaped morphology, similar to mesenchymal cells (109).

During mammary development glandular end buds with a high proliferation rate express E-cadherin, whereas the ductal myoepithelial cells express only P-cadherin (110).

D. Tissue and Glandular Differentiation

Electron microscopy examination of brain tissue sections reveal that E-cadherin is normally maximally localized in the lateral membrane borders in the intermediate (adherens) junctions but is shifted to diffuse cell surface sites in tumors (111). In particular, poorly differentiated tumors have decreased E-cadherin expression, and this is associated with retarded development of the intercellular junctions (112), whereas well-differentiated cancers have homogeneous E-cadherin membrane expression (113). Nontransformed MDCK epithelial cells and well-differentiated colon carcinoma cells acquire a poorly differentiated phenotype when intercellular adhesion is specifically inhibited by anti-E cadherin monoclonal antibodies (61, 114).

The development of a fully differentiated glandular structure may require other adhesion molecules, such as integrins. During renal development there is sequential expression of laminins, E-cadherin, and cytokeratins, suggesting that the order of cell adhesion molecule expression is important. It is possible that growth and branching require separate proteins. This pattern of spatiotemporal organization may be ultimately controlled by the actions of several molecules on E-cadherin, including products of homeobox genes, e.g., MSX-1, MSX-2, and goosecoid. Other molecules, including Wnt-2, the int-1 oncogene-related product, lead to grossly increased glandular branching, which is associated with aberrant cell adhesion.

E. Cell Phenotype/Polarity

Differences in the morphology of colorectal cells correlate with their cadherin expression patterns. Round cells (R) are usually invasive and fail to express α-catenin or E-cadherin, whereas spread cells (S) are noninvasive and express abundant E-cadherin and α-catenin. The interaction with catenins in the lateral membranes may serve two functions: (1) maintain cell–cell contacts and (2) to provide a link between cell adhesion molecules, the cytoskeleton, and other integral membrane proteins (115). Binding of catenins to E-cadherin is required for the appropriate membrane distribution of the several other molecules vital for maintaining epithelial polarity: the Na^+, K^+-ATPase pump via interactions with ankyrin and spectrin, EGFR, integrins, and desmosomes. In particular, rat retinal pigment epithelium (RPE) cells transfected with E-cadherin express a desmoglein and a different isoform of ankyrin that is associated with cytokeratin/desmosome complexes and the polarization of the membrane (115). It appears that the changes in surface polarity of cells involve a hierarchy of sorting mechanisms in the Golgi complex and plasma membrane and that Na^+, K^+-ATPase membrane receptors are sorted in apical pathways, partly due the action of E-cadherin (116). In this regard the MDCK cell line has marked polarity and expresses both P- and E-cadherin (64). Alternatively, the loss of E-cadherin catenin complexes in malignant cells leads to simultaneous loss of polarity and functional receptors (10). As cellular polarization proceeds, spotlike adherens junctions are gradually fused side by side, with concomitant shortening of the associated stress fiberlike actin filament bundles (117).

F. Cytokinesis–Invasion

In many tumors involving the skin, bladder, breast, and stomach, abnormal P- and E-cadherin functions are associated with both invasion and metastasis (118–120). In the colon most reports also confirm this association (121–124), although at least one study has failed to show any association of E-cadherin with tumor spread (125). This may be due in part to the criteria involved in deciding positive immunoreactivity, because only the membranous fraction of E-cadherin contributes to adhesion function.

More direct evidence comes from several sources. First, transfection with E-cadherin reduces the invasive potential of the f-derivative of MDCK cells (deficient in E-cadherin). Furthermore, this action can be reversed through the use of specific antisense E-cadherin RNA (126). E-Cadherin-deficient tumors form increased numbers of dispersed colonies in collagen gels associated with tissue invasion in xenografted tumors (127). Second, in our studies, colonies with detectable E-cadherin mutations and decreased adhesive function also have the fastest growth in xenografts. Other experimental evidence for E-cadherin as a invasive suppressor gene includes the transfection of wild-type p53, DCC, and APC into SW480 cells, which leads to suppressed tumorigenicity; however, these lines still retained anchorage independence, suggesting cell–cell adhesion defects (128).

Evidence implicating E-cadherin in formation of metastases is weak and is based on the association between high-metastasizing cell lines and low E-cadherin immunoreactivity. One recent study has shown *in vitro* that E-cadherin-negative tumor cells are more likely than E-cadherin-positive cells to be dislodged from the primary tumor by low shear forces, such as those found in venules or lymphatic systems (129). It has been postulated, however, that the aberrant adhesion of cells to the extracellular matrix as a result of aberrant carcinoembryonic antigen (CEA), CD44 expression, E-cadherin mutations, or E-cadherin down-regulation may dictate sites of metastasis, such as in the liver (9, 122). Recurrence of cancer may be indicative of field carcinogen effects in that residual cells may be left *in situ* following surgery and invasion may follow after a lag period (130). A report has also suggested that E-cadherin proteolysis and subsequent shedding (turnover) may be accelerated in patients with epithelial tumors, perhaps indicating increased endogenous breakdown (131).

Intriguingly, increased E-cadherin expression is associated with intravascular components of adenocarcinomas rather than with the extravascular compartment (132). E-Cadherin also decreases the expression of gelatinase secretion from colorectal COKFu cells, thereby decreasing invasion (133). The production of mucus by epithelial cancers is also antiadhesive and this can inhibit both cadherin-mediated cell–cell contact and laminin-mediated ECM adhesion (134).

G. Epithelial/Leukocyte Interactions

E-Cadherin may also be important in the generation of epithelial leukocyte interactions because $\alpha E\beta 7$ integrins of T lymphocytes can bind E-cadherin in mucosae (43). Furthermore, Langerhans cells express E-cadherin and this may be important in establishing and maintaining interactions between Langerhans cells and squamous epithelial cells (135).

H. Cell Adhesion and Apoptosis

Almost all epithelial cells undergo apoptosis [programmed cell death (PCD)] in certain benign ulcerative lesions, e.g., non-steroidal antiinflammatory drug (NSAID)-induced enteropathy PCD (136, 137). We have also found that in colitis there is an increased level of apoptosis in active ulceration compared with normal colon (J. Jankowski, F. Bedford, and Y. Kim, personal observations).

ECM cell adhesion molecules have been proved to be of major importance in generating cell survival signals (138, 139). Cell–cell adhesion, however, has also been postulated to regulate apoptosis in epithelial tissues. In this regard, treatment of confluent MDCK cells with either scatter factor or the phorbol ester TPA (both of which break adherens junctions and the cytoskeleton) increases apoptosis. Furthermore, Hermiston and Gordon have developed a murine chimeric model derived from 129/Sv embryonic stem cells stably transfected with a dominant-negative N-cadherin mutant under the control of a promoter that functions only in postmitotic enterocytes in normal C57BL/6 (B6) blastocysts (140, 141). Polyclonal villi containing the dominant-negative cadherin have disrupted cell adhesion in addition to causing precocious entry into a cell death program. Intriguingly, increased bcl-2 expression in mammary epithelial cells has been shown to decrease E-cadherin expression while increasing ECM molecules (142).

IV. Decreased or Defective E-Cadherin Expression in Colorectal Disease

A. Defects in Cell Adhesion during Colonic Inflammation—Spatial Changes of E-Cadherin in Vivo

It has been reported that intercellular adhesion molecules are down-regulated in UC (143). Similarly, there is up-regulation of nonfunctional cytoplasmic E-cadherin expression in UC during the inflammatory and granulation phases of epithelial adaptation, whereas the membranous expression decreases. As the disease activity progresses and enters the late remodeling phase, both cytoplasmic and membranous E-cadherin expression is lost altogether, reflecting in part the poor degree of cellular differentiation and maturity. The cytoplasmic distribution may be necessary to allow cells to migrate, become effete, or proliferate as necessary. Inhibition of adhesion is known to occur in vitro as a result of the cytokines HGF, IL-1, and TGF-β. Tumor necrosis factor-α (TNF-α) expression is associated with disorganization of the both desmosomal and classical cadherins at the adherens junctions during menstruation (144). Alternatively, IL-12 has been shown to increase expression of E-cadherin in colorectal cancer cells (145).

B. Structural Changes of E-Cadherin during Tumorigenesis

1. E-CADHERIN AS A CANDIDATE ONCOSUPPRESSOR GENE AND INTEGRITY OF THE DNA TEMPLATE

The allelic loss of heterozygosity of the E-cadherin gene on chromosome 16q has been reported in 40% of cervical and prostrate carcinomas (120), in 50% of breast adenocarcinomas (146), hepatocellular carcinoma (147), and Wilms' tumors (148), and in 75% of gastric adenocarcinomas. In a recent study of cytogenetic abnormalities in colorectal tumors and cell lines (149) loss of 16q (E-cadherin) as well as 6q, 1p, 7p, 8p, 17p, and 18p is a frequent occurrence (150). It has been postulated that as a consequence of the LOH decreased E-cadherin transcription may result in major morphospatial deregulation. Barrett's adenocarcinomas (151) and diffuse gastric adenocarcinomas (150) also have exon skipping of the E-cadherin gene, which results in truncated mRNA and protein.

2. Point Mutations

To date, point mutations of E-cadherin have been shown in clinical cases of gynecologic cancer and in lung, renal, gastric, and colorectal cell lines (149, 152, 153). In most cases of somatic point mutations of E-cadherin the wild-type allele is maintained (152). These point mutations are most frequent in the extreme 5' end of the mature peptide coding sequence or the Ca^{2+} domain of EC1 and lead to a loss of adhesive activity, suggesting that this may be a mutation cluster region (MCR) similar to those in codons 428–504 of the APC gene as well as codons 12 and 13 of the c-ras gene (Fig. 4). It seems possible, however, that different mutations of E-cadherin could result in a spectrum of biological alterations and as such the variety of mutations could explain the different phenotypes and biology.

3. Transcriptional Regulation

The E-cadherin palindromic promoter region (E-pal) sequence binds several nuclear binding factors (trans-acting elements) that are present in well-differentiated cancers and normal tissue but are down-regulated in several poorly differentiated cancers (154). Nuclear factors that stimulate c-erbB2 transcription may be inhibitory to the E-cadherin promoter gene (20), whereas estrogen can up-regulate E-cadherin expression. Mutations in the E-pal site also result in E-cadherin down-regulation. Retinoic acid and tamoxifen have both been shown to inhibit breast cancer in animal models in vivo in part by increasing E-cadherin expression (155). Recent evidence also suggests that E-cadherin mRNA stability may also be decreased in cancer (114, 156).

4. Posttranscriptional Processing

The phenomenon of exon skipping in the E-cadherin gene has been associated with several poorly differentiated carcinomas, including those of the gastric mucosa. Exons 8 and 9 are frequently skipped in the cis-acting elements, which include the putative calcium-binding domains. Differentiated intestinal tumors, on the other hand, have silent mutations. Interestingly, all patients reported to date expressed abundant transcripts, albeit truncated, and had immunoreactivity because loss of either exon 8 or 9 does not cause a frame shift. It has been speculated that these exon-skipping defects occur due to dietary polycyclic aromatic hydrocarbon derivatives (16, 150, 157).

5. Posttranslational Level

Internalization, conformation changes, or proteolysis may alter the expression of extracellular epitopes of E-cadherin on cells. These changes result in the failure of E-cadherin to become polarized on the cell membrane, remaining instead in the intracellular space, which is especially in low Ca^{2+}. In this regard E-cadherin is truncated in Barrett's esophageal mucosa and has a molecular mass less than 90 kDa (151). Furthermore, immunoreactivity is predominately cytoplasmic due to loss of cytoskeletal binding elements. In the stomach, proteolysis may be important because truncated protein bands are also seen (158). Minor anomalies of the N-terminus can also reduce E-cadherin function, such as incorrect cleavage of the E-cadherin polypeptide (60, 159). Increased levels of soluble E-cadherin have been reported in diseases associated with increased turnover and proteolysis and cancer (160, 161).

6. Changes in Catenin Expression

In gastric and breast cancer there is no evidence for mutations in the α- and β-catenin genes (162). These components of the cadherin/catenin complex are, however, disrupted in other gastrointestinal lesions; α-catenin is lost in 75% of colorectal cancers whereas β-catenin is lost by partial deletion in scirrhous gastric cancer (48, 49, 163). α-Catenin loss is associated with increased invasiveness of human colon cancer cells such as HCT-8 and DLD-1, whereas COLO320M, SW480, and SW620 were defective in E-cadherin expression (19, 164). β-Catenin is down-regulated in over 50% of esophageal, gastric, and colonic tumors, but this is not associated with mutation of the gene (165).

C. Role of Other Cadherins in Generation of a "Cancer Phenotype"

P-Cadherin is found predominantly in multilayered extraembryonic tissue and normal stratified epithelium, including the skin, esophagus, and anus, but is expressed paradoxically at the ZA in epithelial cancers (166). P-Cadherin is therefore not present in the normal colon but is increasingly expressed in adenomatous polyps and well-differentiated cancers (167; J. Jankowski, F. Bedford, and Y. Kim, personal observations). In several tumors, including stomach and basal cell carcinoma, P-cadherin is up-regulated dur-

---

FIG. 5. The qualitative and quantitative mechanisms of cadherin dysfunction in epithelial tumorigenesis. The complexity of dysfunction during tumorigenesis is represented in this figure, including decreased transcription, gene mutation, aberrant function, and the expression of inappropriate cadherins.

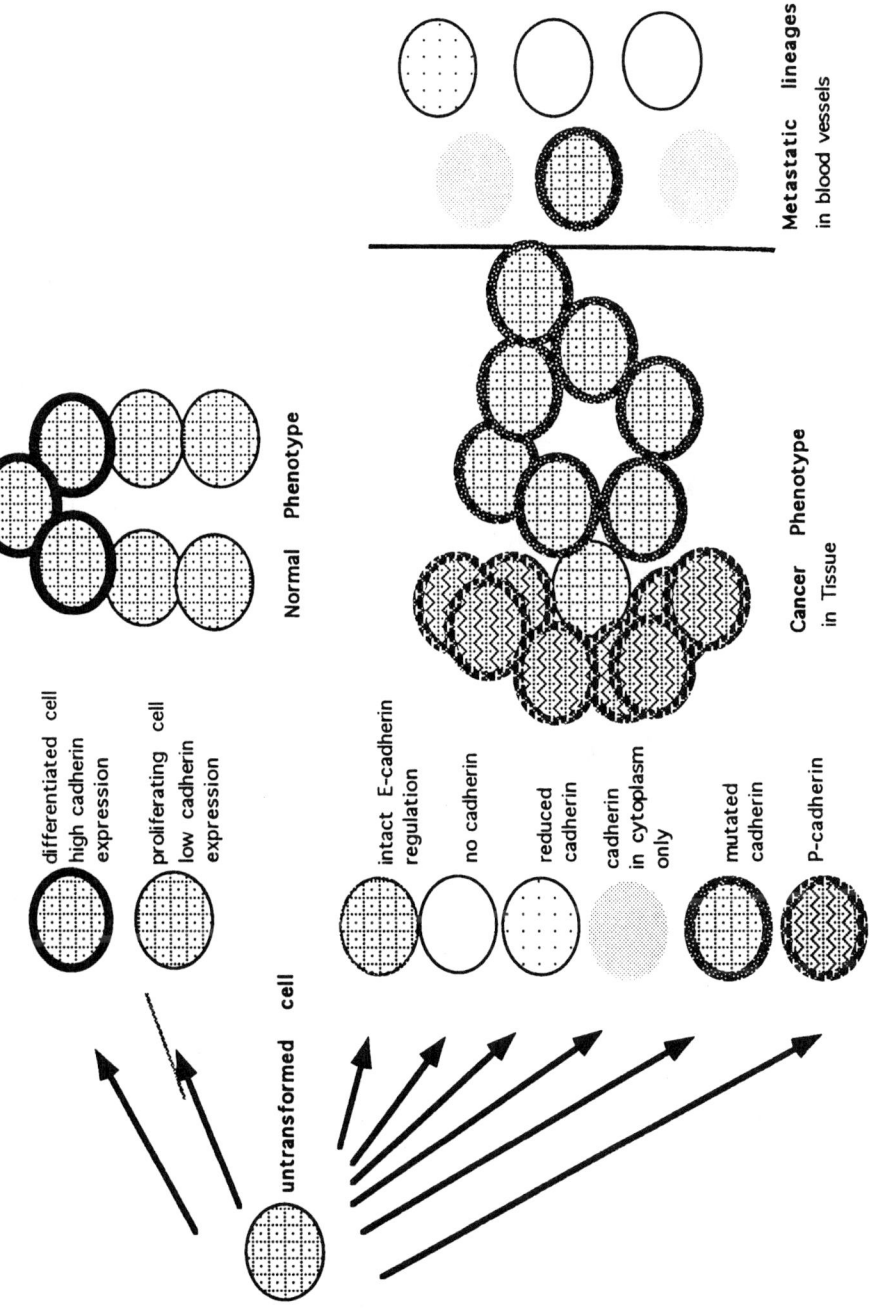

ing tumorigenesis whereas E-cadherin is down-regulated (*168*). In this regard P-cadherin may be responsible for the generation of a multilayered hyperproliferative phenotype.

P-Cadherin may have differential roles in tumorigenesis because it is increased in well-differentiated cancers but not in poorly differentiated cancers. It seems, however, that P-cadherin not only has a role in the development of cancer phenotype but also in cancer biology, and its subsequent loss leads to increased tumor infiltration (*169*). These observations are also confirmed *in vitro* because P-cadherin is expressed less in the poorly differentiated gastric cancer cell line TMK-1 compared with the well-differentiated MKN-28 cell line (*170*). In general, when a tissue normally expresses P-cadherin its loss is associated with increased invasion (*171*), but in mammary tissue, where P-cadherin is not expressed, there is a paradoxical expression of P-cadherin in invasive cancer (*172*).

In addition to P-cadherin other nonnative cadherins have been identified in colorectal tumors, including OB-1 cadherin and desmosomal cadherins (*173*, *174*), but their significance in cancer progression is uncertain. It is unclear whether these alternate cadherins are up-regulated as a result of decreased E-cadherin or if their transcription decreases E-cadherin or competes with the native cadherin binding and/or adhesive signals (*175*, *176*).

It also cannot be excluded that the multiple changes in the expression of basement membrane and extracellular matrix molecules that occur in colorectal cancer may also result in alternations in cadherin expression (*177*, *178*).

V. Conclusions

Cadherins have highly conserved cytoplasmic and extracellular domains between species and among other members of the cadherin family. Each tissue has distinct cadherins and within the tissue; each cell subtype has specific quantities of particular cadherins.

There is qualitative, spatiotemporal expression of cadherins during renal development and subsequently there is quantitative expression in the final stages of organogenesis (*106*, *179*). E-Cadherin is a reliable marker of differentiation when expressed on cell membranes. Tumorigenesis correlates with decreased E-cadherin expression or increased intracytoplasmic localization, both of which correlate well with altered biological actions *in vivo* and *in vitro*, including invasion and cell survival (*126*, *175*, *180*). There is correlation between epithelial morphology, noninvasiveness, and E-cadherin expression. In advanced tumorigenesis, however, reattachment of tumor cells to metastasizing sites may rely on multiple adhesion molecules, including dysfunctional cadherins (*62*). It may also be possible that other cadherins, such as P-cadherin, cadherin-6, and N-cadherin, interact with the cytoskeleton and may paradoxically cause invasive lesions or metastases (*118, 167, 181*) (Fig. 5).

The manipulation of cadherin expression offers several therapeutic possibilities because 12-prostaglandin and retinoic acid increase E-cadherin expression in the adherens junctions and thereby decrease proliferation (*182*).

Certain facts are apparent from the study of E-cadherin in colorectal tissue. First, E-cadherin is an important marker of differentiation in colorectal tissue *in vivo* and *in vitro*. Second, E-cadherin mutations are associated with decreased cellular adhesion and altered phenotype. Third, cell adhesion changes over time with complex interactions with other molecules. As a result, a conceptual framework can be made as to the controlling mechanisms of E-cadherin expression. First, cells express specific regulatory elements that dictate both the coarse and fine tuning of cell–cell contact. Second, during carcinogenesis these regulatory elements fail to coordinate and compensate for downstream changes of E-cadherin (Fig. 5). Our attempts to improve the understanding of this work will focus on deducing a research hypothesis so as to predict the relationships between the regulatory elements that maintain intact cell adhesion.

In conclusion, small differences in specific combinations, concentrations, and distributions of cell adhesion molecules and matrix receptors present in each cell type and can function as the required "morphogenetic" code, i.e., dictate a specific differentiation pattern.

Acknowledgments

We would like to thank Dr. Inke Nathke, Cell and Molecular Biology Laboratory, University of Stanford, California, and Dr. Marion Bussemakers, Department of Urology, University of Nejmegen, The Netherlands, for participation in helpful discussions during the preparation of this manuscript. Without their comments, the completion of this article would not have been possible due to the complexity and rapid advances in the area of cadherin biology.

References

1. J. Jankowski and N. A. Wright, *Sem. Cell. Biol.* **3**, 445 (1992).
2. M. Takeichi, *Science* **251**, 1451 (1991).
3. M. Overduin, T. S. Harvey, S. Bagby, K. I. Tong, P. Yau, M. Takeichi, and M. Ikura, *Science* **267**, 386 (1995).
4. J. Arnemann, K. H. Sullivan, A. I. Magee, I. A. King, and R. S. Buxton, *J. Cell. Sci.* **104**, 741 (1993).
5. I. A. King, K. H. Sullivan, R. Bennett, and R. S. Buxton, *J. Invest. Dermatol.* **105**, 314 (1995).

6. P. J. M. D. Koch, R. Goldschimidt, R. Zimbelman, R. Troyanovsky, and W. W. Franke, *Proc. Natl. Acad. Sci. U.S.A.* **89,** 353 (1992).
7. K. H. Sano, R. L. Tanihara, S. Heimark, M. Obata, T. Davidson, S. St. john, and S. Suzuki, *EMBO J.* **12,** 2249 (1993).
8. B. Ranscht and M. T. Dours-Zimmerman, *Neuron* **7,** 391 (1991).
9. J. Behrens, L. Vakaet, R. Friis, E. Winterhager, F. Van Roy, M. M. Mareel, and W. Birchmeier, *J. Cell. Biol.* **120,** 757 (1993).
10. U. B. Frixen, J. Behrens, M. Sachs, G. Eberle, B. Voss, A. Warda, D. Lochner, and W. Birchmeier, *J. Cell. Biol.* **113,** 173 (1991).
11. J. M. Staddon, K. Herrenknecht, C. Smales, and L. L. Rubin, *J. Cell. Sci.* **108,** 609 (1995).
12. G. Berx, K. Staes, J. V. Hengel, F. Molemans, M. J. G. Bussemakers, A. Van Bokhoven, and F. Van Roy, *Genomics* **26,** 281 (1995).
13. S. Miyatani, N. G. Copeland, D. J. Gilbert, N. A. Jenkins, and M. Takeichi, *Proc. Natl. Acad. Sci. U.S.A.* **89,** 8443 (1992).
14. K. Hatta, S. Miyatani, N. G. Cpeland, D. J. Gilbert, N. A. Jenkins, and M. Takeichi, *Nucleic Acids Res.* **19,** 4437 (1991).
15. D. L. Rimm and J. S. Morrow, *Mol. Biol.* **12,** 34 (1991).
16. K. F. Becker, U. Reich, C. Schott, and H. Hofler, *Hum. Genet.* **96,** 739 (1995).
17. J. Behrens, O. Lowrich, L. Klein-Hitglass, and W. Birchmeier, *Proc. Natl. Acad. Sci. U.S.A.* **88,** 11495 (1991).
18. M. L. Faraldo and A. Cano, *J. Mol. Biol.* **231,** 935 (1993).
19. S. J. Vermeulen, E. A. Bruyneel, F. M. van-Roy, M. M. Mareel, and M. E. Bracke, *Br. J. Cancer* **72,** 1447–53 (1995).
20. B. D'souza and J. Taylor-Papadimitiou, *Proc. Natl. Acad. Sci. U.S.A.* **91,** 7202 (1994).
21. J. Palacios, N. Benito, A. Pizarro, M. A. Limeres, A. Suarez, A. Cano, and C. Gamallo, *Virchows Arch.* **427,** 259 (1995).
22. G. Carruba, D. Miceli, D. A'Amico, R. Farruggio, L. Comito, A. Montesanti, L. Polito, and L. A. Castagnetta, *Biochem. Biophys. Res. Commun.* **212,** 624 (1995).
23. A. M. Fenyves, J. Behrens, and K. Spanel-Borowski, *J. Cell. Biol.* **61,** 172 (1993).
24. W. G. Jiang, S. Hiscox, M. B. Hallett, D. F. Horrobin, R. E. Mansel, and M. C. Puntis, *Cancer Res.* **55,** 5043 (1995).
25. S. Gopalakrishnan and M. P. Quinlan, *Cell Growth Differen.* **6,** 985 (1995).
26. G. Hennig, J. Behrens, M. Truss, S. Frisch, E. Reichman, and W. Birchmeier, *Oncogene* **11,** 475 (1995).
27. G. Hennig, O. Lowrick, W. Birchmeier, and J. Behrens, *J. Biol. Chem.* **271,** 595 (1996).
28. K. Yoshiura, Y. Kanai, A. Ochiai, Y. Shimoyama, T. Sugimura, and S. Hirohashi, *Proc. Natl. Acad. Sci. U.S.A.* **92,** 7416 (1995).
29. C. Caulin, F. G. Scholl, P. Frontelo, C. Gamallo, and M. Quintanilla, *Cell Growth Differen.* **6,** 1027 (1995).
30. W. I. Weis, *Structure* **3,** 425 (1995).
31. L. Shapiro, P. D. Kwong, A. M. Fannon, and W. A. Colman Hendrickson, *Proc. Natl. Acad. Sci. U.S.A.* **92,** 6793 (1995).
32. J. J. Deman, N. A. Van Larebeke, E. A. Bruyneel, M. E. Bracke, S. J. Vermeulen, K. M. Vennekens, and M. M. Mareel, *In Vitro Cell Dev. Biol. Anim.* **31,** 633 (1995).
33. S. H. Jaffe, *Proc. Natl. Acad. Sci. U.S.A.* **87,** 3589 (1990).
34. T. Fujimori and M. Takeichi, *Mol. Cell. Biol.* **14,** 37 (1993).
35. C. Kintner, *Cell* **69,** 225 (1992).
36. A. Nagafuchi, S. Ishihara, and S. Tsukita, *J. Cell. Biol.* **127,** 235 (1994).
37. P. Navarro, L. Caveda, F. Breviario, I. Mandoteanu, M. G. Lampugnani, and E. Dejana, *J. Biol. Chem.* **270,** 30965 (1995).

38. R. L. Juliano and S. Haskill, *J. Cell. Biol.* **120,** 577 (1993).
39. K. J. Hodivala and F. M. Watt, *J. Cell. Biol.* **124,** 589 (1994).
40. P. Navarro, E. Lozano, and A. Cano, *J. Cell Sci.* **105,** 923 (1993).
41. S. Finneman, M. Kihl, G. Otto, and D. Wedlich, *Mol. Cell. Biol.* **15,** 5082 (1994).
42. T. C. Klinowska, G. W. Ireland, and S. J. Kimber, *Differentiation* **57,** 7 (1994).
43. P. I. Karecla, S. J. Bowden, S. J. Green, and P. J. Kilshaw, *Eur. J. Immunol.* **25,** 852 (1995).
44. C. Murphy-Erdosh, C. K. Yoshida, N. Paradies, and L. F. Reichardt, *J. Cell. Biol.* **129,** 1379 (1994).
45. L. K. Su, B. Vogelstein, and K. W. Kinzler, *Science* **262,** 1734 (1993).
46. B. Rubinfeld, B. Souza, I. Albert, O. Muller, S. H. Chamberlain, F. R. Masiarz, S. Munemitsu, and P. Polakis, *Science* **262,** 1731 (1993).
47. L. Hinck, I. S. Nathke, J. Papkoff, and W. J. Nelson, *J. Cell. Biol.* **125,** 1327 (1994).
48. J. Hulsken, W. Birchmeier, and J. Behrens, *J. Cell. Biol.* **127,** 2961 (1994).
49. J. Hulsken, J. Behrens, and W. Birchmeier, *Curr. Biol.* **6,** 711 (1994).
50. B. Rubinfeld, B. Souza, I. Albert, S. Munemitsu, and P. Polakis, *J. Biol. Chem.* **270,** 5549 (1995).
51. J. P. Doyle, J. G. Stempak, P. Cowin, D. R. Colman, and D. Urso, *J. Cell. Biol.* **131,** 465 (1995).
52. M. Kato, S. Saunders, H. Nguyen, and M. Bernfield, *Mol. Biol. Cell* **6,** 559 (1995).
53. M. Glukhova, V. Koteliansky, X. Sastre, and J. P. Thiery, *Am. J. Pathol.* **146,** 706 (1995).
54. K. Herrenknecht, M. Ozawa, A. Erkershorn, A. Loittspeich, P. Lenter, and R. Kemler, *Proc. Natl. Acad. Sci. U.S.A.* **88,** 9156 (1991).
55. S. Hirano, N. Kimoto, Y. Shimoyama, S. Hirohashi, and M. Takeichi, *Cell* **70,** 283 (1992).
56. A. Nagafuchi, M. Takeichi, and S. Tsukita, *Cell* **65,** 849 (1991).
57. P. D. McCrea, C. W. Turck, and B. Gumbiner, *Science* **254,** 1359 (1991).
58. J. M. Trent, R. Wiltshire, L. K. Su, N. C. Nicolaides, Vogelstein, and K. W. Kinzler, *Cytogenet. Cell. Genet.* **71,** 343 (1995).
59. K. A. Knudsen and M. J. Wheelock, *J. Cell. Biol.* **118,** 671 (1992).
60. M. Ozawa, M. Ringwald, and R. Kemler, *Proc. Natl. Acad. Sci. U.S.A.* **87,** 4246 (1990).
61. B. M. Gumbiner and P. D. McCrea, *J. Cell. Sci.* **17,** 155 (1993).
62. M. Peifer, *Science* **262,** 1667 (1993).
63. J. D. McPherson, R. A. Morton, C. M. Ewing, J. J. Wasmuth, J. Overhauser, A. Nagafuchi, S. Tsukita, and W. B. Isaacs, *Genomics* **19,** 188 (1994).
64. J. C. Wu, C. W. Gregory, and R. M. DePhilip, *Biochem. Biophys. Res. Commun.* **195,** 1329 (1993).
65. K. R. Johnson, J. E. Lewis, J. Wahl, A. P. Soler, K. A. Knudsen, and M. J. Wheelock, *Exp. Cell Res.* **207,** 252 (1993).
66. I. S. Nathke, L. Hinck, J. R. Swedlow, J. Papkoff, and W. J. Nelson, *J. Cell. Biol.* **125,** 1341 (1994).
67. L. Hinck, W. J. Nelson, and J. Papkoff, *J. Cell. Biol.* **124,** 729 (1994).
68. S. Tsukita, M. Itoh, S. Nagafuchi, and S. Tsukita, *J. Cell. Biol.* **123,** 1049 (1993).
69. I. Miyashiro, T. Senda, A. Matsumine, G. H. Baeg, T. Kuroda, T. Shimano, S. Miura, T. Noda, S. Kobayashi, and M. Monden, *Oncogene* **11,** 89 (1995).
70. S. Munemitsu, B. Souza, O. Muller, I. Albert, B. Rubinfeld, and P. Polakis, *Cancer Res.* **54,** 3676 (1994).
71. K. J. Smith, D. B. Levy, P. Maupin, T. D. Pollard, B. Vogelstein, and K. W. Kinzler, *Cancer Res.* **54,** 3672 (1994).
72. M. S. Kinch, G. J. Clark, C. J. Der, and K. Burridge, *J. Cell. Biol.* **129,** 507 (1995).
73. J. Stappert and R. Kemler, *Cell. Adhes. Commun.* **2,** 319 (1994).

74. N. Matsuyoshi, M. Hamaguchi, S. Taniguchi, A. Nagafuchi, S. Tsukita, and M. Takeichi, *J. Cell. Biol.* **118,** 703 (1992).
75. J. L. Guan and D. Shalloway, *Nature (London)* **358,** 690 (1992).
76. L. Naldini, E. Vigna, R. P. Narsimhan, G. Gaudino, R. Zarnegar, G. K. Michalopoulos, and P. M. Comoglio, *Oncogene* **6,** 501 (1991).
77. M. Watabe, K. Matsumoto, T. Nakamura, and M. Takeichi, *Cell. Struct. Funct.* **18,** 117 (1993).
78. G. Hartmann, K. M. Weidner, H. Schwatz, and W. Birchmeier, *J. Biol. Chem.* **269,** 21936 (1994).
79. B. Boyer, S. Dufour, and J. P. Thiery, *Exp. Cell. Res.* **201,** 347 (1992).
80. M. Watabe, A. Nagafuchi, S. Tsukita, and M. Takeichi, *J. Cell. Biol.* **127,** 247 (1994).
81. G. M. Edelman and K. L. Crossin, *Annu. Rev. Biochem.* **60,** 155 (1991).
82. W. E. Pullman and W. F. Bodmer, *Nature (London)* **356,** 529 (1993).
83. H. Durbin, M. Novelli, and W. Bodmer, *Genomics* **19,** 181 (1994).
84. J. M. Daniel and A. B. Reynolds, *Mol. Cell. Biol.* **15,** 4819 (1995).
85. J. M. Staddon, C. Smales, C. Schulze, F. S. Esch, and L. L. Rubin, *J. Cell. Sci.* **108,** 369 (1995).
86. S. Shibamoto, M. Haakawa, K. Takeuchi, T. Hori, K. Miyazawa, N. Kitamura, K. R. Johnson, M. J. Wheelock, N. Matsuyoshi, and M. Takeichi, *J. Cell. Biol.* **128,** 949 (1995).
87. Y. Kanai, A. Ochiai, T. Shibata, T. Oyama, S. Ushijima, S. Akimoto, and S. Hirohashi, *Biochem. Biophys. Res. Commun.* **208,** 1067 (1995).
88. D. F. Newgreen and J. Minichiello, *Dev. Biol.* **170,** 91 (1995).
89. A. Skoudy and A. Garcia de Herreros, *FEBS Lett.* **374,** 415 (1995).
90. M. Nilsson and L. E. Ericson, *Exp. Cell. Res.* **219,** 626 (1995).
91. M. Pasdar and Z. Li, *Cell Motil. Cytoskeleton* **26,** 163 (1993).
92. D. L. Rimm, E. R. Koslov, P. Kebriaei, C. D. Cianci, and J. S. Morrow, *Proc. Natl. Acad. Sci. U.S.A.* **92,** 8813 (1995).
93. K. A. Knudsen, A. P. Soler, K. R. Johnson, and M. J. Wheelock, *J. Cell. Biol.* **130,** 67 (1995).
94. T. Oda, Y. Kanai, Y. Shimoyama, A. Nagafuchi, S. Tsukita, and S. Hiroshashi, *Cancer Res.* **53,** 1696 (1993).
95. H. Shiozaki, K. Iihara, H. Oka, T. Kadowaki, S. Matsui, J. Gofuku, M. Inoue, A. Nagafuchi, S. Tsukita, and T. Mori, *Am. J. Pathol.* **144,** 667 (1994).
96. D. Berndorff, R. Gessner, B. Kreft, N. Schnoy, A. M. Lajous-Petter, N. Loch, W. Reutter, M. Hortsch, and R. Tauber, *J. Cell. Biol.* **125,** 1353 (1994).
97. A. Nose, A. Nagafuchi, and M. Takeichi, *Cell* **54,** 993 (1988).
98. M. J. Wheelock and P. J. Jensen, *J. Cell. Biol.* **117,** 415 (1992).
99. J. E. Lewis, P. J. Jensen, and M. J. Wheelock, *Invest. Dermatol.* **102,** 870 (1994).
100. H. Sasaki, A. Tada, M. Takeichi, M. Obata, and Y. Terashima, *Proc. Annu. Meet. Am. Assoc. Cancer Res.* **33,** A191 (1992).
101. G. Levi, B. Gumbiner, and J. P. Tiery, *Development* **111,** 159 (1991).
102. Y. Hirai, A. Nose, S. Kobayashi, and M. Takeichi, *Development* **105,** 263 (1989).
103. D. Riethmacher, V. Brinkmann, and C. Birchmeier, *Proc. Natl. Acad. Sci. U.S.A.* **92,** 855 (1995).
104. L. Clayton, J. M. McConnell, and M. H. Johnston, *Zygote* **3,** 177 (1995).
105. M. S. Steinberg and M. Takeichi, *Proc. Natl. Acad. Sci. U.S.A.* **91,** 206 (1994).
106. M. T. Tassin, A. Beziau, M. C. Gubler, and B. Boyer, *Int. J. Develop. Biol.* **38,** 45 (1994).
107. B. Geiger and O. Ayalon, *Annu. Rev. Cell. Biol.* **8,** 307 (1992).
108. K. L. Lutz and T. J. Soiahaan, *Biochem. Biophys. Res. Commun.* **211,** 21 (1995).
109. C. A. Burdsal, C. H. Damsky, and R. A. Pedersen, *Development* **118,** 829 (1993).
110. C. W. Daniel, P. Strickland, and Y. Friedmann, *Dev. Biol.* **169,** 511 (1995).

111. Y. Tohma, T. Yamashima, and J. Yamashita, *Cancer Res.* **52**, 1981 (1992).
112. H. Gabbart, *Cancer Metastasis Rev.* **4**, 293 (1985).
113. Y. Shimoyama and S. Hirohashi, *Cancer Res.* **51**, 2185 (1993).
114. M. Mareel, K. Vleminckx, S. Vermeulen, M. Bracke, and F. Van Roy, *Bull. Cancer* **79**, 347 (1992).
115. J. A. Marrs, C. Andersson-Fisone, M. C. Jeong, L. Cohen-Gould, C. Zurzolo, I. R. Nabi, E. Rodriguez-Boulan, and W. J. Nelson, *J. Cell. Biol.* **129**, 507 (1995).
116. R. W. Mays, K. A. Siemers, B. A. Fritz, A. W. Lowe, G. van Meer, and W. J. Nelson, *J. Cell. Biol.* **130**, 1105 (1995).
117. S. Yonemura, M. Itoh, A. Nagafuchi, and S. Tsukita, *J. Cell. Sci.* **108**, 127–42 (1995).
118. A. Tang, M. S. Eller, M. Hara, M. Yaar, S. Hirohashi, and B. A. Gilchrest, *J. Cell. Sci.* **107**, 983 (1994).
119. W. Birchmeieer, K. Weidner, J. Hulsken, and J. Behrens, *Semin. Cancer Biol.* **4**, 231 (1993).
120. W. Birchmeier, J. Hulsken, and J. Behrens, *Cancer Surv.* **24**, 129 (1995).
121. A. A. M. Vanderwurff, J. Tenkate, E. P. M. Vanderlinden, W. N. M. Dinjens, J. W. Arends, and F. T. Bosman, *J. Pathol.* **168**, 287 (1992).
122. S. Dorudi, A. M. Hanby, R. Poulsom, J. Northover, and J. R. Hart, *Br. J. Cancer* **71**, 614 (1995).
123. S. Dorudi, J. Sheffield, R. Poulsom, J. M. A. Northover, and J. R. Hart, *Am. J. Pathol.* **142**, 981 (1993).
124. M. Pignatelli, D. Liu, M. M. Nasim, G. W. H. Stamp, S. Hirano, and M. Takeichi, *Br. J. Cancer* **66**, 629 (1992).
125. A. R. Kinsella, B. Green, G. C. Leptss, C. L. Hill, G. Bowie, and B. A. Taylor, *Br. J. Cancer* **67**, 904 (1993).
126. K. Vleminckx, L. Vakaet, M. Mareel, W. Fiers, and F. Van Roy, *Cell* **66**, 107 (1991).
127. J. E. de Vries, W. N. Dinjens, G. K. De Bruyne, H. W. Verspaget, E. P. van der Linden, A. P. de Bruine, M. M. Mareel, F. T. Bosman, and J. ten Kate, *Br. J. Cancer* **71**, 271 (1995).
128. M. C. Goyette, K. Cho, and C. L. Fashing, *Mol. Cell. Biol.* **12**, 1387 (1992).
129. S. W. Byers, C. L. Sommers, B. Hoxter, A. M. Mercurio, and A. Tozeren, *J. Cell. Sci.* **108**, 2053 (1995).
130. H. Oka, H. Shiozaki, K. Kobayashi, M. Inque, H. Tahara, T. Kobayashi, Y. Takatsuka, N. Matsuyoshi, M. Takeichi, and T. Mori, *Cancer Res.* **153**, 1696 1701 (1993).
131. M. Katayama, S. Hirai, M. Yasumoto, K. Nishikawa, S. Nagata, M. Otsuka, K. Kamihagi, and I. Kato, *Int. J. Oncol.* **5**, 1049 (1994).
132. G. P. Cowley and M. E. Smith, *Int. J. Cancer* **60**, 325 (1995).
133. M. Miyaki, K. Tanaka, R. Kikuchi-Yanoshita, M. Muraoka, M. Konishi, and M. Takeichi, *Oncogene* **11**, 2547 (1995).
134. H. Kemperman, Y. Wijnands, J. Wesseling, C. M. Niessen, A. Sonnenberg, and E. Roos, *J. Cell. Biol.* **127**, 2971 (1994).
135. A. Blauvelt, S. I. Katz, and M. C. Udey, *J. Invest. Dermatol.* **104**, 293 (1995).
136. F. D. Lee, *J. Clin. Pathol.* **46**, 118 (1993).
137. M. Tsuji and R. N. Dubois, *Cell* **83**, 493 (1995).
138. S. M. Frisch and H. Francis, *J. Cell. Biol.* **124**, 619 (1994).
139. N. Boudreau, C. J. Sympson, Z. Werb, and M. J. Bissell, *Science* **267**, 891 (1995).
140. M. L. Hermiston and J. I. Gordon, *J. Cell. Biol.* **129**, 489 (1995).
141. M. L. Hermiston and J. I. Gordon, *Science* **270**, 1203 (1995).
142. P. J. Lu, Q. L. Lu, A. Rughetti, and J. Taylor-Papadimitiou, *J. Cell. Biol.* **129**, 1363 (1995).
143. M. Balsitis, K. Morrell, Y. R. Mahida, and C. J. Hawkey, *Eur. J. Gastroenterol. Hepatol.* **6**, 352 (1994).

144. S. Tabibzadeh, Q. F. Kong, S. Kapur, P. G. Satyaswaroop, and K. Aktories, *Hum. Reprod.* **10**, 994 (1995).
145. S. Hiscox, M. B. Ballett, M. C. Puntis, and W. G. Jiang, *Clin. Exp. Metastasis* **13**, 396 (1995).
146. T. Sato, A. Tanigami, K. Yamakawa, F. Akiyama, F. Kasumi, G. Sakamoto, and Y. Nakamura, *Cancer Res.* **50**, 7184 (1990).
147. H. Tsuda, Y. Zhang, Y. Shimosato, Y. Yokota, M. Terada, T. Sugimura, T. Miyamura, and S. Hiroshashi, *Proc. Natl. Acad. Sci. U.S.A.* **87**, 6791 (1990).
148. M. A. Maw, P. E. Grundy, L. J. Millow, M. R. Eccles, R. S. Dunn, P. J. Smith, A. P. Feinberg, D. J. Law, M. C. Paterson, P. E. Telzerow, D. F. Callen, A. D. Thompson, R. I. Richards, and A. E. Reeve, *Cancer Res.* **52**, 3094 (1992).
149. F. K. Bedford and J. Jankowski, *J. Clin. Pathol. (Mol. Pathol.)* **48**, M6 (1995).
150. K. F. Becker, M. J. Atkinson, U. Teich, H. H. Huang, H. Nekarda, J. R. Siewert, and H. Hofler, *Human Mol. Genet.* **2**, 803 (1993).
151. J. Jankowski, P. Newham, O. Kandemir, S. Hirano, M. Takeichi, and M. Pignatelli, *Int. J. Oncol.* **4**, 441 (1994).
152. J. I. Risinger, A. Berchuck, M. E. Kohler, and J. Boyd, *Nature Genet.* **7**, 98 (1994).
153. T. Oda, Y. Kanai, T. Oyama, J. Yoshiura, Y. Shimoyama, W. Birchmeier, T. Sgimura, and S. Hirohashi, *Proc. Natl. Acad. Sci. U.S.A.* **91**, 1854 (1994).
154. B. Mayer, J. P. Johnson, F. Leitl, K. W. Jauch, M. M. Heiss, F. W. Schildberg, W. Birchmeier, and I. Funke, *Cancer Res.* **53**, 1690 (1993).
155. M. R. Anzano, S. W. Byers, J. M. Smith, C. W. Peer, L. T. Mullen, C. C. Brown, A. B. Roberts, and M. B. Sporn, *Cancer Res.* **54**, 4614 (1994).
156. M. Mareel, M. Bracke, and F. Van-Roy, *Cancer Detect. Prev.* **19**, 451 (1995).
157. K. F. Becker, M. J. Atkinson, U. Reich, I. Becker, H. Nekarda, J. R. Stewart, and H. Hofler, *Cancer Res.* **54**, 3845 (1994).
158. K. Matsuura, J. Kawanishi, S. Fuji, M. Imamura, S. Hirano, M. Takeichi, and Y. Nitsu, *Br. J. Cancer* **66**, 1122 (1992).
159. P. F. Bongiorno, M. Alkasspooles, S. W. Lee, W. J. Rachwal, J. H. Moore, R. I. Whyte, M. B. Orringer, and D. G. Beer, *Br. J. Cancer* **71**, 166 (1995).
160. R. E. Banks, W. H. Porter, P. Whelan, P. H. Smith, and P. J. Selby, *J. Clin. Pathol.* **48**, 179 (1995).
161. N. Matsuyoshi, T. Tanaka, K. Toda, H. Okamoto, F. Furukawa, and S. Imamura, *Br. J. Dermatol.* **132**, 745 (1995).
162. S. Candidus, P. Bischoff, K. F. Becker, and H. Hofler, *Cancer Res.* **56**, 49 (1996).
163. J. Kawanishi, J. Kato, K. Sasaki, S. Fujii, N. Watanabe, and Y. Niitsu, *Nippon Rinsho* **53**, 1590 (1995).
164. S. J. Vermeulen, E. A. Bruyneel, M. E. Bracke, G. K. De-Bruyne, K. M. Vennekens, K. L. Vleminckx, G. J. Berx, E. M. van-Roy, and M. M. Mareel, *Cancer Res.* **55**, 4722 (1995).
165. T. Takayama, H. Shiozaki, S. Shibamoto, H. Oka, Y. Kimura, S. Tamura, M. Inoue, T. Ito, and M. Monden, *Am. J. Pathol.* **148**, 39 (1996).
166. A. Nose and M. Takeichi, *J. Cell. Biol.* **103**, 2649 (1986).
167. Y. Shimoyama, M. Gotoh, T. Terasaki, M. Kitajima, and S. Hirohashi, *Cancer Res.* **55**, 2206 (1995).
168. A. Pizarro, C. Gamallo, N. Benito, J. Palacios, M. Quintanilla, A. Cano, and F. Contreres, *Br. J. Cancer* **72**, 327 (1995).
169. W. Yasui, T. Sano, K. Nishimura, Y. Kitadai, Z. Q. Ji, H. Yokozaki, H. Ito, and E. Tahara, *Int. J. Cancer* **54**, 49 (1993).
170. A. Tannapfel, C. Wittekind, and E. Tahara, *Z. Gastroenterol.* **32**, 91 (1994).
171. J. Vanaken, C. A. Cuvelier, N. Dewever, J. Roels, Y. Gao, and M. M. Mareel, *Pathol. Res. Pract.* **189**, 975 (1993).

172. S. A. Rasbridge, C. E. Gillett, S. A. Sampson, F. S. Walsh, and R. R. Millis, *J. Pathol.* **169,** 245 (1993).
173. S. B. Munro, I. M. Turner, R. Farookhi, O. W. Blaschuk, and S. Jothy, *Exp. Mol. Pathol.* **62,** 118 (1995).
174. D. R. Garrod, *Curr. Opin. Cell. Biol.* **5,** 30 (1993).
175. M. Takeichi, *Curr. Opin. Cell. Biol.* **5,** 806 (1993).
176. K. Vleminckx, *Cancer Res.* **19,** 873 (1994).
177. A. K. Nigam, F. J. Savage, P. B. Boulos, G. W. H. Stamp, D. Liu, and M. Pignatelli, *Br. J. Cancer* **68,** 507 (1993).
178. R. Visser, J. W. Arends, I. M. Leigh, and F. T. Bosman, *J. Pathol.* **179,** 285 (1993).
179. B. Ranscht, *Curr. Biol.* **6,** 740 (1994).
180. J. C. Reed, *J. Cell. Biol.* **124,** 1 (1994).
181. T. Tani, L. Laitman, L. Kangas, V. P. Lehto, and I. Virtanen, *Int. J. Cancer* **64,** 407 (1995).
182. K. Ikai, M. Yamamoto, N. Matauyosji, and M. Kukushima, *Prost. Leukot. Essent. Fatty Acids* **52,** 303 (1995).
183. M. J. G. Bussemakers, A. van Bokhoven, S. G. M. Mees, R. Kemler, and J. A. Schalken, *Mol. Biol. Rep.* **17,** 123 (1993).
184. A. Nose, A. Nagafuchi, and M. Takeichi, *EMBO J.* **6,** 3655 (1987).
185. K. Hatta, A. Nose, A. Nagafauchi, and M. Takeichi, *J. Cell. Biol.* **106,** 873 (1988).
186. S. Nakagawa and M. Takeichi, *Development* **121,** 1321 (1995).
187. A. Sjodin, U. Dahl, and H. Semb, *Exp. Cell. Res.* **221,** 413 (1995).
188. B. J. Fredette and B. Ranscht, *J. Neurosci.* **14,** 7331 (1994).
189. M. Zeschnigk, D. Kozian, C. Kuch, M. Schmoll, and A. Starzinski-Powitz, *J. Cell. Sci.* **108,** 2973 (1995).
190. D. Simrak, C. M. Cowley, R. S. Buxton, and J. Arnemann, *Genomics* **25,** 591 (1995).
191. F. Breviario, L. Caveda, M. Corada, I. Martin-Padura, P. Navarro, J. Golay, M. Introna, D. Gulino, M. G. Lampugnani, and E. Dejana, *Arterioscler. Thromb. Vasc. Biol.* **15,** 1229 (1995).
192. L. Simmoneau, M. Kitagawa, S. Suzuki, and J. P. Thiery, *Cell Adhes. Commun.* **3,** 115 (1995).

The Formation of DNA Methylation Patterns and the Silencing of Genes

JEAN-PIERRE JOST*,[1]
AND ALAIN BRUHAT[†]

*Friedrich Miescher Institute
CH-4002 Basel, Switzerland
†Unité de Nutrition Cellulaire et
    Moléculaire
INRA de Theix
Champanelle, France

I. The Formation of DNA Methylation Patterns 217
   A. DNA Methyltransferases 219
   B. Demethylation of DNA 222
   C. Determination Factors 228
   D. Models for Formation of Specific DNA Methylation Patterns 229
   E. DNA Hypermethylation: An Aberrant Form of DNA
      Methylation or a Defense Mechanism 233
II. DNA Methylation and the Silencing of Genes 234
   A. Methylation of CpGs Can Change DNA Structure 234
   B. Many Transcription Factors Do Not Bind to Methylated DNA 234
   C. DNA Methylation Favors Binding of Specific Repressor Proteins ... 235
III. Conclusions 243
    References 244

I. The Formation of DNA Methylation Patterns

Several lines of evidence strongly suggest that the pattern of DNA methylation is vital for the normal development of vertebrates (*1–4*). For example, targeted mutations of the DNA methyltransferase gene in transgenic mice have resulted in embryonic death (*5*). In the same experiments it was shown that the level of 5-methylcytosine (m^5C) in the DNA of embryonic stem cells and of embryos homozygous for the DNA methyltransferase mutation was reduced to about one-third of that found in heterozygous or wild-type cells or embryos. Though no obvious phenotype difference was apparent in homozygous embryonic stem cells in culture, homozygous embryos displayed severe stunting, developmental delay and death at midgestation. It is known

[1]To whom correspondence should be addressed

that too much or too little DNA methylation can result in abnormal growth and development (5–14). Because m^{5}C can undergo spontaneous hydrolytic deamination leading to base transition, an overmethylation of specific genes can lead to their inactivation and at the same time create hot spots of mutations. From 30 to 40% of all human germ-line mutations are thought to be methylation-induced even though the CpG dinucleotide is underrepresented and efficient cellular repair systems exist (9). For example, the tumor suppressor gene *p53* contains methylated CpGs that serve as hot spots of mutations in some but not all human cancers (9, 15). The role of critical levels of DNA methylation during tumorgenesis has been investigated in the murine adrenocortical tumor cell line system (11). In this system, an inhibition of DNA methyltransferase causes a demethylation of DNA and a reversal of the tumorgenic phenotype. Similarly, DNA hypomethylation in "Min" mice causes a suppression of intestinal neoplasia (14). Both hyper- and hypomethylation of DNA can be involved in the genesis of cancer (see Fig. 1) depending on which gene, and where on the gene, hyper- or hypomethylation has taken place. Hypermethylation of a tumor suppressor gene, for example, or hypomethylation of an oncogene could have the same final effect in generating cancer. This raises the important questions about how much methylation or

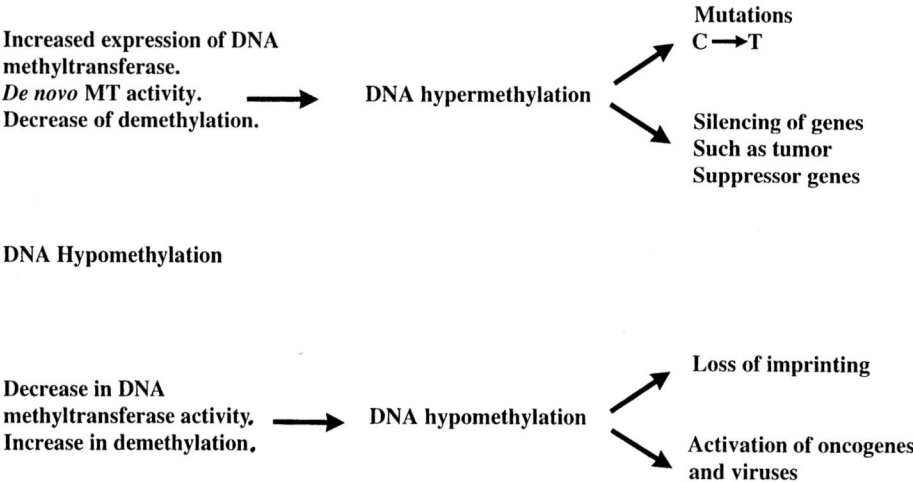

FIG. 1. Possible biological implications of hypermethylation and hypomethylation of DNA. MT, DNA methyltransferase; C, cytosine; T, thymine.

demethylation of the genome is needed for normal cell function and embryonic development, and of the importance of a stringent control of methylation pattern formation of specific genes. How is a specific methylation pattern of a given gene maintained or changed during development? It is believed that at least three different components are needed for the establishment of a specific DNA methylation pattern: DNA methyltransferases, the demethylation system of DNA (passive and/or active), and the determination factors (cis and trans). DNA methyltransferases and the demethylation system are CpG or CpXpG specific but not strictly sequence specific, whereas the determination factors are sequence and developmental stage specific. In this paper we present these three important components and discuss their possible interactions during the formation of specific DNA methylation patterns.

A. DNA Methyltransferases

In contrast to prokaryotic DNA methyltransferases, the vertebrate enzyme exhibits a relaxed sequence specificity for only a three-nucleotide motif ($m^5CG \cdot C$) within the CG dinucleotide pair (16). In vertebrates, there are at least two different DNA methyltransferases specific for CpGs. One is expressed throughout the development and life of vertebrates and a second is only expressed in early embryonic development (17–21). In plants, there are also two distinct DNA methyltransferases: one is specific for CpGs and the other is specific for CpXpG (22–25). For vertebrate and probably plant DNA methyltransferases three different modes of DNA methylation are distinguishable: methyl-directed maintenance methylation, *de novo* methylation, and DNA structure-induced methylation.

1. METHYL-DIRECTED MAINTENANCE DNA METHYLATION

It has been shown that a hemimethylated DNA substrate is methylated about 100 times faster than the same unmethylated DNA substrate (16). The very strong stimulation of DNA methyltransferase by hemimethylated DNA attests to its important role in the somatic inheritance of methylation patterns in vertebrates (26, 27). In preimplantation mouse embryos there are very high levels of DNA methyltransferase activity. However, one form of DNA methyltransferase in early embryos has a slightly higher mobility on gel electrophoresis than the form found in other cells and tissues (17). In spite of the very high level of DNA methyltransferase, there is a substantial decrease in the level of DNA methylation throughout the preimplantation period. This apparent paradox could be explained by a difference in the cellular distribution of the enzyme. From the oocyte stage to the four-cell stage, most DNA methyltransferase is concentrated in peripheral cytoplasm and the nuclei do not contain detectable DNA methyltransferase. In four- and eight-cell embryos, however, DNA methyltransferase is present in large amounts in the nuclei. In addition to the selective distribution of DNA methyltransferase between cytoplasm and nuclei, the activity of the enzyme could possibly be regulated by protein factors (28–30).

Because of its importance in maintaining DNA methylation, DNA methyltransferase is associated with replicating foci during the S phase of replication and it has a diffuse nucleoplasmic distribution in non-S phase cells (31). The association of the enzyme with the replication foci is mediated by a special sequence located near the N terminus of the enzyme (31). This sequence is not needed for enzyme activity, and if deleted there is no proper nuclear targeting. When this sequence is fused to β-galactosidase, it causes the fusion protein to associate with the replication foci in a cell cycle dependent manner (31). The rise in activity of DNA methyltransferase in late G_1 and early S phase is mainly due to an increase in the synthesis of DNA methyltransferase mRNA (32–34). It has been shown that DNA methyltransferase is controlled by the Ras signaling pathway (35). The first observation was the down regulation of Ras activity on 9-Jun (a transdominant negative mutant of Jun) in Y1 cells, which reversed the transformed morphology of the cells and resulted in a reduction in the level of DNA methyltransferase mRNA. Introduction of an oncogene, Ha-*Ras*, into the above cells resulted in a reversion to a transformed morphology and an increase in DNA methyltransferase activity (35).

A different series of experiments, carried out with transient transfections, strongly suggested that the promoter region of DNA methyltransferase is under the positive control of Ras and AP-1 (Fos/Jun) (35, 36) (see Fig. 2). There is also evidence that DNA methyltransferase is regulated posttranscriptionally. For example, at the onset of myoblast differentiation there is a very sharp decrease in DNA methyltransferase activity (37) that could be explained by a threefold increase in the turnover of its mRNA and protein.

2. *De Novo* DNA Methylation

When compared with DNA maintenance methylation, the initial velocity of the reaction of *de novo* DNA methylation with DNA methyltransferase from vertebrates is exceedingly low (1). The kinetics of the reaction with respect to DNA is hyperbolic, indicating either a single DNA binding site on the enzyme or, alternatively, multiple noncooperative binding sites (16). *In vitro*, the enzyme can normally only gain access to the duplex DNA at the ends or at regions rendered single-stranded by breathing (38, 39), a reaction strongly inhibited by salt. The phenomenon of *de novo* methylation can, under special extreme circumstances, lead to a hypermethylation of the genome. This phenomenon has been observed in vertebrate cells (40, 41) and in plants (107) under the selective pressure of specific drugs. *De novo* methy-

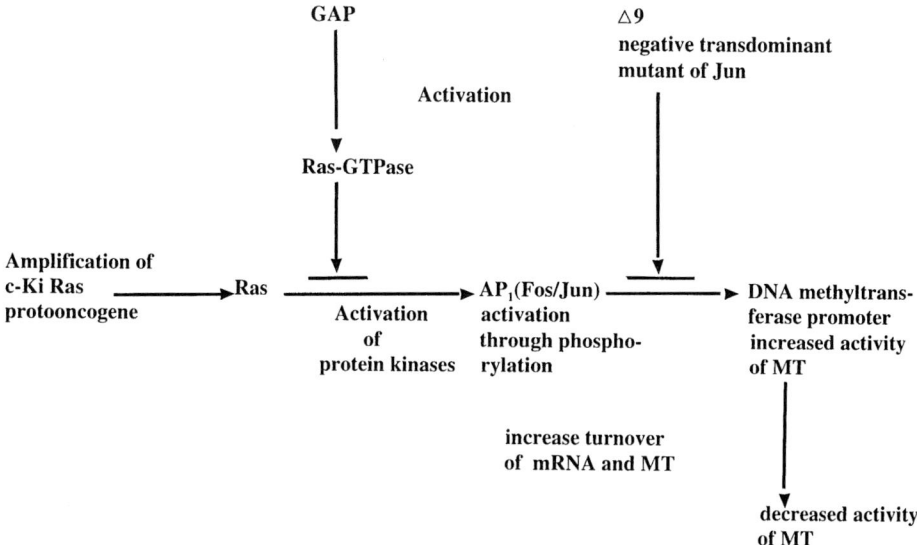

FIG. 2. Regulation of DNA methyltransferase by the Ras signaling pathway (model according to Ref. 35). GAP is a human Ras-GTPase activating protein; Ras-GTPase is a downmodulator of Ras; MT, DNA methyltransferase.

lase activity accounts for the inactivation of the major part of the adenoviral genome in transfected cells and may also account for the site-specific *de novo* methylation that occurs when a gene is switched off during development (4). In the case of adenoviruses, *de novo* methylation starts in the internal part of the colinearly integrated Ad 12 DNA molecule. The initiation site for *de novo* methylation appears to spread across a certain area and does not seem to be confined to a narrow region (42, 43).

The control of *de novo* methylation resides in the N-terminal region of the DNA methyltransferase (44). The N-terminal domain contains a zinc binding site and the N and C terminal domains can be separated by cleavage with either trypsin or *Staphylococcus aureus* V8 protease. Though intact enzyme has little effect on unmethylated DNA substrate, cleavage between the C and the N terminal domain with trypsin causes a large stimulation of *de novo* methylation of DNA. Thus the N-terminal domain of DNA methyltransferase ensures the clonal propagation of the methylation pattern through inhibition of the *de novo* activity of the C-terminal domain (17). It has been suggested that ectopic DNA methylation, as observed in certain tumors, aging animals (45, 46), and established cell lines (48), could result from the activation of the *de novo* activity of DNA methyltransferase by proteolytic cleavage (17, 56). *De novo* methylation also depends on the DNA sequence in general and the spacing between CpGs (48). In the 1.6-kb repetitive DNA sequence in *Petunia hybrida*, for example, there is a specific signal in the DNA that favors *de novo* methylation of DNA (50). In an attempt to identify sequences capable of enhancing *de novo* DNA methylation *in vitro*, Christman et al. (51) have shown that some m^5C residues in single-stranded DNA can stimulate *de novo* methylation of adjacent sites. It was suggested that double-stranded DNA may not be the primary substrate for *de novo* methylation and that looped single-stranded structures formed during the normal course of DNA replication or repair may serve as nucleation sites for the *de novo* methylation of adjacent DNA regions (51).

3. Methylation Induced by Specific DNA Structures and Sequences

Any situation mimicking the transition state of the DNA methyltransferase reaction, such as CG sites in which cytosine is unstacked or protonated, should facilitate the reaction of DNA methyltransferase (16). Heteroduplexes affecting the structure of CG dinucleotides are known to be excellent substrates for *de novo* and methyl-directed methylation by vertebrate enzymes (52–55). For example, when a thymine replaces one of the cytosines in what would otherwise be a CG dinucleotide, a twofold increase is observed in the rate of methylation (53). If a guanine in the same CG dinucleotide is replaced by O^6-methylguanine, the paired cytosine in the CG on the opposite strand is methylated at four times the normal rate (55). Similarly, the replacement of a guanine residue in the CG dinucleotide by an adenine residue results in a 12- to 14-fold increase in the rate of cytosine methylation on the opposite strand in the A·C mispair. According to Smith (16), a lesion at the CG site that generates both a C·C mispair and an adjacent A·C mispair can result in up to a 100-fold stimulation of methylation and a precise selectivity for only one of the cytosines in the C·C mispair.

B. Demethylation of DNA

DNA can be demethylated by either a passive or an active mechanism. As shown in Fig. 3, in passive DNA demethylation there is a site-specific inhibition of the maintenance methylase activity during DNA replication, thus preventing the methylation of the newly synthesized daughter strand. This inhibition may occur through competitive binding of a more abundant protein with the same binding affinity as DNA methyltransferase for a specific hemimethylated DNA sequence. The inhibition could also be due to a delay in the appearance of DNA methyltransferase during replication (57). After the first cycle of replication, the inhibition of DNA methyltransferase will result in a 50% demethylation of DNA, and 25% DNA methylation will remain after the second cycle of replication (1, 2).

Passive demethylation

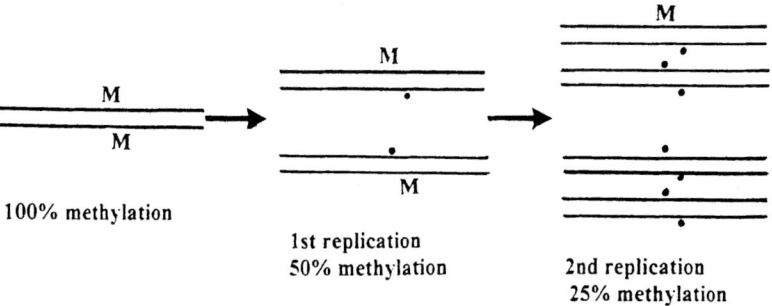

Active demethylation

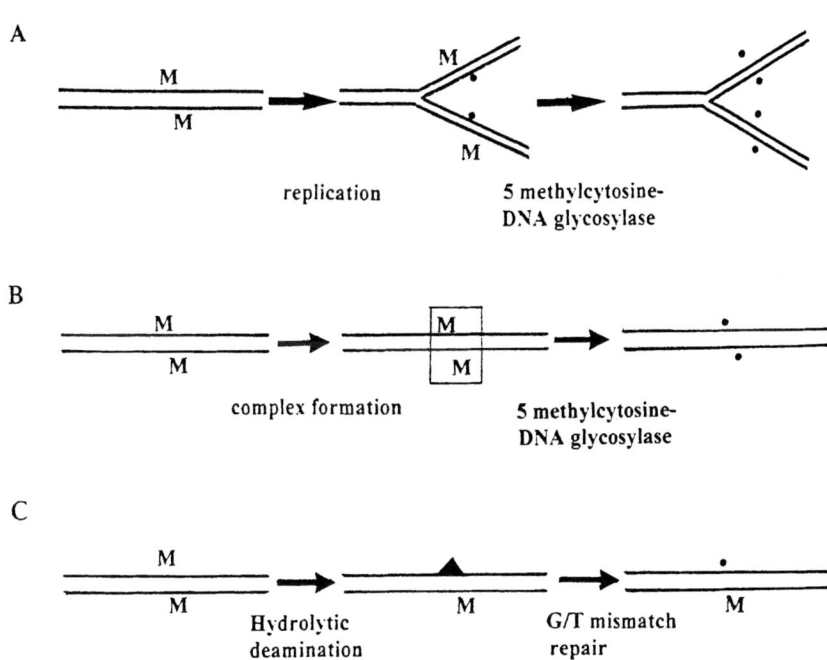

FIG. 3. Various ways by which DNA can be demethylated. M, m^{5}C; •, unmethylated cytosine. (Reprinted, with permission, from Ref. 157.)

There are four possible ways by which DNA can be actively demethylated. As shown in Fig. 3A, DNA demethylation may involve an m^5C-DNA glycosylase that removes m^5C from hemimethylated DNA during the S phase of replication (58), or a glycosylase may remove m^5C from symmetrically methylated nonreplicating DNA (59), as shown in Fig. 3B. In the first case, there is evidence that an m^5C-DNA glycosylase is present in developing chick and mouse embryos and in differentiating mouse myoblasts (60, 61). The enzyme is at least 10 times more potent on hemimethylated DNA than on symmetrically methylated DNA (58). In the second case, a similar enzyme present in HeLa cells was shown to remove m^5C from symmetrically methylated DNA in the presence of Mg^{2+} (59). However, these results were challenged by Steinberg (62), who showed that the enzyme present in HeLa cells could have been a nuclease. The third possibility (Fig. 3C) involves the hydrolytic deamination of m^5C and its conversion to thymine, creating a G/T mismatch, for which an enzyme called mismatch-specific thymine-DNA glycosylase has been identified and purified (63). However, it is doubtful that this reaction would be efficient enough to account for the active demethylation of DNA, which occurs very rapidly during embryonic development. It is known that the rate constants for the spontaneous hydrolytic deamination of m^5C and cytosine in double-stranded DNA at 37°C are 5.8×10^{-13} and 2.6×10^{-13} sec^{-1}, respectively (64). Such rate constants are efficient enough to account for the appearance of spontaneous mutations but they are not sufficient to explain the active demethylation of DNA occurring during differentiation. The fourth possibility (not shown in Fig. 3) for DNA demethylation could be a small patch DNA repair pathway. So far there is no evidence for the existence of such a mechanism.

1. Site-Specific versus Genome-Wide Demethylation

The methylation patterns of several genes that are expressed or not expressed in various tissues have already been published (65, 66). From the results obtained in hundred of maps from over 70 different genes, it has been shown that there is an inverse correlation between gene methylation and gene expression. Not all sites on a given gene are demethylated at the same time. Usually the promoter region and/or the 3' end-flanking regions may be demethylated first, followed later on by the demethylation of the structural gene. In many cases the demethylation of the structural gene occurs long after the start of transcription of the gene.

Genome-wide demethylation is a phenomenon whereby a large portion of the genome becomes transiently demethylated during differentiation—for example, differentiating mouse myoblasts (61), teratocarcinomas (67–69), and erythroleukemia cells (70). When mouse teratocarcinoma cells are induced to differentiate *in vitro*, there is a genome-wide demethylation of 30% of all CpGs (*68, 69*). Similarly, in murine erythroleukemia cells 4–30% of m^{5}CpGs are replaced by CpGs during the early differentiation triggered by different chemical substances (*70*). In the case of erythroleukemia cells, the demethylation is transient and precedes differentiation, whereas in teratocarcinoma cells the level of demethylation is proportional to the length of treatment with retinoic acid. Genome-wide demethylation has also been observed during the induction of Epstein–Barr virus in producer cells (*71*). In these cells, up to 30% of all CpGs become demethylated in the absence of DNA synthesis, suggesting that an active mechanism of DNA demethylation was responsible for the replacement of m^{5}C by C. Genome-wide demethylation has also been observed *in vivo* during spermatogenesis (*72*). In this system, demethylation in meiotic and postmeiotic cells was not only a property of repetitive sequences but was also found in the middle repetitive and unique sequences. It has been shown more recently that in G8 mouse myoblasts differentiating into myotubes, the transient genome-wide demethylation (~30–40% of all CCGG) was accompanied first by a decrease in DNA methyltransferase activity, and then by an increase in m^{5}C-DNA glycosylase activity (*61*). It is possible that the drop in DNA methyltransferase activity generated hemimethylated DNA, which is the favored substrate for m^{5}C-DNA glycosylase (*58*). In the mouse genome, which contains 5×10^9 base pairs, there are approx. 2.5×10^7 CpGs, of which 70–90% are normally methylated. A 1% methylation change therefore reflects a change in methylation of about 2.5×10^5 CpGs. This means that a 30% demethylation of CpGs in mouse myoblasts corresponds to 7.5×10^6 methylated CpG sites per haploid genome. If we assume an average of 500 methylated CpGs per gene, approximately 15,000 different genes could be demethylated per haploid genome. This value probably exceeds the number of genes that are turned on during the differentiation of myoblasts (*157*). In this case, what is the biological meaning of genome-wide demethylation? Because genome-wide demethylation in differentiating myoblasts is rather a late event, it may be a consequence and not the cause of differentiation. It is possible that genome-wide demethylation may help DNA repair, or the survey of entire gene domains that have recently undergone the transition from silent to active chromatin. There is little experimental evidence to support this idea, but it has been observed that the repair of a UV-damaged dihydrofolate reductase gene in hypomethylated Chinese hamster cells was twice as fast as in normally methylated cells (*73*). Furthermore, it is also possible that genome-wide demethylation could be related to periodic changes in chromatin organization associated with a general rearrangement of DNA repair patches (*74*).

2. Evidence for the Presence of an Active Demethylating System

The first observation made by Gjerset and Martin (75) indicated that in a cell-free extract prepared from differentiating erythroleukemia cells there is an enzymatic removal of acid-soluble labeled material from DNA double labeled with ^{32}P and methylated with *Hpa*II methyltransferase in the presence of tritiated S-adenosylmethionine. Unfortunately, the product of the reaction was not analyzed and no enzyme has been purified from this system. Using the same system, but a different approach, it has been shown that in differentiating mouse erythroleukemia cells the replacement of m^{5}C by cytosine is achieved by an enzymatic reaction (4, 70). Similarly, active DNA demethylation has been demonstrated in teratocarcinoma cells transfected with DNA containing methylated CpG islands (76). More recently, *in vitro* nuclear cell free systems have been reconstituted from chick embryos (60) and differentiating mouse myoblasts (61), where the m^{5}C present in a synthetic oligonucleotide was replaced by cytosine. One of the enzymes responsible for the reaction is m^{5}C-DNA glycosylase. This enzyme has been purified 30,000-fold from 12-day-old chicken embryos (58), and some of its properties are listed below.

3. Properties of m^{5}C-DNA Glycosylase

m^{5}C-DNA glycosylase was first detected in developing chicken embryos and differentiating mouse myoblasts (60, 61). However, this enzyme could not be detected in HeLa cells (J. P. Jost, unpublished results). In developing chicken embryos, the activity of m^{5}C-DNA glycosylase was high during the first 16 days of development, after which it declined to base line levels. The activity of the enzyme is highest in developing brain and muscles and is very low in these organs in adults. These results were confirmed by Western blots using polyclonal antibodies directed against synthetic peptides. The reaction products of the highly purified enzyme incubated with hemimethylated oligonucleotides are m^{5}C and the apyrimidic sugar, which are cleaved from the DNA by alkaline hydrolysis (58). The purified enzyme is approximately 52.5 kDa on an SDS-polyacrylamide gel. The optimal pH is between 6.5 and 7.5 and the p*I* is 5.5–7.5. The catalytic properties for hemimethylated DNA are 8×10^{-8} M with a V_{max} of 4×10^{-11} moles hr^{-1} mg^{-1} protein. Both the crude and the highly purified enzymes are very specific for m^{5}CpG and do not cleave m^{5}CpC, m^{5}CpT, m^{5}CpA, m^{5}ApG, or m^{5}CpApG. Furthermore, the purified enzyme reacts with hemimethylated DNA about 10 times faster than with symmetrically methylated DNA. This suggests that, *in vivo*, m^{5}C-DNA glycosylase may have a preference for newly synthesized DNA that is still in the hemimethylated form. The small but significant activity of m^{5}C-DNA glycosylase in the presence of symmetrically methylated DNA indicates that the same enzyme may, under specific conditions, demethylate symmetrically methylated DNA. In this case, the simultaneous demethylation of symmetrical CpGs may result in double-strand breaks of DNA, indicating that additional factors may be needed to prevent the breaking of the chromosome. In this context it is interesting to note that the up-regulation of a protein called the nonhistone protein 1 (NHP-1) is always observed during the active DNA demethylation of the avian vitellogenin II gene promoter. As previously shown, NHP-1 binds in a sequence-independent manner, and, with very high affinity, DNA-containing CpG base pairs (77, 78). The binding activity of NHP-1 increases about twofold in the liver of estradiol-treated immature roosters, where it coincides with the active demethylation of CpGs situated in the promoter region of the vitellogenin gene (79). Similarly, it was recently shown that an increase in NHP-1 subunits p75 and p85 occurred during the differentiation of mouse G8 myoblasts at a time when there was a genome-wide demethylation (80). In the same system there was also a very narrow correlation between the kinetics of NHP-1 and m^{5}C-DNA glycosylase induction. NHP-1 was purified to homogeneity and the peptides derived from its p75 and p85 subunits were microsequenced. The sequences of these peptides had between 64 and 100% homology to the human Ku autoimmune antigen (80). It is already known that Ku proteins play a very important role in DNA repair and recombination (81–93). It is thought that one of the functions of the Ku protein is to hold the ends of DNA together during DNA recombination (87, 93). Circumstantial evidence indicates that NHP-1 (alias Ku) could be the protein needed to avoid double-strand breaks at a symmetrical CpG during the active demethylation of symmetrically methylated DNA. The m^{5}C-DNA glycosylase does not use single-stranded methylated DNA as a substrate and, like DNA methyltransferase, its action is distributive. m^{5}C-DNA glycosylase was shown to copurify with a mismatch-specific thymine-DNA (and uridine-DNA) glycosylase activity. A single peptide band eluted from an SDS-polyacrylamide gel, when denatured and slowly renatured, gave activity for both enzymes (58). However, the enzyme has no true uracyl-DNA glycosylase activity, because it does not use single-stranded DNA containing uracyl as a substrate. It remains now to show whether one or two different peptides of the same size perform these different enzymatic reaction. Recently, a mismatch-specific thymine-DNA glycosylase that has no apparent m^{5}C-DNA glycosylase activity (63) has been purified from HeLa cells. In addition, there are major differences in the chromatographic behavior of the two enzymes. On affinity columns, m^{5}C-DNA glycosylase has no selectivity for methylated or unmethylated DNA, whereas the mismatch-specific thymine DNA glycosylase binds only DNA containing a G/T mismatch.

C. Determination Factors

The determination factors are DNA sequences and protein factors that can influence the methylation/demethylation of specific CpG or CpXpG sites in a specific manner.

1. Cis-Acting Regulatory Elements of Methylation

Most housekeeping genes have CpG islands that are not normally methylated. For example, in the mouse adenine phosphoribosyl transferase (*aprt*) gene the CpG islands are totally free of methylation, whereas the flanking regions have methylated CpGs (*94*). In a transgenic mouse assay, deletion mutagenesis of the Sp1 sites flanking the CpG islands of the *aprt* gene initiated a *de novo* methylation of the CpG islands. These experiments demonstrated that the peripherally located Sp1 sites were necessary to keep the *aprt* CpG islands methylation free (*94, 95*). Though certain cis-acting elements can effectively prevent the methylation of DNA, other sequences can enhance DNA methylation. For example, Mummanemi *et al.* (*96*) showed conclusively that a cis-acting element located 1.3 kb upstream of the mouse *aprt* gene acted as a signal for the methylation and epigenetic inactivation of the gene. During X chromosome inactivation, specific sequences in the X inactivation center (XIC) may also play a fundamental role for the spreading of methylation from the XIC in both directions (*97*). In the above cases it is clear that cis control elements can prevent or give the signal for *de novo* methylation. It is, however, not certain by which mechanism they function. Two possibilities, which are not mutually exclusive, should be considered: the structure of the DNA could either enhance or prevent the action of DNA methyltransferase, or there could be an indirect effect mediated by some specific RNA and proteins.

2. Trans-Acting Regulatory Elements of DNA Methylation

Penny *et al.* (*98*) provided crucial evidence that the transcription of the X-inactive specific transcript (XIST) gene was necessary for the inactivation of the X chromosome that bore it. By removing 7 kb of DNA in the first exon of XIST, the X chromosome was not inactivated, whereas inactivation did occur in the X chromosome with the intact XIST gene (*98*).

We can also consider any protein or RNA that could, at a critical concentration or activity, compete with either DNA methyltransferase or m^{5}C-DNA glycosylase as possible trans-acting regulatory elements of DNA methylation.

D. Models for Formation of Specific DNA Methylation Patterns

Our current working hypothesis is that a specific DNA methylation pattern is the result of an interplay between DNA methyltransferase, the demethylating mechanism(s), and the sequence-specific determination factors (cis and trans). We are also going to consider an alternative hypothesis that could explain the making of a specific DNA methylation pattern.

1. Interplay of Methyltransferase, m^{5}C-DNA Glycosylase, the Determination Factors

The combinatorial interaction of DNA methyltransferase, m^{5}C DNA glycosylase, and the trans determination factors is shown in Fig. 4. Because both DNA methyltransferase and m^{5}C-DNA glycosylase prefer hemimethylated DNA as a substrate, it is most likely that the profile of DNA methylation is established during DNA replication, shortly after DNA synthesis. In our model it is also assumed that, before replication, a specific CpG site is either 100% methylated (Fig. 4A–F) or 0% methylated (Fig. 4G). In order to simplify the model, it is assumed that DNA methyltransferase (m), m^{5}C-DNA glycosylase (d), and the determination factors (f) have comparable DNA physical binding constants and that the only variable to be considered here is the relative number of active molecules. From Fig. 4 it can be seen that any change in the relative ratio of the number of active molecules of m, d, and f (represented by the bar diagrams) can profoundly influence the state of methylation of the CpG site present on the replicating DNA. The same results could also be obtained by having the same number of molecules of m, d, and f, but with a variable binding constant for the DNA. A change in the binding constant could be achieved by the covalent modification of m, d, and f or by cofactors. Assuming there is one single cell containing one single symmetrically methylated CpG, after one cycle of replication, several combinations of m, d, and f could result in the methylation of CpG in the two daughter cells, ranging from 0 to 100% (Fig. 4). In the cases presented in Fig. 4A–F, it is possible to explain how a fully methylated CpG site becomes partially or fully demethylated. It is interesting to note that in Fig. 4 there are two ways by which DNA becomes demethylated: passive and active. For example, in Fig. 4D there is a relatively small number of DNA methyltransferase and m^{5}C-DNA glycosylase molecules in the presence of a large excess of determination factor that could inhibit the methylation of the CpG site, leading to a 50% passive demethylation after one cycle of replication. On the other hand, in the case presented in Fig. 4F, where there is an excess of m^{5}C-DNA glycosylase over the other two components, full demethylation of the

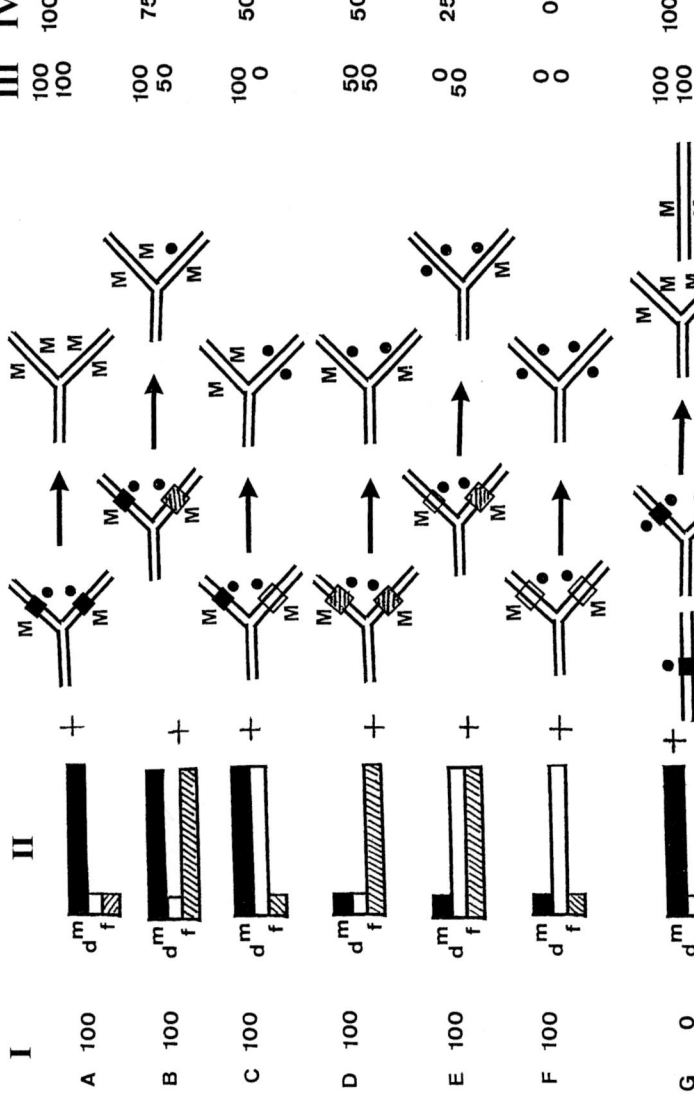

FIG. 4. Formation of DNA methylation patterns. A model based on the interaction of DNA methyltransferase (m), m5C-DNA glycosylase (d), and the determination factors (f). The lengths of the bars are proportional to the "activity" of m, d, and f. M, m5C; •, unmethylated cytosine. This model represents the interaction of m, d, and f on DNA at the replication fork. I, Initial percentage of DNA methylation; II, relative activity of m, d, and f at the time of DNA modification; III, percentage of methylation in each daughter cell; IV, average percentage of methylation.

CpG site is achieved after only one cycle of DNA replication (active demethylation).

During the formation of the DNA methylation pattern, *de novo* methylation of specific CpG sites was also observed. As has already been discussed, the *in vitro* activity of DNA methyltransferase for *de novo* methylation is much lower than for maintenance DNA methylation. In this case, one cannot exclude the presence of other factor(s) that could stimulate *de novo* DNA methylation *in vivo*. As symbolically shown in Fig. 4G, *de novo* methylation could occur on replicating or nonreplicating DNA. The results presented in Fig. 4B–E indicate that the state of methylation of alleles may differ among daughter cells.

From the preceding model, four predictions can be made: (1) the DNA methylation pattern is established during replication or shortly after DNA synthesis; (2) the state of methylation of a given CpG site will depend on the relative ratio of the DNA methyltransferase, m^5C-DNA glycosylase, and the sequence specific factors binding to the DNA containing the CpG site; (3) in a homogeneous population of cells, differences in the state of methylation of the alleles of a given gene should be observed in the daughter cells after replication; (4) in a homogeneous population of cells, the average percentage of DNA methylation after replication should not be a clear-cut 0 or 100%, but in specific cases there should be values ranging between 0 and 100%.

These predictions may be verified by collecting sufficient experimental data from known systems that have been studied in detail or by experimental verification, whereby the three components of the model could be selectively tested one at a time. According to the model, the methylation pattern of many sites may be different in the alleles of cells derived from a single clone. This is the case for the human c-Ha *ras 1* gene (*158*) and for the clone-derived cells from leiomyoma (*159*), where despite the clonality of these tumors, there is a heterogeneity in the methylation at a number of genomic sites. So far, only in one case is there evidence for a change in the ratio of DNA methyltransferase to m^5C-DNA glycosylase leading to a genome-wide demethylation (*61*). During the transient genome-wide demethylation in differentiating myoblasts, there is a sharp decrease in DNA methyltransferase activity followed by an increase in m^5C-DNA glycosylase activity (*61*). These changes could have two consequences: a decrease in DNA methyltransferase may favor the formation of hemimethylated DNA, which is the best substrate for m^5C-DNA glycosylase, and/or the change in the mole ratio of the two enzymes may facilitate the access of m^5C-DNA glycosylase to the hemimethylated DNA.

Should the synthesis of a sequence specific DNA binding protein be prevented, the state of methylation of the CpG present in the protein binding site would solely depend on the relative activities of DNA methyltransferase and m⁵C-DNA glycosylase. Evidence for an *in vivo* titration of a rate-limiting and sequence-specific DNA binding protein is strongly indicated by experiments where an increase in the copy number of transgenes resulted in their silencing by DNA methylation (*99*). A related phenomenon is repeat induced gene silencing (RIGS), where it has been shown in several plant and animal systems that the expression of structurally intact genes may be silenced epigenetically when a transgenic construct increases the copy number of DNA sequences (*99*). The silencing is reversible and correlates with a decrease in steady-state mRNA and an increase in DNA methylation. Usually a single gene insert continued to be expressed, but multigene inserts at the identical locus often segregated silenced progeny. This phenomenon has been observed in *Petunia* (*99–101*), tobacco (*102, 103*), *Arabidopsis* (*99, 104*), fungi (*105*), and insects and vertebrates (*106*). An alternative explanation of the phenomenon is that the repeated sequence-induced silencing could involve the pairing of homologous DNA sequences into local three- or four-stranded structures (*99*). Such structures could be ideal substrates for *de novo* methylation.

2. Transient Expression of Multiple Sequence-Specific *de Novo* DNA Methyltransferase Activities

This hypothesis states that the transient expression of a series of sequence-specific *de novo* DNA methyltransferase activities could mediate *de novo* methylation of selected sequences at a given developmental stage of an organism (*16*). Up to now only two different DNA methyltransferases have been found in vertebrates and in plants. It seems unlikely that the presence of these two enzymes alone could account for the diversity of DNA methylation patterns.

3. Transient Expression of DNA Methyltransferase Modulators

According to Smith and Tolbey (*16*), the expression of a series of DNA methyltransferase modulators could confer sequence specificity to a single DNA methyltransferase. To date no systematic search for such factors has been carried out and in the interim we have to await experimental evidence for the existence of such factors.

4. Transient Formation of Unusual DNA Structures

It has already been shown that oligonucleotides containing heteroduplex sites of DNA damage, Watson–Crick paired foldback structures, foldback structures containing non-Watson–Crick pairs, or gaps are all actively methylated *in vitro* (*16*). These results raise the possibility that DNA methylation has evolved as a biological response to the transient formation of these unusual

DNA structures (16). Some of these unusual structures could be introduced by the binding of specific proteins to the DNA.

E. DNA Hypermethylation: An Aberrant Form of DNA Methylation or a Defense Mechanism

The silencing of certain genes can occur through a random hypermethylation of the whole genome under the influence of certain drugs (40, 41). For example, Nyce et al. (40) have observed a genome-wide DNA hypermethylation, a low-frequency silencing of thymidine kinase (TK) gene expression, and a progressive resistance to the DNA synthesis inhibitor 3-azido-3'-deoxythymidine (AZT) in Chinese hamster cells treated with AZT. Within 24 hr of exposure to the drug, they observed a twofold enhancement of genome DNA methylation and a production of TK^- epimutants at a rate 40-fold above the background. As expected, the AZT-induced TK^- epimutants were shown to be severely reduced in their capacity to activate AZT to AZT-5'-monophosphate. Such TK^- epimutants were also shown to have a collateral resistance to 5-fluoro-2'-deoxyuridine. The TK^- epimutation could be reverted within 24 hr at a high frequency (8–73%) by exposing the cells to the demethylating agent 5-azacytidine. Concomitantly these revertants regained the ability to metabolize AZT to its 5'-monophosphate. These experiments clearly showed a potential mechanism of drug resistance involving drug-induced DNA hypermethylation that resulted in the transcriptional inactivation of cellular genes whose product was required for the activation of the drugs (40). Other drugs such as cytosine arabinoside, hydrozyurea, or aphidicolin (inhibitors of DNA synthesis) also had a similar effect on DNA hypermethylation in hamster fibrosarcoma cells (41). However, cycloheximide (an inhibitor of protein synthesis) and actinomycin D (an inhibitor of RNA polymerase) or serum deprivation had no effect on DNA hypermethylation. In addition, DNA hypermethylation occurred predominantly in the fraction of DNA synthesized in the presence of the inhibitor. Hypermethylation remained in the absence of the drug. Similar observations have also been made in transgenic plants that were under the selective pressure of hygromycin and Claforan. In this case, the hypermethylation was dose and time dependent. After 1 month of growth on a hygromycin- or Claforan-containing medium, an increase of up to 2×10^7 methylated CpGs per haploid genome was observed (107). The DNA hypermethylation occurred in high, moderate, and low repeated DNA levels. In the absence of the antibiotics the DNA hypermethylation was reversible. Analysis of the promoter region of the auxin binding protein gene showed a heterogeneity in its hypermethylation that was biased toward CpGs. As a possible explanation of the phenomenon, there may be a prolonged G_1–S phase in the presence of drugs, that opening a wider window for the DNA methyltransferase. Another possible explanation is that a proteolytic cleavage of the DNA methyltransferase, induced by the drug, could change the substrate specificity of the enzyme, which would cause massive *de novo* methylation of the DNA, or the enzyme could possibly be activated by an unknown factor. As a last possible explanation, drugs could induce the DNA amplification of certain gene domains (*108–111*), which would serve as substrates for *de novo* DNA methylation.

The hypermethylation of plant DNA in response to antibiotics could be regarded as part of the plant defense mechanism against invading organisms. An increase in DNA methylation potential triggered by antibiotics (or its analogs) could neutralize the invading DNA and/or it could render the host DNA less susceptible to the restriction enzymes produced by the invading organism.

II. DNA Methylation and the Silencing of Genes

The methylation of cytosine at position 5 might silence genes by changing the structure of DNA, which may directly influence the interaction of proteins with DNA or it could interfere with the binding of transcription factors or enhance the binding of potential repressors.

A. Methylation of CpGs Can Change DNA Structure

Methylation of CpGs in a stretch of alternating purines and pyrimidines can lead to a transition from right-handed B-DNA to left handed Z-DNA. Cytosine methylation at position 5 stabilizes the Z helix at a drastically lower ionic strength than is required by the unmethylated polymer (0.7 versus 2.5 M NaCl, or 0.6 versus 700 mM MgCl$_2$, respectively). The B form of DNA has a helix repeat of 10.5 bp per helical turn, whereas Z-DNA has a helix repeat of approximately 13.6 bp per turn (*112, 113*). B-DNA has two grooves, one slightly broader than the other, whereas Z-DNA has only one deep minor groove and an outer convex surface that is analogous to the major groove of DNA (*112, 113*). It is to be expected that a Z structure of DNA could drastically change the binding properties of the DNA to proteins. However, there is to date no solid evidence that *in vivo* such structures stabilized by DNA methylation play a role on gene expression.

B. Many Transcription Factors Do Not Bind to Methylated DNA

One of the most direct ways that methylation of DNA could inhibit transcription is by preventing the binding of trans-acting factors to the DNA. Many systems have been described whereby the binding of transcription factors to their cognate region in the promoter was inhibited by the methylation of CpGs, and a list compiling examples taken from vertebrates and plant systems has been published (114). However, there are also protein factors for which the binding properties are affected very little by DNA methylation. For example, site-directed methylation of the CpG on both strands of an Sp1 recognition site gave little (115) or no (116) change in the extent of specific DNA–protein complex formation with Sp1 from HeLa cells.

C. DNA Methylation Favors Binding of Specific Repressor Proteins

Just as there are bacterial restriction endonucleases that recognize their substrates sequence only if they are methylated (114), so there are also some eukaryotic DNA binding proteins (in vertebrates and plants) that are specific for methylated DNA. These proteins have been subdivided into sequence-specific and sequence-nonspecific proteins (114). Only the proteins binding to nonspecific methylated DNA sequences have been extensively studied.

In vivo and *in vitro* transcription competition experiments with methylated promoters have shown that methylation of DNA may silence a gene through the binding of a repressor that can be displaced by an exogenous excess of methylated DNA sequences (117). For example, in transfection experiments with methylated promoters and methylated DNA competitors the repression caused by a protein could be relieved (117). Similarly, the repression of a methylated promoter of mouse metallothionein I gene could be overcome by the cotransfection with an excess of untranscribable DNA methylated with *Sss*I methyltransferase (118). In a different system, it was shown that the *in vitro* transcription of a methylated template was inhibited by a variant of histone H1 and that the transcription inhibition could also be relieved by competition with methylated DNA (119). More recently, it has been demonstrated (120) that the repression of methylated avian vitellogenin promoter by the methylated DNA binding protein 2 H1 (MDBP-2-H1), a member of the histone H1 family, could be relieved by *in vitro* competition with a methylated oligonucleotide but not with the same concentration of the unmethylated oligonucleotide (120). The above results strongly indicate that DNA methylation of a promoter inhibits transcription by an indirect mechanism involving, in part, the selective binding of proteins having a stronger affinity for methylated than for unmethylated DNA. Table I lists properties of three well-characterized proteins that bind preferentially to methylated DNA.

1. M^{5}CPG BINDING PROTEINS MECP1 AND MECP2

Two proteins that bind preferentially to methylated DNA have been purified and characterized. The first of these, MeCp1, needs 15 or more symmetrically methylated CpGs for a strong binding to DNA. Cross-linking ex-

TABLE I
SOME PROPERTIES OF PROTEINS BINDING PREFERENTIALLY TO METHYLATED DNA

Property	MeCP1	MeCP2	MDBP-2-H1
Methylated CpG pairs need for binding to DNA	12	1	1
Minimal length of oligo required for binding	—	—	30 bp
Size by gel filtration	400–800 kDa	100 kDa	—
Size on SDS-PAGE	120 kDa	84 kDa	21 kDa
Abundance of molecules per cell	5×10^3	1.5×10^5	n.d.
K_d M	n.d.^{a}	$10^{-9}\,M$	2.5×10^{-9}
Binding to m^{5}CpG controlled by hormones	—	—	Yes
Preferential repression of methylated DNA *in vitro*	Yes	No	Yes

^{a}n.d., Not determined.

periments detected a 120-kda protein that correlated with the binding of MeCp1. MeCp1 is widely distributed in mammals except in embryonic carcinoma cell lines, which have very low levels of MeCp1 (*121*). Clusters of methylated CpGs are found at the CpG island-like sequences in the X chromosomes of eutherian mammals or in the CpG-rich promoter regions of LINE elements in mammalian genomes (*122*) and in the silenced retroviral genomes of mice infected early in development (*123*). All of these sequences are potential substrates for MeCp1.

MeCp2 is more abundant than MeCp1 (see Table I) and is also able to bind to an oligonucleotide containing only a single methylated CpG pair. Experiments carried out with antibodies directed against MeCp2 have shown a uniform distribution of the protein within the interphase nuclei. The antibody also stained metaphase chromosomes, indicating that MeCp2 is associated with chromosomes. The staining was dispersed throughout the chromosome arms (*124*). Interphase nuclei of mouse L cells stained most strongly over the prominent heterochromatic bodies (*124*). MeCp2 is barely detectable in early embryonic cells. Nan *et al.* (*126*) designed further experiments to find out whether the heterochromatic localization of MeCp2 was dependent on DNA methylation. For this purpose they transiently expressed MeCp2–LacZ fusion protein in cells containing normal or low levels of DNA methylation. The fusion protein was targeted to heterochromatin in the wild-type methylated cells but was inefficiently localized in mutant cells with low levels of DNA methylation. In addition the results showed that the selective binding of MeCp2 to heterochromatin containing methylated DNA was an exclusive property of the 85 amino acids of the m^5CpG-binding domain (MBD). These experiments strongly indicate that MeCp2 is indeed an important factor for the heterochromatization of methylated DNA.

Purified MeCp2 protein inhibits transcription from both methylated and unmethylated DNA templates *in vitro* (*127*). The primary structure of MeCp2 contains two kinds of short motifs that have been found in proteins that interact with the minor groove of B-DNA. The motif (R)GRP(K) that occurs twice is a feature of mammalian HMG-I/Y proteins (*124*). The second set of motifs belongs to the SPKK consensus sequence. Eight copies of this motif occur in MeCp2. This motif has been found to bind to the minor groove of AT-rich DNA sequences (*128*). A minimal MBD isolated from MeCp2 (P8) is 85 amino acids in length and binds exclusively to DNA that contains one or more symmetrically methylated CpGs. The minimal MBD has negligible nonspecific affinity for DNA, confirming that the nonspecific and methyl-specific binding domains of MeCp2 are distinct. *In vitro* footprinting and binding kinetics indicate that MBD binding can protect a 12-nucleotide region surrounding an m^5CpG pair, where it binds with a dissociation constant of approximately 10^{-9} M (*129*). From these observations it is clear that MeCp2 could serve as one of the universal repressors for methylated DNA, whereas MeCp1, which is about 100 times less abundant than MeCp2, could have only special repressor functions in CpG-rich regions of DNA.

a. In Vivo MeCp2 Is Required for Normal Embryonic Development. It has been shown that mutant embryonic stem cells lacking MeCp2 were growing in the same way as the parental embryonic stem cell line and were capable of differentiation. However, chimeric embryos derived from a different mutant cell line showed obvious developmental defects (*155*).

The severity of the developmental defects correlated with the contribution of the mutant cells. Embryos with the highest level of chimerism failed to gastrulate and died between days 8.5 and 12.5 of gestation (*155*). There is a striking similarity between MeCp2 and mouse DNA methyltransferase, which was also shown to be essential for embryonic development but not for embryonic stem (ES) cells. In addition, experiments involving MeCp2, an abundant protein whose chromosomal distribution depends on DNA methylation (*126*), strongly emphasize the importance of DNA methylation for normal embryogenesis.

2. METHYLATED DNA BINDING PROTEIN-2-H1

Methylated DNA binding protein-2-H1 (MDBP-2-H1) is the only protein described so far that selectively binds to methylated DNA under the control of estradiol (*130*).

a. MDBP-2-H1 Is a Member of the Histone H1 Family. The CpG situated at position +10 of the vitellogenin II gene was shown to be methylated in the liver of roosters, where the gene was not active, and demethylated in hen liver, where the gene was expressed (*131*). In rooster liver, a nuclear protein fraction was shown to bind preferentially to this methylated region of the promoter. Binding activity was absent in the liver of egg-laying hens but, during the egg-laying pause and during the molting period, binding activity appeared in the liver, suggesting the presence of a cyclic hormonal control. This was tested by injecting estradiol into immature roosters: the binding activity of MDBP-2-H1 disappeared under the influence of the hormone (*79*). MDBP-2-H1 was purified and peptides derived from a tryptic digestion were microsequenced. Two peptides, KPAGPSVTELITK and ALLAGGYDVEK, were found to be identical to chicken histone H1 core protein (*132*). MDBP-2-H1, a variant of the histone H1, represented at the most 5–8% of the total population of histone H1 in rooster liver. On SDS-polyacrylamide gels, MDBP-2-H1 has an apparent molecular mass of 21 kDa, and antibodies directed against calf thymus total histone H1 cross-reacted with it. Several different sequence variants of histone H1 have been described for most eukaryotic systems that have been analyzed carefully (*133–142*). The profiles of histone H1 variants differ qualitatively among different animal species and quantitatively among different tissues of the same species (*133–142*). The proportions of these different subtypes also vary with the progress of differentiation as seen in embryonic development, hormonally induced changes, spermatogenesis, and terminal differentiation (*137*). The expression of most histone H1 variants is replication-independent, except in mice, where two of the seven nonallelic forms of histone H1 are expressed in a replication-dependent manner (*141*). At present, nothing is known about the regulation of MDBP-2-H1 synthesis and it is not clear whether the histone H1 variant c tested by Johnson *et al.* (*119*) is related to MDBP-2-H1. In sharp contrast to MDBP-2-H1, the selective binding of other histone H1s to methylated DNA could not be demonstrated. For example, it has been shown that methylation at CpG sequences does not influence total histone H1 binding to a nucleosome present on *Xenopus borealis* 5S rRNA gene (*143*). More recently, no preferential binding of total calf histone H1 to methylated DNA could be demonstrated (*144;* J. P. Jost, unpublished results). Earlier, while using an immunochemical approach, it was shown that at least 80% of m^5C is located in nucleosomes that contained histone H1 and it was concluded that, because DNA methylation is symmetrical and semiconservative, replication coupled with maintenance methylation could ensure the faithful inheritance of methyl groups by daughter cells (*145*). Methylation of DNA could, therefore, be the signal for selective deposition of histone H1 (*145*). It was also argued that the fidelity of histone H1 propagation associated with chromatin structure could be guaranteed in this way. Although this concept is very attractive, it is not supported by the evidence presented by two other groups (143, 144). However, more recently, while using a more sensitive assay, it was shown that the histone H1 had a preferential binding for methylated DNA. The cooperative binding of histone H1 to methylated DNA was dependent on the level of DNA methylation and the presence of the globular domain of the histone H1 (160).

b. DNA Binding Properties of MDBP-2-H1. Affinity-purified MDBP-2-H1 binds preferentially to methylated DNA. The binding is sequence nonspecific and requires a minimum length of 30 base pairs and one pair of symmetrically methylated CpG dinucleotides. A UV cross-linkage experiment showed that the molecular mass of the protein binding to methylated DNA was about 40 kDa, indicating that a dimer of the 21-kDa MDBP-2-H1 may possibly have been formed (146). The contact points of the protein complex with methylated DNA were further studied by means of DNA binding interference experiments with partially depurinated and depyrimidinated end-labeled oligonucleotides (146). Strong contact points between MDBP-2-H1 and the methylated oligonucleotide are located symmetrically on one strand of the DNA duplex, giving two distinct footprints of equal length. However, there is no binding of MDBP-2-H1 to the shorter oligonucleotides containing each individual footprint, suggesting that there is probably a very strong cooperation between the two monomers binding to the 30-nucleotide-long methylated oligonucleotides (132). How m^5C enhances the binding of MDBP-2-H1 to the oligonucleotide is not known. One possibility is that methylation of the oligonucleotide causes a distortion of the double helix, making the minor groove more accessible to MDBP-2-H1. In the liver, MDBP-2-H1 associates with another, larger protein that is found mainly in roosters (132). This new protein does not bind directly to DNA, but forms a larger complex with MDBP-2-H1 prior to its binding to methylated DNA. This larger complex, as measured by real-time kinetics (surface plasmon resonance), binds faster to methylated DNA than MDBP-2-H1 alone (147). The important question of whether MDBP-2-H1 also binds selectively to methylated DNA *in vivo* was addressed by UV cross-linkage experiments and selective immunoprecipitation with antibodies directed against histone H1 (132). The results showed that, *in vivo*, MDBP-2-H1 preferentially binds to the silent, methylated vitellogenin promoter (132).

c. In Vivo Estradiol Down-Regulates the Binding Capacity of MDBP-2-H1 to Methylated DNA. In estradiol-treated roosters the kinetics of DNA demethylation does not precede the onset of vitellogenin gene transcription. At a few selected sites, demethylation of DNA is parallel to the onset of transcription. In sharp contrast, the ability of MDBP-2-H1 to bind methylated DNA decreases very sharply on estradiol injection and the loss of binding ability precedes the onset of vitellogenin II gene transcription (79). These results indicate that in the absence of MDBP-2-H1 binding to the methylated promoter region, there is an initiation of transcription, suggesting that the inhibitory effect of methylated DNA on transcription is most probably indirect. The indirect effect of MDBP-2-H1 on transcription was also demonstrated in *in vitro* transcription competition assays in homologous (148) and heterologous transcription assays (120). The results clearly showed that MDBP-2-H1 acted as a repressor, because a competition carried out with a methylated oligonucleotide of critical length, covering the region of the first exon of the vitellogenin gene, was able to relieve the repression caused by MDBP-2-H1 (120). The decrease in the binding of MDBP-2-H1 to methylated DNA observed *in vivo* could either reflect a decrease in the total amount of MDBP-2-H1 protein present in the nuclei or could be a consequence of a covalent modification of the protein. This question was addressed by immunoblots with antibodies directed against MDBP-2-H1. The results clearly showed that there were no significant differences in MDBP-2-H1 antigen concentrations when tissues with a high MDBP-2-H1 binding activity (rooster liver) were compared to those with a low binding activity (estradiol-treated roosters and egg-laying hens). In this case, the results clearly indicated that estradiol was responsible for a covalent modification of MDBP-2-H1 that influenced the capacity of the protein to bind methylated DNA (132). Though a down-regulation of the binding of MDBP-2-H1 to methylated DNA was observed, a reciprocal increase in binding activity of NHP-1 was also observed. NHP-1 is a member of the Ku protein family and, as mentioned previously, this protein may be involved in the active demethylation of symmetrically methylated DNA.

d. Phosphorylation and Dephosphorylation of MDBP-2-H1 Is Important for the Modulation of Its Binding to Methylated DNA. From the experiments just described, the loss of binding activity between MDBP-2-H1 and methylated DNA strongly suggests that estradiol may have caused a covalent modification of the repressor. It has been shown that MDBP-2-H1 purified from rooster liver is a phosphoprotein and that phosphorylation of this protein is essential for the binding to methylated DNA (149). Phospho-amino acid analysis indicates that phosphorylation occurs exclusively on serine residues. As shown by two-dimensional gel electrophoresis and tryptic phosphopeptide analysis, phosphorylation occurs at several sites. A treatment of mature rooster with estradiol triggered dephosphorylation of at least two sites in MDBP-2-H1 and resulted in the loss of capacity to bind to methylated DNA. In egg-laying hens, MDBP-2-H1 was found to be in the dephosphorylated form, which bound very poorly to methylated DNA. These results suggested that dephosphorylation of MDBP-2-H1 resulting from estradiol treatment may represent a rapid way to relieve the repression consequent to the binding of the repressor to methylated DNA (see Fig. 5). In this case, on release of MDBP-2-H1 the methylated gene can be transcribed without delay and demethylation can proceed independently of the initiation of transcription. It is possible that the chromatin in its repressed form is inaccessible for the demethylating enzymes and MDBP-2-H1 has to be removed before active demethylation can start. At present the role of phosphorylation of histone H1 (total) in the formation of mitotic chromosomes is at best confusing, and contradictory results have been published (*150, 151*). It is rather surprising that in the vast majority of published work on histone H1, no mention is ever made of the multiple variants of this molecule or their differences.

e. The Phosphorylated Form of MDBP-2-H1 Is a Repressor for the in Vitro Transcription of Methylated DNA. As with other methylated DNA binding proteins, MDBP-2-H1 has a repressor function, whereby methylated DNA plays an indirect role. In an *in vitro* transcription system (homologous or heterologous) lacking MDBP-2-H1, transcription proceeded even in the presence of methylated CpGs (in this case, the methylated CpG was not situated in the binding site of an important transcription factor). The addition of purified MDBP-2-H1 from rooster liver repressed transcription from a methylated template more efficiently than from unmethylated DNA. Dephosphorylation of rooster liver MDBP-2-H1 by phosphatase or estradiol treatment of roosters resulted in the loss of the inhibitory activity of MDBP-2-H1 when used in a cell-free transcription system (*120*). These results establish a relationship between the dephosphorylation of MDBP-2-H1 caused by estradiol, the down-regulation of its binding activity to methylated DNA, and the derepression of the avian vitellogenin II gene (see Fig. 5).

In the case of the avian vitellogenin gene, there is obviously a complex signal transduction system that leads to the dephosphorylation of MDBP-2-H1 (and possibly other related proteins) on estradiol treatment. *In vivo*, the covalent modification of MDBP-2-H1 is obviously reversible because during the egg-laying pause there is a form of MDBP-2-H1 in hen liver that binds to methylated DNA with high affinity. Temporal changes in DNA methylation, modification of MDBP-2-H1, and the level of active estradiol receptors may explain the transition from the silent to the active form of the gene and its modulation during the ovulation cycle of the hen. In immature chickens or adult roosters the vitellogenin gene is silent and the concentration of estradiol in the blood is low (*153*), most of the CpGs in the promoter region of

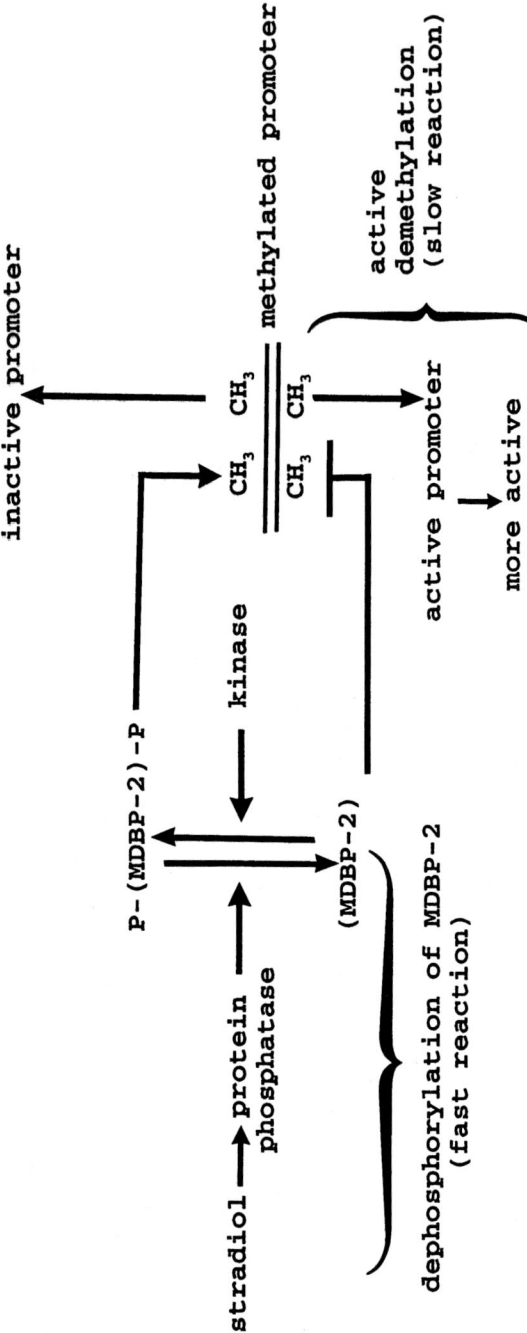

FIG. 5. Activation of the methylated DNA binding protein 2 (MDBP-2-H1) in avian liver and its binding to methylated DNA. Estradiol triggers a dephosphorylation of MDBP-2-H1. The dephosphorylated form of MDBP-2-H1 binds poorly to methylated DNA. The dephosphorylation is a fast reaction. In the absence of high concentrations of estradiol, MDBP-2-H1 is in the phosphorylated form and binds tightly to methylated DNA on the silenced gene. Demethylation of DNA in the absence of bound MDBP-2-H1 is a slow reaction that proceeds long after the onset of gene transcription. In this case, the effect of DNA methylation is indirect and is mediated by the phosphorylated form of MDBP-2-H1.

vitellogenin gene are methylated, and the binding activity of MDBP-2-H1 is very high, ensuring the silencing of the gene. In maturing hens, 2 weeks before the start of egg-laying, there is a very large and transient increase in the concentration of blood estradiol (from 50 to 350 pg ml^{-1}) (153). Similarly, in immature roosters a single injection of estradiol results in a transient and very high level of blood estradiol (154), sufficient to trigger a cascade of reactions leading to the dephosphorylation of MDBP-2-H1 and the active demethylation of DNA that was observed during the process of avian vitellogenin gene activation.

III. Conclusions

m^{5}C was discovered in 1948 (156), but only very recently was its vital function for normal embryonic development of vertebrates demonstrated (5, 155). Two lines of work involving transgenic mice have clearly shown that mice with mutations in the DNA methyltransferase gene die during midgestation (5) and mice with mutations in the MeCp2 gene also fail to develop past midgestation (155). These very important findings strongly support the hypothesis that the methylation of DNA and, most probably, the methylation patterns of specific genes are of vital importance for the normal development of embryos. Moreover, the effect of DNA methylation in this case is clearly indirect and mediated in part by MeCp2. However, these findings do not exclude the possibility that other proteins have a similar role. As has been discussed in this article, other proteins, such as MeCp1, MDBP1, and MDBP-2-H1, also bind selectively to methylated DNA. In the case of MDBP-2-H1 can also be shown that its binding to methylated DNA depends on the phosphorylation of serine residues which is controlled by estradiol (149). The biological importance of all other proteins that bind selectively to methylated DNA has not yet been demonstrated.

Because specific DNA methylation patterns are vital for normal development, it is now of paramount importance to establish firmly how such methylation profiles are formed. We have presented the possible interplay of DNA methyltransferase and m^{5}C-DNA glycosylase with site-specific cis- and trans-acting determination factors as a working hypothesis. It is clear that new experiments testing the validity of this will have to be designed and carried out.

ACKNOWLEDGMENTS

We would like to thank Y.-C. Jost for typing the manuscript and S. Oakeley for editing it. We are also grateful to E. Oakeley and F. Schmitt for their constructive discussions.

References

1. R. L. P. Adams and R. H. Burdon, eds., "Molecular Biology of DNA Methylation." Springer-Verlag, Berlin and New York, 1985.
2. A. Razin, H. Cedar, and A. D. Riggs, eds., "DNA Methylation: Biochemistry and Biological Significance." Springer-Verlag, Berlin and New York, 1984.
3. J. P. Jost and H. P. Saluz, eds., "DNA Methylation: Molecular Biology and Biological Significance." Birkhäuser-Verlag, Basel, Switzerland, 1993.
4. A. Razin and T. Kafri, *This Series* **48**, 53 (1994).
5. E. Li, T. H. Bestor, and R. Jaenisch, *Cell* **69**, 915 (1992).
6. J. L. Counts and J. I. Goodman, *Cell* **83**, 13 (1995).
7. A. Razin and R. Shemer, *Human Mol. Genet.* **4**, 1751 (1995).
8. R. Holliday and G. W. Grigg, *Mutat. Res.* **285**, 61 (1993).
9. P. A. Jones, W. M. Rideout III, J. C. Shen, D. H. Spruck, and Y. C. Tsai, *BioEssays* **14**, 33 (1992).
10. A. P. Feinberg and B. Vogelstein, *Nature (London)* **301**, 89 (1983).
11. A. R. MacLeod and M. Szyf, *J. Biol. Chem.* **270**, 8037 (1995).
12. T. L. Kautiainen and P. A. Jones, *J. Biol. Chem.* **261**, 1594 (1986).
13. M. A. Gama-Sosa, V. A. Slagel, R. W. Trewyn, R. Oxenhandler, K. C. Kuo, C. W. Gehrke, and M. Ehrlich, *Nucleic Acids Res.* **11**, 6883 (1983).
14. P. W. Laird, L. Jackson-Grusby, A. Fazeli, S. L. Dickinson, W. E. Jung, E. Li, R. A. Weinberg, and R. Jaenisch, *Cell* **81**, 197 (1995).
15. W. M. Rideout III, G. A. Coetzee, A. F. Olumi, and P. A. Jones, *Science* **249**, 1288 (1990).
16. S. S. Smith, *This Series* **49**, 65 (1994).
17. L. L. Carlson, A. W. Page, and T. H. Bestor, *Genes Dev.* **6**, 2536 (1992).
18. T. H. Bestor and G. L. Verdine, *Curr. Opin. Cell Biol.* **6**, 380 (1994).
19. T. H. Bestor, A. Landano, R. Mattaliano, and V. Ingram, *J. Mol. Biol.* **203**, 971 (1988).
20. R. L. P. Adams, *BioEssays* **17**, 139 (1995).
21. J. M. Trasler, A. A. Alcivar, L. E. Hake, T. H. Bestor, and N. B. Hecht, *Nucleic Acids Res.* **20**, 2541 (1992).
22. C. E. Houlston, H. Lindsay, S. Pradhan, and R. L. P. Adams, *Biochem. J.* **293**, 617 (1993).
23. M. Bezdek, B. Koukalova, V. Kuhrova, and B. Vysket, *FEBS Lett.* **300**, 268 (1992).
24. S. Pradhan and R. L. P. Adams. *Plant J.* **7**, 471 (1995).
25. E. J. Finnegan and E. S. Dennis, *Nucleic Acids Res.* **21**, 2383 (1993).
26. R. Holliday and J. E. Pugh, *Science* **187**, 226 (1975).
27. A. D. Riggs, *Cytogenet. Cell Genet.* **14**, 9 (1975).
28. A. Tomasetti, P. Hernaiz-Driever, G. P. Pfeifer, and D. Drahovsky, *Biochim. Biophys. Acta* **951**, 201 (1988).
29. E. S. Lyon, P. McPhie, and W. B. Jakoby, *Biochem. Biophys. Res. Commun.* **108**, 846 (1982).
30. E. S. Lyon, L. Buonocore, and M. Miller, *Mol. Cell. Biol.* **7**, 1759 (1987).
31. H. Leonhardt, A. W. Page, H. U. Weier, and T. H. Bestor, *Cell* **71**, 865 (1992).
32. M. Szyf, V. Bozovic, and G. Tanigawa, *J. Biol. Chem.* **266**, 10027 (1991).
33. R. L. P. Adams, H. Lindsay, A. Reale, C. Seivwright, S. Kass, M. Cummings, and C. Houlston, *in* "DNA Methylation: Molecular Biology and Biological Significance" (J. P. Jost and H. P. Saluz, eds.), p. 120. Birkhäuser-Verlag, Basel, Switzerland, 1993.
34. M. C. Vogel, T. Papadopoulos, H. K. Müller-Hermelink, D. Drahovsky, and G. P. Pfeifer, *FEBS Lett.* **236**, 9 (1988).
35. A. R. MacLeod, J. Rouleau, and M. Szyf, *J. Biol. Chem.* **270**, 11327 (1995).
36. J. Rouleau, A. R. MacLeod, and M. Szyf, *J. Biol. Chem.* **270**, 1595 (1995).
37. Y. Liu, L. Sun, and J. P. Jost, *Nucleic Acids Res.* **24**, 2718 (1996).

38. J. Turnbull and R. L. P. Adams, *Nucleic Acids Res.* **3**, 677 (1976).
39. G. P. Pfeifer, E. B. Speiss, S. Grunwald, T. L. J. Boehm, and D. Drahovsky, *EMBO J.* **4**, 2879 (1985).
40. J. Nyce, S. Leonard, D. Canupp, S. Schulz, and S. Wong, *Proc. Natl. Acad. Sci. U.S.A.* **90**, 2960 (1993).
41. J. Nyce, L. Liu, and P. A. Jones, *Nucl. Acids. Res.* **14**, 4353 (1986).
42. W. Doerfler, *in* "DNA Methylation: Molecular Biology and Biological Significance" (J. P. Jost and H. P. Saluz, eds.), p. 262. Birkhäuser-Verlag, Basel, Switzerland, 1993.
43. M. Toth, U. Müller, and W. Doerfler, *J. Mol. Biol.* **214**, 673 (1990).
44. T. H. Bestor, *EMBO J.* **11**, 2611 (1992).
45. S. B. Baylin, J. W. Hoppener, A. de Bustros, P. H. Steenbergh, C. J. Lips, and B. D. Nelkin, *Cancer Res.* **46**, 2917 (1986).
46. A. L. Silverman, J. G. Park, S. R. Hamilton, A. F. Gazdar, G. D. Luk, and S. B. Baylin, *Cancer Res.* **49**, 3468 (1989).
47. Y. Uehara, T. Ono, A. Kurishita, H. Kokuryu, and S. Okada, *Oncogene* **4**, 1023 (1989).
48. F. Antequera, J. Boyes, and A. Bird, *Cell* **62**, 503 (1990).
49. A. H. Bolden, C. M. Nalin, C. A. Ward, M. S. Poonian, and A. Weissbach, *Mol. Cell. Biol.* **6**, 1135 (1986).
50. M. ten Lohuis, A. Müller, I. Heidmann, I. Niedenhof, and P. Mayer, *Plant J.* **8**, 919 (1995).
51. J. K. Christman, G. Sheikhnejad, C. J. Marasco, and J. R. Sufrin, *Proc. Natl. Acad. Sci. U.S.A.* **92**, 7347 (1995).
52. S. S. Smith, T. A. Hardy, and D. J. Baker, *Nucleic Acids Res.* **15**, 6899 (1987).
53. D. J. Baker, T. A. Hardy, and S. S. Smith, *Biochem. Biophys. Res. Commun.* **146**, 596 (1987).
54. D. J. Baker, J. L. C. Kan, and S. S. Smith, *Gene* **74**, 207 (1988).
55. S. S. Smith, J. L. C. Kan, D. J. Baker, B. E. Kaplan, and P. Dembek, *J. Mol. Biol.* **217**, 39 (1991).
56. R. L. P. Adams, R. H. Burdon, K. McKinnon, and A. Rinaldi, *FEBS Lett.* **163**, 194 (1983).
57. R. L. P. Adams, A. Hanley, and A. Rinaldi, *FEBS Lett.* **269**, 29 (1990).
58. J. P. Jost, M. Siegmann, L. Sun, and R. Leung, *J. Biol. Chem.* **270**, 9734 (1995).
59. M. Vairapandi and N. J. Duker, *Nucleic Acids Res.* **21**, 5323 (1993).
60. J. P. Jost, *Proc. Natl. Acad. Sci. U.S.A.* **90**, 4684 (1993).
61. J. P. Jost and Y. C. Jost, *J. Biol. Chem.* **269**, 10040 (1994).
62. R. A. Steinberg, *Nucleic Acids Res.* **23**, 1621 (1995).
63. P. Neddermann and J. Jiricny, *J. Biol. Chem.* **268**, 21218 (1993).
64. J. C. Shen, W. M. Rideout III, and P. A. Jones, *Nucleic Acids Res.* **22**, 972 (1994).
65. J. Yisraeli and M. Szyf, *in* "DNA Methylation: Biochemistry and Biological Significance" (A. Razin, H. Cedar, and A. D. Riggs, eds.), p. 353. Springer-Verlag, Berlin and New York, 1984.
66. A. Yeivin and A. Razin, *in* "DNA Methylation: Molecular Biology and Biological Significance" (J. P. Jost and H. P. Saluz, eds.), p. 523. Birkhäuser-Verlag, Basel, Switzerland, 1993.
67. B. Teubner and W. A. Schulz, *J. Cell Phys.* **165**, 284 (1995).
68. T. H. Bestor, S. B. Hellewell, and V. M. Ingram, *Mol. Cell. Biol.* **4**, 1800 (1984).
69. P. Y. Young and S. M. Tilgham, *Mol. Cell. Biol.* **4**, 898 (1984).
70. A. Razin, C. Webb, M. Szyf, J. Yisraeli, A. Rosenthal, T. Naveh-Many, and N. Sciaky-Gallili, *Proc. Natl. Acad. Sci. U.S.A.* **81**, 2275 (1984).
71. M. Szyf, L. Eliasson, V. Mann, G. Klein and A. Razin, *Proc. Natl. Acad. Sci. U.S.A.* **82**, 8090 (1985).
72. N. Rocamora and C. Mezquita, *FEBS Lett.* **247**, 415 (1989).
73. L. Ho, V. A. Bohr, and P. C. Hanawalt, *Mol. Cell. Biol.* **9**, 1594 (1989).
74. G. Mathis and F. R. Althaus, *J. Biol. Chem.* **261**, 5758 (1986).
75. R. A. Gjerset and D. W. Martin, *J. Biol. Chem.* **257**, 8581 (1982).

76. D. Frank, N. Keshet, M. Shani, A. Levine, A. Razin, and H. Cedar, *Nature (London)* **351**, 239 (1991).
77. M. J. Hughes, H. Liang, J. Jiricny, and J. P. Jost, *Biochemistry* **28**, 9137 (1989).
78. M. J. Hughes and J. P. Jost, *Nucleic Acids Res.* **17**, 8511 (1989).
79. J. P. Jost, H. P. Saluz, and A. Pawlak, *Nucleic Acids Res.* **19**, 5771 (1991).
80. H. Oderwald and J. P. Jost, *FEBS Lett.* **382**, 313 (1996).
81. T. Mimori and J. A. Hardin, *J. Biol. Chem.* **261**, 10375 (1986).
82. T. Mimori, J. A. Hardin, and J. A. Steitz, *J. Biol. Chem.* **261**, 2274 (1986).
83. E. deVries, W. van Driel, W. G. Bergsma, A. C. Arnberg, and P. C. van der Vliet, *J. Mol. Biol.* **208**, 65 (1989).
84. M. H. Stuiver, F. E. J. Coenjaerts, and P. C. van der Vliet, *J. Exp. Med.* **172**, 1049 (1990).
85. P. R. Blier, A. J. Griffith, J. Craft, and J. Hardin, *J. Biol. Chem.* **268**, 7594 (1995).
86. N. V. Boubnov, K.T. Hall, Z. Wills, S. E. Lee, D. M. He, D. M. Benjamin, C. R. Pulaski, H. Band, W. Reeves, E. A. Hendrickson and D. T. Weaver, *Proc. Natl. Acad. Sci. U.S.A.* **92**, 890 (1995).
87. W. K. Rathmell and G. Chu, *Proc. Natl. Acad. Sci. U.S.A.* **91**, 7623 (1994).
88. G. E. Taccioli, T. M. Gottlieb, T. Blunt, A. Priestly, J. Demengeot, R. Mizuta, A. R. Lehmann, F. W. Alt, S. P. Jackson, and P. A. Jeggo, *Science* **265**, 1442 (1994).
89. V. Smider, W. K. Rathmell, M. R. Lieber, and G. Chu, *Science* **266**, 288 (1994).
90. R. C. Getts and T. D. Stamato, *J. Biol. Chem.* **269**, 15981 (1994).
91. M. Falzon, J. W. Fewell, and E. L. Kuff, *J. Biol. Chem.* **268**, 10546 (1993).
92. S. Paillard and F. Strauss, *Nucleic Acids Res.* **19**, 5619 (1991).
93. W. K. Rathmell and G. Chu, *Mol. Cell Biol.* **14**, 4741 (1994).
94. D. MacLeod, J. Charlton, J. Mullins, and A. P. Bird, *Genes Dev.* **8**, 2282 (1994).
95. M. Brandeis, D. Frank, I. Keshet, Z. Siegfried, M. Mendelsohn, A. Nemes, V. Temper, A. Razin and H. Cedar, *Nature (London)* **371**, 435 (1994).
96. P. Mummaneni, K. A. Walker, P. L. Bishop, and M. S. Turker, *J. Biol. Chem.* **270**, 788 (1995).
97. M. F. Lyon, *Nature (London)* **379**, 116 (1996).
98. G. D. Penny, G. F. Kay, S. A. Sheardown, S. Rastan, and N. Brockdorff, *Nature (London)* **379**, 131 (1996).
99. F. F. Assaad, K. L. Tucker, and E. R. Signer, *Plant Mol. Biol.* **22**, 1067 (1993).
100. J. D. G. Jones, D. E. Gilbert, K. L. Grady, and R. A. Jorgensen, *Mol. Gen. Genet.* **207**, 478 (1987).
101. F. Linn, I. Heidmann, H. Saedler, and P. Meyer, *Mol. Gen. Genet.* **222**, 329 (1990).
102. S. L. A. Hobbs, P. Kpodar, and C. M. O. DeLong, *Plant Mol. Biol.* **15**, 851 (1990).
103. I. Ingelbrecht, H. Van Houdt, M. Van Montagu, and A. Depicker, *Proc. Natl. Acad. Sci. U.S.A.* **91**, 10502 (1994).
104. O. Mittelsten Scheid, J. Paszkowski, and I. Potrykus, *Mol. Gen. Genet.* **228**, 104 (1991).
105. E. U. Selker, *Annu. Rev. Genet.* **24**, 579 (1990).
106. M. C. Kricker, J. W. Drake, and M. Radman, *Proc. Natl. Acad. Sci. U.S.A.* **89**, 1075 (1992).
107. F. Schmidt, E. J. Oakeley, and J. P. Jost, *J. Biol. Chem.* **272**, 1534 (1997).
108. R. T. Schimke, *Cell* **37**, 705 (1984).
109. R. T. Schimke, *J. Biol. Chem.* **263**, 5989 (1988).
110. G. R. Stark, M. Debatisse, E. Giulotto, and G. M. Wahl, *Cell* **37**, 901 (1989).
111. K. W. Scotto, J. L Biedler, and P. W. Meleva, *Science* **232**, 751 (1986).
112. A. Rich, in "DNA Methylation: Biochemical and Biological Significance" (A. Razin, H. Cedar, and A. D. Riggs, eds.), p. 279. Springer-Verlag, Berlin and New York, 1984.
113. W. Zacharias, in "DNA Methylation: Molecular Biology and Biological Significance" (J. P. Jost and H. P. Saluz, eds.), p. 145. Birkhäuser-Verlag, Basel, Switzerland, 1993.

114. M. Ehrlich and K. C. Ehrlich *in* "DNA Methylation: Molecular Biology and Biological Significance" (J. P. Jost and H. P. Saluz, eds.), p. 145. Birkhäuser-Verlag, Basel, Switzerland, 1993.
115. J. Ben-Hattar, P. Beard, and J. Jiricny, *Nucleic Acids Res.* **17,** 10179 (1989).
116. M. Höller, G. Westin, J. Jiricny, and W. Schaffner, *Genes Dev.* **2,** 1127 (1988).
117. J. Boyes and A. Bird, *Cell* **64,** 1123 (1991).
118. A. Levine, G. L. Cantoni, and A. Razin, *Proc. Natl. Acad. Sci. U.S.A.* **88,** 6515 (1991).
119. C. A. Johnson, J. P. Goddard, and R. L. P. Adams, *Biochem. J.* **305,** 791 (1995).
120. A. Bruhat and J. P. Jost, *Nucleic Acids Res.* **24,** 1816 (1996).
121. R. R. Meehan, J. D. Lewis, S. McKay, E. L. Kleiner, and A. Bird, *Cell* **58,** 499 (1989).
122. D. D. Loeb, R. W. Padgett, S. C. Hardies, W. R. Shehee, M. B. Comer, M. H. Edgell, and C. A. Hutchinson III, *Mol. Cell. Biol.* **6,** 168 (1986).
123. D. Jähner, H. Stuhlmann, C. L. Stewart, K. Harbers, J. Lohler, I. Simon, and R. Jaenisch, *Nature (London)* **298,** 623 (1982).
124. J. D. Lewis, R. R. Meehan, W. J. Henzel, I. Maurer-Foggy, P. Jeppesen, F. Klein, and A. Bird, *Cell* **69,** 905 (1992).
125. R. R. Meehan, J. D. Lewis, S. McKay, E. L. Kleiner, and A. P. Bird, *Cell* **58,** 499 (1989).
126. X. Nan, P. Tate, E. Li, and A. Bird, *Mol. Cell. Biol.* **16,** 414 (1996).
127. R. R. Meehan, J. D. Lewis, and A. Bird, *Nucleic Acids Res.* **20,** 5085 (1992).
128. M. Churchill and M. Susuki, *EMBO J.* **8,** 4189 (1989).
129. X. Nan, R. R. Meehan, and A. Bird, *Nucleic Acids Res.* **21,** 4886 (1993).
130. J. P. Jost and H. P. Saluz, *in* "DNA Methylation: Molecular Biology and Biological Significance" (J. P. Jost, and H. P. Saluz, eds.), p. 425. Birkhäuser-Verlag, Basel, Switzerland, (1993).
131. H. P. Saluz, I. M. Feavers, J. Jiricny, and J. P. Jost, *Proc. Natl. Acad. Sci. U.S.A.* **85,** 6697 (1988).
132. J. P. Jost and J. Hofsteenge, *Proc. Natl. Acad. Sci. U.S.A.* **89,** 9499 (1992).
133. M. Bustin and R. D. Cole, *J. Biol. Chem.* **243,** 4500 (1968).
134. J. M. Kinkade, *J. Biol. Chem.* **244,** 3375 (1969).
135. W. F. Brandt, W. N. Strickland, M. Strickland, L. Carlisle, D. Woods, and C. Von Holt, *Eur. J. Biochem.* **94,** 1 (1979).
136. R. W. Lennox and L. H. Cohen, *J. Biol. Cchem.* **258,** 262 (1983).
137. R. D. Cole, *Analyt. Biochem.* **136,** 24 (1984).
138. R. D. Cole, *Int. J. Pept. Prot. Res.* **30,** 433 (1987).
139. D. Wells and C. McBride, *Nucleic Acids Res.* **17,** r311 (1989).
140. E. Schulze, L. Trieschmann, B. Schulze, E. R. Schmidt, S. Pitzel, K. Zechel, and U. Grossbach, *Proc. Natl. Acad. Sci. U.S.A.* **90,** 2481 (1993).
141. Y. Dong, A. M. Sirotkin, Y. S. Yang, D. T. Brown, D. B. Sittman, and A. I. Skoultchi, *Nucleic Acids Res.* **22,** 1428 (1994).
142. D. T. Brown, B. T. Alexander, and D. Sittman, *Nucleic Acids Res.* **24,** 486 (1986).
143. K. Nightingale and A. P. Wolffe, *J. Biol. Chem.* **270,** 4197 (1995).
144. F. J. Campoy, R. R. Meehan, S. McKay, J. Nixon, and A. Bird, *J. Biol. Chem.* **270,** 27473 (1995).
145. D. J. Ball, D. S. Gross, and W. T. Garrard, *Proc. Natl. Acad. Sci. U.S.A.* **80,** 5490 (1983).
146. A. Pawlak, M. Bryans, and J. P. Jost, *Nucleic Acids Res.* **19,** 1029 (1991).
147. J. P. Jost, O. Munch, and T. Andersson, *Nucleic Acids Res.* **19,** 2788 (1991).
148. M. Vaccaro, A. Pawlak, and J. P. Jost, *Proc. Natl. Acad. Sci. U.S.A.* **87,** 3047 (1990).
149. A. Bruhat and J. P. Jost, *Proc. Natl. Acad. Sci. U.S.A.* **92,** 3678 (1995).
150. S. Y. Roth and C. D. Allis, *Trends Biochem. Sci.* **17,** 93 (1992).

151. T. Hirano, *Trends Biochem. Sci.* **20,** 357 (1995).
152. H. Schröcksnadel, A. Bator, and J. Frick, *Steroids* **6,** 767 (1973).
153. B. E. Senior, *J. Reprod. Fertil.* **41,** 107 (1974).
154. U. Joss, C. Bassand, and C. Ventling, *FEBS Lett.* **66,** 293 (1976).
155. P. Tate, W. Skarnes, and A. Bird, *Nature Genet.* **12,** 205 (1996).
156. R. D. Hotchkiss, *J. Biol. Chem.* **168,** 315 (1948).
157. J. P. Jost *in* "Epigenetics" (V. Russo, R. Martienssen, and A. Riggs, eds.), p. 109, 1996.
158. L. A. Chandler, H. Ghazi, P. A. Jones, P. Boukamp, and N. E. Fusenig, *Cell* **50,** 711 (1987).
159. A. J. Silva, K. Ward, and R. White, *Dev. Biol.* **156,** 391 (1993).
160. M. McArthur and J. O. Thomas, *EMBO J.* **15,** 1705 (1996).

The Role of mRNA Stability in the Control of Globin Gene Expression

J. Eric Russell,
Julia Morales, and
Stephen A. Liebhaber[1]

Departments of Genetics and Medicine
and the Howard Hughes Medical
Institute
University of Pennsylvania
Philadelphia, Pennsylvania 19104

I. General Aspects of Eukaryotic mRNA Stability	250
A. Common Determinants of mRNA Stability	250
B. Specific Determinants Confer Different Stabilities on Individual mRNAs	251
C. Trans-Acting Factors Participate in Determining mRNA Stability	253
II. Specific Aspects of Globin mRNA Stability	253
A. Cytologic Features of Erythroid Differentiation	253
B. Globin mRNAs Accumulate to High Levels in Terminally Differentiating Erythroid Cells	255
C. Nonglobin mRNAs May Be Selectively Destabilized in Terminally Differentiating Erythroid Cells	256
D. All Globin mRNAs Are Not Equally Stable	257
E. Induced Destabilization of Globin mRNAs	260
F. Initial Attempts to Define Mechanisms of Globin mRNA Turnover	261
G. Genetic Evidence for the Importance of mRNA Stability to Globin Gene Expression	261
III. Structural Determinants of α-Globin mRNA Stability	263
A. Experimental Approach to Defining Determinants That Stabilize Human α-Globin mRNA	263
B. Evidence for a Stability Determinant within the 3'-UTR	265
C. Stability Element Function	266
D. The Stability Element Maps to a Series of C-Rich Pyrimidine Tracks in the 3' UTR	267
E. Role of RNP Formation in Stabilization of Human α-Globin mRNA	269
F. Identification of Trans-Acting Factors Involved in Human α-Globin mRNA Stability	271
G. Identification of a 39-kDa Poly(C) Binding Protein, αCP	271
H. Deadenylation of Stable and Unstable Human α-Globin mRNAs	274
I. Structural Basis of α-Globin mRNA Stability: A Model	274

[1] To whom correspondence may be addressed.

IV. Stability of Other Globin mRNAs 276
    A. Determinants of α-Globin mRNA Stability
        Are Highly Conserved 276
    B. Parallel Evolution of Cis- and Trans-Acting Components
        of the Stability-Determining Complex from Mouse to Humans 276
    C. Human β-Globin mRNAs Are Stabilized by One
        or More Elements within the 3′ UTR 280
    D. Human α- and β-Globin mRNAs Are Stabilized by Structurally
        and Mechanistically Distinct Elements 282
    E. Possible Function of RNP Complexes Related to the α Complex
        in Stabilization of Nonglobin Erythroid mRNAs 283
V. Potential Therapeutic Applications 283
    References 285

The hematopoietic stem cell differentiates into eight morphologically and functionally distinct cell lines (1). The developmental controls that underlie this process are based on ordered activation and silencing of numerous genes, which generate mRNAs encoding both general and cell type-specific proteins. Cells of the erythroid lineage display three unique characteristics distinguishing them from other hematopoietic cells. First, erythroid cells contain extraordinarily high levels of the oxygen transport molecule, *hemoglobin*. Hemoglobin, a heterotetramer composed of two α- or α-like and two β- or β-like globin subunits, constitutes up to 95% of soluble erythrocyte protein content. Second, the expression of constituent globin subunits switches at two points during human ontogeny. Separate sets of α- and β-like globin genes are sequentially expressed as the site of erythroid cell proliferation migrates from the embryonic yolk sac to the fetal liver and finally to the adult bone marrow. Finally, the terminal stages of erythroid differentiation occur in cells that have undergone global, irreversible transcriptional arrest. In these cells the contribution of posttranscriptional mechanisms to the control of gene expression is, by definition, absolute. This review describes the role of mRNA stability in the control of globin gene expression during terminal differentiation of erythroid cells, and summarizes recent findings which begin to define the corresponding crucial *cis*- and *trans*-acting determinants. In addition, the contribution of globin mRNA stability to the embryonic-to-fetal and fetal-to-adult switches in globin gene expression will be discussed.

I. General Aspects of Eukaryotic mRNA Stability

A. Common Determinants of mRNA Stability

The cellular concentration of a particular mRNA species is determined by the balance between its rates of synthesis and degradation. Although the rates of transcription, processing, and nuclear-to-cytoplasmic transport are important variables that affect mRNA expression, the level to which an mRNA accumulates in the cytoplasm is determined to a great extent by its relative stability (2, 3). Individual mRNA species degrade at characteristic rates, resulting in a wide spectrum of observed mRNA half-lives ($t_{1/2}$). mRNA stability is determined by a combination of general and specific structures and pathways (2, 4–6). Structures common to all eukaryotic mRNAs, such as the $^{m7}G(5')ppp(5')N$ cap and the 3' polyadenylate [poly(A)] tail, appear to be general determinants of stability. These structures act, at least in part, by inhibiting exonuclease degradation of the 5' and 3' ends of an mRNA. mRNA-specific $t_{1/2}$ values appear to be determined by multiple, frequently overlapping pathways, several of which have been defined most clearly in yeast (6). For prototype yeast mRNAs, the rate-limiting event appears to be decapping, which is followed by $5'\rightarrow 3'$ exonucleolytic degradation (7, 8). In many cases, decapping is linked to a preceding shortening of the poly(A) tail (9–12). Less commonly, specific mRNA degradation is initiated by a rate-limiting endonucleolytic cleavage event (13). In the special category of nonsense-mutation mediated mRNA decay, decapping occurs in the absence of prior poly(A) shortening (14).

B. Specific Determinants Confer Different Stabilities on Individual mRNAs

In higher eukaryotes, mRNA $t_{1/2}$ values vary almost 1000-fold, from a low of several minutes to a high of several days (3). The spectrum of mRNA $t_{1/2}$ values is summarized in Fig. 1. The $t_{1/2}$ of a specific mRNA frequently reflects the function of its translated protein product. For example, oncogenes, cytokines, and cell cycle control factors are generally encoded by highly labile mRNAs (3, 6). Such short half-lives permit changes in gene transcription to be reflected rapidly in the levels of mRNA and subsequent rates of protein expression. A large number of these short-lived mRNAs share one or more repeats of an AU-rich motif (AUUUA) within the 3' untranslated region (UTR) (15, 16). More than 10 proteins have been reported to bind to this AU-rich motif; whether any of these factors actually contributes to the mechanism(s) through which mRNA $t_{1/2}$ is determined is unknown (16).

At the other end of the spectrum is a collection of highly stable mRNAs with $t_{1/2}$ values in the range of several days. These mRNAs encode globins (17, 18), collagens (19, 20), crystallins (21), and other proteins that typically accumulate to unusually high levels, particularly in highly differentiated post-mitotic cells. The expression of "bulk" proteins encoded by these mRNAs does not require rapid response control but favors more economical long-lived mRNAs. The importance of high-level mRNA stability to the expression of these genes is best illustrated by cases in which this characteristic is lost.

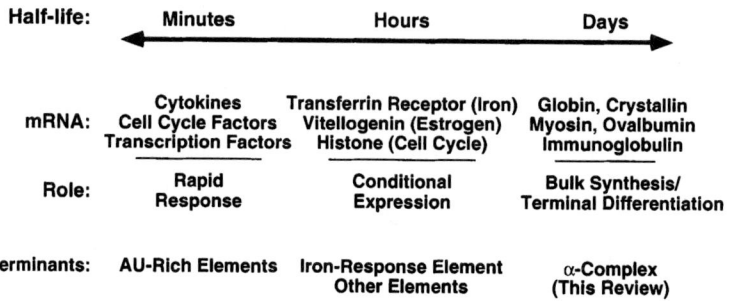

FIG. 1. Eukaryotic mRNAs display significant differences in stability. Representative eukaryotic mRNAs are grouped into three classes according to their structural half-lives. The first category includes very unstable mRNAs, the second includes mRNAs whose stability can vary in response to specific cellular and environmental stimuli (indicated parenthetically for each mRNA), and the third includes highly stable mRNAs. The AUUUA motif, the stem–loop iron response elements, and the α complex typify structural determinants that mediate decay phenotypes in these categories. The control of stability for mRNAs in the third category, and specifically for globin mRNAs, is detailed in this review.

As an example, a single base substitution at the translation termination codon of the human α-globin mRNA (UAA → CAA) permits ribosomes to continue translation into the 3′ UTR (22), destabilizing the α-globin mRNA (23, 24). This mutation, known as hemoglobin $\alpha^{Constant\ Spring}$ (α^{CS}), results in a >95% loss of gene expression. The α^{CS} locus accounts for the majority of nondeletional forms of α-thalassemia throughout the world.

mRNA stability can be a dynamic property that responds to developmental and/or environmental stimuli. The modulation of transferrin receptor (TfR) expression represents a clear example of such control (25). The TfR receptor mediates transport of iron–transferrin complexes from the serum into the cell. The stability of TfR mRNA increases under conditions of low intracellular iron levels, resulting in increased expression of TfR and consequent increase in iron transport. Reciprocally, under conditions of high intracellular iron levels, TfR mRNA stability decreases, resulting in decreased expression of TfR and reduced iron transport. The control of TfR mRNA stability is mediated by interaction of a pentad of highly similar RNA stem-and-loop structures in the 3′ UTR (iron response element, or IRE) (26) with a 90-kDa cytosolic iron response element binding protein (IRE-BP) (27). The IRE-BP contains an iron-binding site that when occupied alters the conformation of the IRE-BP such that its IRE-binding affinity decreases by several orders of magnitude (28). Under these conditions, the deprotected IRE is cleaved by an unidentified endonuclease that initiates rapid degradation of TfR mRNA (13). Conversely, endonucleolytic cleavage of the IRE is prevented by high-affinity binding of the IREbp under conditions of low cellular iron, when the

IRE-BP iron-binding site is unoccupied. Additional systems have been identified in which the alteration of mRNA stability is important to cell function or to developmental control (for examples, see Fig. 1), although details of the mechanisms are not as well defined (29).

C. Trans-Acting Factors Participate in Determining mRNA Stability

The majority of described eukaryotic RNA binding proteins were initially identified as components of heterogeneous nuclear RNA–protein (hnRNP) complexes. Many of these proteins contain one or more shared motifs (30). These motifs tend to be relatively large, forming β-sheet "platforms" on which RNA is bound (31). Because many hnRNP proteins are involved in the general packaging of heterogeneous mRNA, their RNA binding sites exhibit relaxed sequence specificities. In addition, these proteins appear to bind equally well to both RNA and single-stranded DNA *in vitro*. More recently, RNA binding proteins have been identified as essential components of certain mRNA stability determinants. In contrast to hnRNP proteins, RNA binding proteins or protein complexes with clearly defined functions such as the IRE-BP (25), Rev (response element binding protein (32, 33)), and the human α-stability complex (34) are highly sequence specific. Thus, RNA binding proteins may either participate in highly specific interactions with target mRNAs, or may serve more general packaging functions (hnRNPs) that are necessary for efficient mRNA processing, transport, and translation. The mechanisms through which RNA binding proteins interact with mRNA primary and higher order structures to mediate vital cellular functions is the focus of great interest.

II. Specific Aspects of Globin mRNA Stability

A. Cytologic Features of Erythroid Differentiation

The importance of globin mRNA stability to normal erythroid development was first inferred from the study of erythroid terminal differentiation. Erythropoeisis can be segmented into six stages on the basis of conventional histologic staining and light microscopy. These stages are shown schematically in Fig. 2. The basophilic erythroblast (BE), the first cytologically distinct committed erythroid precursor, contains a large nucleus and fine nuclear chromatin. This cell is transcriptionally active: the deep blue color of the Wright/Giemsa-stained cytoplasm reflects the accumulation of large amounts of RNA. The polychromatophilic erythroblast (PCE), the second stage of erythroid maturation, is smaller than the BE, with coarsening of the

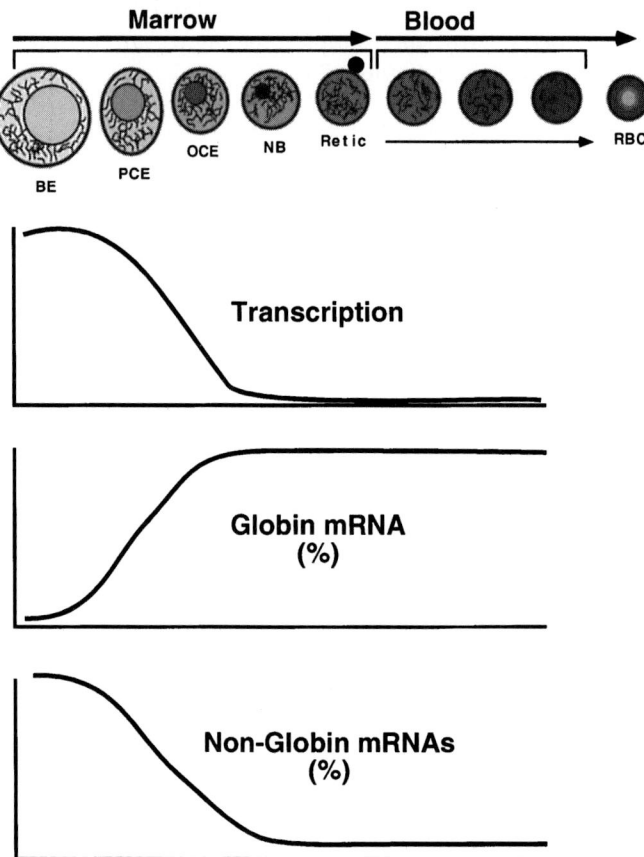

FIG. 2. Globin mRNAs are enriched in terminally differentiated erythroid cells. Cytologically distinct stages of erythroid terminal differentiation are illustrated. The bone marrow phase of differentiation is characterized morphologically by progressive decrease in cell size, change in cytoplasmic staining from the blue basophilic erythroblast (BE) to the pink reticulocyte (Retic), and nuclear condensation and chromatin clumping, which terminates in extrusion of the nucleus from the normoblast (NB). The silencing in overall transcription (upper curve) parallels morphological nuclear changes, whereas the change in cytoplasmic staining reflects progressive accumulation of hemoglobin protein. Levels of globin and nonglobin mRNA at each stage are illustrated by the middle and bottom curves, respectively. PCE, Polychromatophilic erythroblast; OCE, orthochromatophilic erythroblast; RBC, red blood cell (mature erythrocyte).

nuclear chromatin and a cytoplasm that appears lighter and slightly purple in color. The change in cytoplasmic staining reflects initial accumulation of small quantities of hemoglobin. At the orthochromatophilic erythroblast stage (OCE) the fully condensed chromatin and the contracted, opacified nucleus indicate progressive transcriptional silencing. The cytoplasm of the OCE stains a distinct pink due to continued accumulation of hemoglobin.

The final stages of erythroid terminal differentiation are characterized by a cellular program of differentiation that operates despite transcriptional inactivation. The normoblast (NB), which represents the fourth stage of erythroid maturation, has a nucleus that is fully condensed and is eventually extruded from the cell. Continued hemoglobin synthesis and accumulation during this stage results in a distinctly red cytoplasm. The reticulocyte (Retic) represents the fifth stage of erythroid maturation. This cell is slightly larger than a mature red cell (erythrocyte), has no nucleus, and appears to be fully hemoglobinized. Residual RNA and cellular organelles appear as dark speckles (reticulin) when reticulocytes are stained using special techniques. Once formed, reticulocytes migrate from the bone marrow into the peripheral circulation where they mature into fully differentiated erythrocytes in 2 to 3 days. During this period residual RNA is degraded, cytoplasmic organelle fragments are destroyed, and the cellular membrane is trimmed by reticuloendothelial cells in the spleen and liver. The fully matured erythrocyte is a biconcave discoid cell, 8 µm in diameter, with a hemoglobin content of 34 g/dl. Mature erythrocytes circulate for approximately 100 days, transporting O_2, CO_2, and perhaps other ligands (35) between the lungs and the peripheral tissues.

B. Globin mRNAs Accumulate to High Levels in Terminally Differentiating Erythroid Cells

The cytologic stages of erythroid differentiation are paralleled by dramatic changes in cell function (36). The changes in transcription rates and in mRNA accumulation are summarized in Fig. 2. The bulk of globin mRNA appears to be synthesized in basophilic erythroblasts, although small amounts may also be transcribed in proerythroblast precursor cells (37, 38). Globin mRNA and protein appear to accumulate at equal rates, suggesting that there are no substantial translational controls on globin gene expression (39). A wide variety of nonglobin genes that encode cytoskeletal components, cell surface cytokine receptors, enzymes essential for energy production, and factors that enhance the function of the hemoglobin tetramer are also transcribed in the erythroblast.

Global silencing of gene transcription commences in late polychromatophilic and early orthochromatophilic erythroblasts and is complete by the normoblast stage. The mechanisms that trigger this event are unknown. Further, it has not been established whether transcriptionally active genes are silenced simultaneously or in a specific order. It is clear, however, that terminal differentiation through the normoblast and reticulocyte stages is accompanied by a marked reduction in mRNA complexity. In late reticulocytes, as much as 95–98% of cellular mRNA encodes globin (40).

Two factors contribute to the accumulation of highly enriched populations of globin mRNAs. First, globin mRNAs are highly stable. Half-life determinations have been carried out by *in vivo* (whole animal) studies in anemic mice, and in cell culture studies using mouse erythroleukemia (MEL) cells (18, 38, 40–42). Globin mRNA half-lives determined in most of these studies are in the range of 17–24 hr. This agreement among multiple $t_{1/2}$ determinations is remarkable considering differences in technique, the exact species of globin mRNA being studied, and the conditions under which different cell lines were evaluated. In one particularly careful determination, where artifactual lysis of mature cultured cells was avoided by using high osmotic 'stabilizing' media (43), the globin mRNA $t_{1/2}$ was found to be greater than 60 hr, or roughly equivalent to the $t_{1/2}$ of ribosomal RNA (42). Direct measurements of globin mRNA in primary bone marrow preparations confirm its high-level stability (18).

C. Nonglobin mRNAs May Be Selectively Destabilized in Terminally Differentiating Erythroid Cells

Although the stability of globin mRNA during terminal erythroid differentiation is critical to its high-level accumulation, a second factor, selective destabilization of nonglobin mRNAs, is also thought to contribute to this process. As in all cells, erythroblasts contain mRNAs with a full spectrum of stabilities (17, 40, 44). Although the bulk of these mRNAs have $t_{1/2}$ values of several hours, as is characteristic of mRNA populations in higher eukaryotes, there is also a substantial fraction of highly stable nonglobin mRNAs. The stability of these nonglobin mRNAs predicts that they should be retained during terminal differentiation and constitute a large fraction of the total cellular mRNA in reticulocytes (40); in fact, this does not occur. This paradox was resolved by mathematical models that hypothesized selective destabilization of the long-lived nonglobin mRNAs in maturing erythroid cells (44), while the high stability of globin mRNA is maintained. The predictions of this kinetic model have been generally supported. A number of mRNAs that are stable in early erythroid cells, such as tubulin and actin, are actively destabilized during terminal differentiation of induced MEL cells (45). This destabilization is attributed to the activation of a specific degradation pathway and does not appear to correlate with shortening of the target mRNA poly(A) tail. Although early reports carried out in MEL cells suggested that globin mRNA stability might also decrease during terminal differentiation (46), subsequent studies have demonstrated that globin mRNA stability is maintained (42).

Thus, the accumulation of globin mRNA in the terminally differentiated erythroid cell reflects its stability and the selective destabilization and clearance of most other mRNAs.

D. All Globin mRNAs Are Not Equally Stable

With few exceptions, studies of globin mRNA stability have been limited to analysis of adult α- and β-globin mRNAs. Other human globin mRNAs, such as the embryonic ζ- and ε-globin mRNAs, are difficult to study because they are expressed exclusively in the embryonic yolk sac (Fig. 3). As erythropoiesis shifts from the yolk sac to the fetal liver early in gestation, the ex-

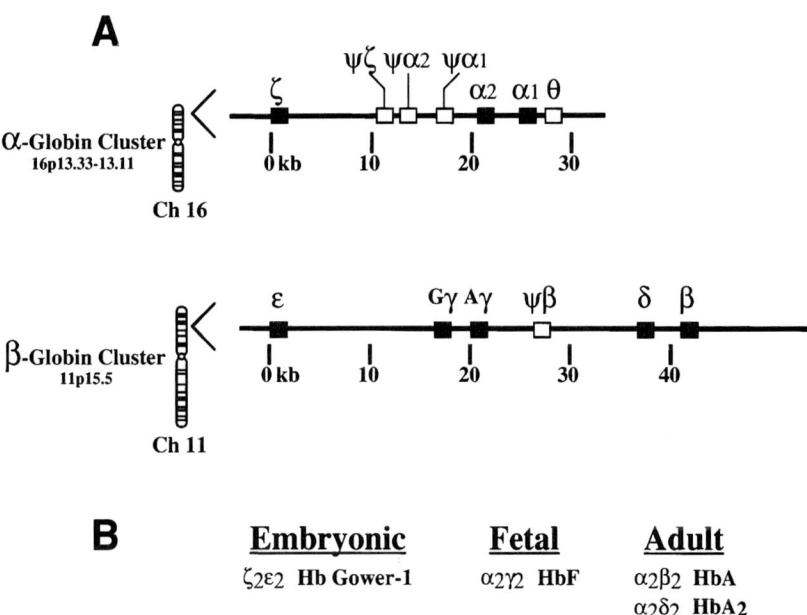

FIG. 3. The human α-like and β-like globin genes are organized in separate clusters and are expressed in a developmental stage-specific pattern. (A) Structure of the human α- and β-globin gene clusters. The α-globin cluster, located at the telomere of the short arm of chromosome 16, contains three genes of functional significance: the embryonic ζ-globin gene and two coexpressed fetal/adult α-globin genes. The β-globin gene cluster, located on chromosome 11, contains five functional genes: the embryonic ε-globin gene, two coexpressed γ-globin genes, the minor adult δ-globin gene, and the major adult β-globin gene. Functional genes are indicated as black boxes; pseudogenes within each cluster are indicated as open boxes. (B) Specific hemoglobin heterotetramers characterize different stages of human development. Globin genes are differentially expressed during embryonic (ζ, ε), fetal (α, γ), and adult (α, β) developmental stages. Globin monomers assemble into hemoglobin heterotetramers characteristic of each stage of human development: embryonic (Hb Gower-1, $\zeta_2\epsilon_2$), fetal (Hb F $\alpha_2\gamma_2$), and adult (Hb A, $\alpha_2\beta_2$). Hb A2 ($\alpha_2\delta_2$) is a minor hemoglobin that constitutes 2–3% of total adult hemoglobin.

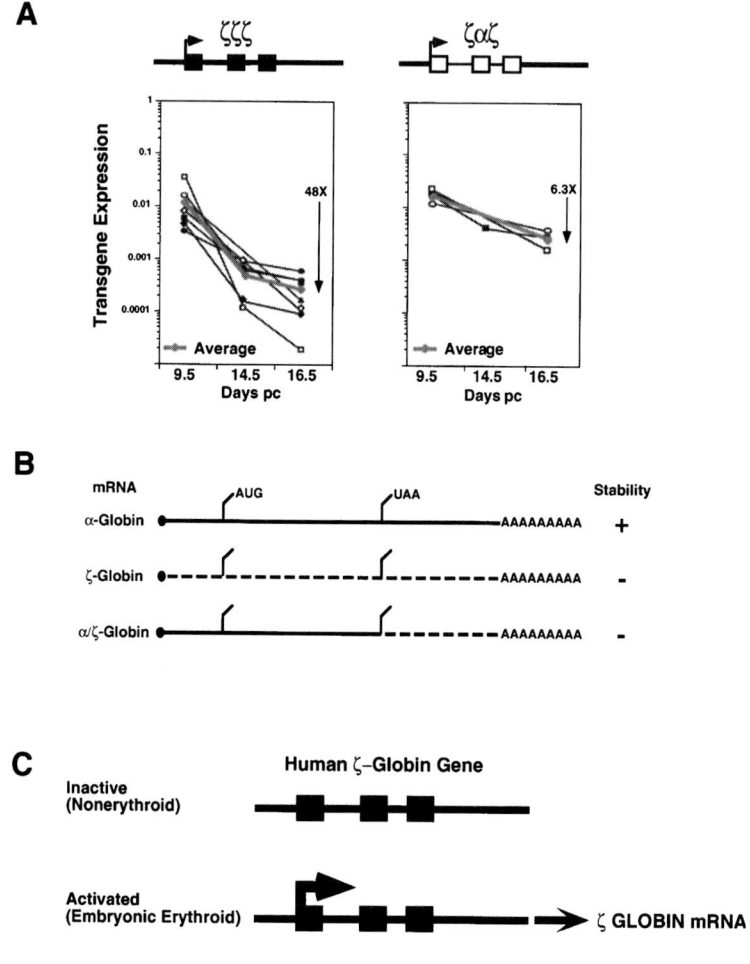

FIG. 4. The instability of human ζ-globin mRNA contributes to its silencing in adult erythroid cells. (A) Contribution of the ζ-globin transcribed region to ζ-globin gene silencing during the embryonic-to-fetal switch. The expression of two human transgenes was determined from transgenic mouse 9.5-day postcoital (pc) embryonic yolk sacs and 14.5- and 16.5-day pc fetal livers. Both transgenes contain the ζ-globin gene promoter and 3′ flanking region elements (heavy lines), and differ only in the identity of their transcribed regions. The ζζζ (native human ζ-globin gene) and ζαζ transgenes contain coding regions from the human ζ- and α-globin tran-

pression of these two genes is silenced and replaced by the fetally expressed α- and γ-globin genes.

Studies of the human ζ-globin gene in transgenic mice suggest that its normal silencing is a complex process that requires interaction of multiple determinants, both transcriptional and posttranscriptional (47). The intact human ζ-globin transgene with 500 bp of contiguous 5' flanking region and 3 kb of contiguous 3' flanking sequences is appropriately silenced during the embryonic-to-fetal switch (47–52). However, when the transcribed region of the α-globin gene is substituted for the corresponding segment of the ζ-globin gene, the chimeric gene fails to be silenced fully in the adult mouse (Fig. 4A) (47). This unexpected result suggests that one or more determinants in the ζ-globin transcribed region contribute to silencing of its expression.

The possibility that ζ-globin mRNA destabilization contributes to its silencing was tested by directly comparing the stabilities of the human α- and ζ-globin mRNAs in the adult mouse. To achieve high-level expression of human ζ-globin mRNA in adult mice, the ζ-globin gene was linked to the α-globin gene promoter and a transcriptional silencing element located 3' to the ζ-globin gene was deleted (47). The ζ-globin mRNA expressed in these adult mice was unstable relative to endogenous mouse α-globin mRNA. In contrast, the human α-globin mRNA expressed in transgenic mice was equally stable to mouse α-globin mRNA (53). The determinant for this difference in stability was mapped to the 3' UTRs of the respective mRNAs. When the 3' UTR of the α-globin mRNA was replaced by the corresponding segment of the ζ-globin mRNA, the resulting chimeric mRNA was as unstable as the native ζ-globin mRNA (Fig. 4B) (53). These data suggest that the embryonic

---

scribed regions (dark and open boxes), respectively. The developmentally regulated expression profiles of seven ζζζ lines and three ζαζ lines are indicated. All values are normalized to transgene copy number and are expressed as a percentage of the expression of the endogenous mouse α-globin genes. There is an average 48-fold decrease in the expression of the ζζζ transgene during this period, versus 6.3-fold for the ζαζ transgene. These results indicate that the identity of the transcribed region influences the effectiveness of ζ-globin gene silencing. (B) Differences in the α- and ζ-globin mRNA 3' UTRs mediate their different stabilities. The stabilities of human α-globin mRNA (solid line) and human ζ-globin mRNA (dashed line) were determined in transgenic mice. Whereas the α-globin mRNA was equally stable (+) to endogenous mouse α-globin mRNA, ζ-globin mRNA was significantly destabilized (−). Substitution of the ζ-globin 3' UTR for the corresponding α-globin segment (α/ζ-globin mRNA) resulted in destabilization of the chimeric mRNA. For each mRNA the cap (•), initiation and termination codons (tick marks), and poly(A) tail are indicated. (C) Model for ζ-globin gene silencing. The human ζ-globin gene is transcriptionally silent in nonerythroid cells, but is transcriptionally active in embryonic erythroid cells (bold bent arrow). In fetal and adult erythroid cells, the interaction of determinants within the 5' and 3' flanking regions incompletely silence ζ-globin gene transcription (light bent arrow). The small quantity of transcribed ζ-globin mRNA that continues to be produced is rapidly degraded, before measurable quantities of ζ-globin protein can be translated.

ζ-globin mRNA is unstable when expressed in adult cells and that this instability is an important factor in the overall silencing of ζ-globin gene expression (Fig. 4C). This model is consistent with prior observations that ζ-globin gene is transcriptionally active in adult bone marrow cells but that its mRNA cannot be detected in circulating reticulocytes (54).

The observed instability of ζ-globin mRNA may be quite consistent with its function in embryonic (primitive) erythroblasts. Unlike adult erythroblasts in the bone marrow, primitive erythroblasts in the yolk sac of the developing embryo retain transcriptionally active nuclei. This permits continual generation of nascent globin mRNA transcripts, even after release into the embryonic circulation. In addition, primitive erythroblasts survive for only a brief period in the embryonic circulation. Even marginal transcriptional activity is likely to ensure sufficient levels of embryonic ζ-globin mRNA and its encoded protein. Thus the high-level stability observed in globin mRNAs expressed in nonnucleated terminally differentiating fetal/adult cells may not be biologically relevant to globin mRNAs expressed in primitive erythroblasts.

Human δ-globin mRNA is another species whose stability has been established. The δ-globin gene is expressed in adult human erythrocytes at only 2–3% of that of β-globin (Fig. 3). The inefficiency of δ-globin gene expression reflects both transcriptional and posttranscriptional components. A mutation within the proximal promoter region (55, 56) results in poor transcriptional efficiency. This defect does not, however, fully explain the relatively low level of δ-globin expression. Although δ-globin synthesis is detectable in the adult bone marrow, it is absent in peripheral reticulocytes (57). Direct measurement of β- and δ-globin mRNA $t_{1/2}$ in nucleated bone marrow cells (16.5 and 4.5 hr, respectively) indicated that the relative instability of the δ-globin mRNA contributes to its poor expression (58). The substantially lower stability of the δ-globin mRNA did not correlate with differences in its poly(A) tail length. However, the β- and δ-globin mRNA sequences diverge most dramatically in the 3' UTR (59% divergence, versus ~8% divergence in the 5' UTR and coding region). Once the stability determinants within the β-globin mRNA are identified, the particular divergent sequence(s) that mediate δ-globin mRNA destabilization can be determined.

E. Induced Destabilization of Globin mRNAs

The stability of adult globin mRNAs appears to be a constitutive property. Globin mRNAs are stable in both early and late erythroblasts as well as in a variety of nonerythroid tissues in which they have been artificially introduced and expressed (see below). However, at least one line of investigation suggests that stability may be lost in response to specific developmental signals (59). K562 cells are multipotential hematopoietic cells that can be induced to differentiate along either the megakaryocytic or erythrocytic pathway. Under basal culture conditions these cells express high levels of fetal γ-globin mRNA. Exposure of K562 cells to phorbol esters [such as phorbol myristate acetate (PMA)] induces differentiation along the megakaryocytic pathway. Cytoplasmic levels of γ-globin mRNA drop substantially in PMA-induced K562 cells. Changes in transcription only partially account for the decrease in γ-globin mRNA levels, suggesting that γ-globin mRNA may also be destabilized in these induced cells. Subsequent work has demonstrated that induced destabilization of γ-globin mRNA is paralleled by shortening of its poly(A) tail and by the appearance of 3' cleavage products (60). The mechanistic link between poly(A) tail shortening and γ-globin destabilization remains unclear. Because the pattern of the 3' cleavage products is not altered by PMA induction, it is unlikely that they result from activation of a specific mRNA degradation pathway. It is also not clear whether the 3' cleavage events are rate limiting either in induced or uninduced cells; nor is it understood how these observations relate to normal physiologic control of γ-globin gene expression.

F. Initial Attempts to Define Mechanisms of Globin mRNA Turnover

Early studies of globin mRNA turnover suggested that decay of globin mRNA might be initiated through endonucleolytic cleavage within the 3' UTR and regulated by the length of the poly(A) tail (61). More recently, the turnover of globin mRNA has been studied in rabbit reticulocytes (the site of globin mRNA degradation during the final stages of erythroid maturation), where it appears to be mediated by 3' UTR site-specific cleavage (62). This is in general agreement with studies in K562 cells induced to megakaryocytic differentiation which demonstrate a predominance of 3' UTR cleavages in degrading γ-globin mRNA (see above) (60). These studies begin to define the mechanism(s) through which globin mRNAs are ultimately metabolized, though they do not address the basis for globin mRNA stabilization.

G. Genetic Evidence for the Importance of mRNA Stability to Globin Gene Expression

A wide variety of naturally occurring mutations have been described that adversely affect the expression of human globin genes and result in the clinical syndrome of *thalassemia* (reviewed in 63). Many thalassemic mutations inhibit efficient mRNA transcription, resulting in decreased mRNA levels *via* a production defect. Other mutations that block normal polyadenylation or interfere with normal mRNA processing by favoring alternative splice sites can destabilize the nascent or abnormally processed mRNA transcript. Mutations that introduce premature translation termination codons can also trigger accelerated mRNA decay (*14*). Although of interest, none of these mutations have been informative in understanding the basis for the stability of normal globin mRNAs.

Of the many thalassemic mutations whose mechanisms have been studied in depth, only $\alpha^{\text{Constant Spring}}$ appears to destabilize globin mRNA as a primary pathophysiologic event. First described in 1971 (*22, 64*), α^{CS} is the most frequent cause of nondeletion α-thalassemia worldwide. Under normal conditions, an intact termination codon blocks ribosomes from entering the 109-nt α-globin mRNA 3′ UTR. The α^{CS} mutation is a single base substitution at the translation termination codon (UAA → CAA) of the major fetal/adult α-globin locus (α2 gene locus), which permits ribosomes to translate for an additional 31 codons into the ordinarily ribosome-free 3′ UTR (Fig. 5). α^{CS} heterozygotes express the C-terminally extended globin protein at only 3% of the wild-type α-globin level. Because α^{CS} mRNA is transcribed at normal rates, and because the α^{CS} protein is stable (*22*), this loss of expression appears to reflect a defect in the accumulation of cytoplasmic α^{CS} mRNA. The observation that the synthesis of α^{CS} protein [relative to the synthesis of wild-type (α^{wt}) α-globin protein] is higher in bone marrow than in reticulocytes is consistent with this mechanism (*65*). Direct assessment of α^{CS} mRNA levels in an α^{CS} heterozygote confirmed this conclusion: α^{CS} mRNA was present in bone marrow erythroid cells at a level comparable to that of α^{wt} mRNA, but was undetectable in peripheral reticulocytes (*23*) (Fig. 5). The fall in the level of α^{CS} mRNA relative to that of α^{wt} mRNA from the transcriptionally active bone marrow to the transcriptionally silent reticulo-

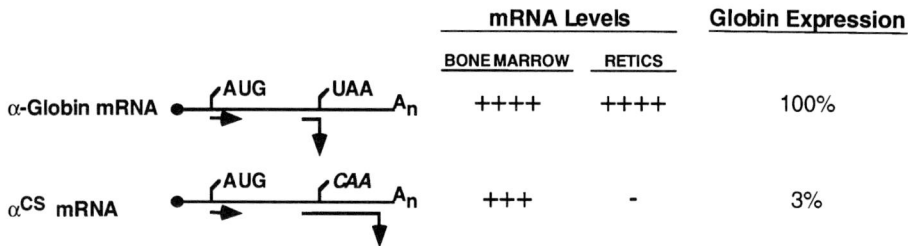

FIG. 5. $\alpha^{\text{Constant Spring}}$ mRNA is unstable *in vivo*. The human α and α^{CS} mRNAs are illustrated (as described in Fig. 4B). The two mRNAs differ by a single U→C mutation, which changes the UAA termination codon into a CAA (glutamine) codon. The sites of translation initiation (horizontal arrow) and translation termination (horizontal arrow with downward deflection) are indicated below each mRNA. Individuals who inherit a wild-type α and an α^{CS} gene express the two mRNAs at nearly equal levels in marrow erythroid cells, but at dramatically different levels in peripheral reticulocytes. This effect results from different stabilities of the two mRNAs, because peripheral reticulocytes (which lack a nucleus) are transcriptionally silent. The difference in globin protein accumulation (100% α versus 3% α^{CS}) is a direct consequence of the loss of α^{CS} mRNA stability.

cytes points to instability of the α^{CS} mRNA. The relative instability of α^{CS} mRNA has been confirmed in experiments utilizing mice containing human α^{wt} and α^{CS} transgenes (65a). Three other less common α-globin gene antitermination mutations produce a phenotype clinically identical to that of the α^{CS} mutation (66–68). Of particular interest is the observation that none of the more than 150 different nondeletional thalassemic mutations that occur at the β-globin gene locus, including several antitermination mutations, appears to destabilize the β-globin mRNA as a primary defect (69–71).

III. Structural Determinants of α-Globin mRNA Stability

A. Experimental Approach to Defining Determinants That Stabilize Human α-Globin mRNA

The observation that α^{CS} suppresses α-globin mRNA levels suggested that entry of the ribosome into the 3′ UTR might be linked mechanistically to the loss of α-globin mRNA stability. We predicted that careful study of the defect in the expression of these mRNAs might permit identification of the determinants crucial to normal α-globin mRNA stability. Several experimental systems were considered for these studies. Because destabilization of α^{CS} mRNA was likely to be linked to ribosomal entry into the 3′ UTR, α^{CS} mRNA destabilization might be poorly modeled in a translationally silent system. Hence, previously established translationally inactive systems that followed the degradation of cytoplasmic mRNAs *in vitro* were inappropriate in this instance (72). An intact erythroid cell in which the α-globin mRNA was normally expressed was felt to be a superior model system in which to study its stability.

The system that was developed to evaluate α^{CS} mRNA stability under conditions of active translation is illustrated in Fig. 6A (73). Equimolar amounts of the α^{wt} and α^{CS} genes are cotransfected into MEL cells, a transformed erythroid cell line arrested at the proerythroblast stage (74). The two genes are identical except for the single nucleotide change defining the α^{CS} mutation. Any differences in the expression of these two genes would thus be a direct consequence of the α^{CS} mutation. The two α-globin genes are transcriptionally active during the initial 18 to 24 hr posttransfection, at which time transcriptional rates fall precipitously. During the subsequent posttranscriptional "chase" period, any difference in the stability of the α^{CS}- and α^{wt}-globin mRNAs would progressively alter their relative concentrations. In this system the $\alpha^{CS}:\alpha^{wt}$ ratios falls from approximately 1.0 at 24 hr

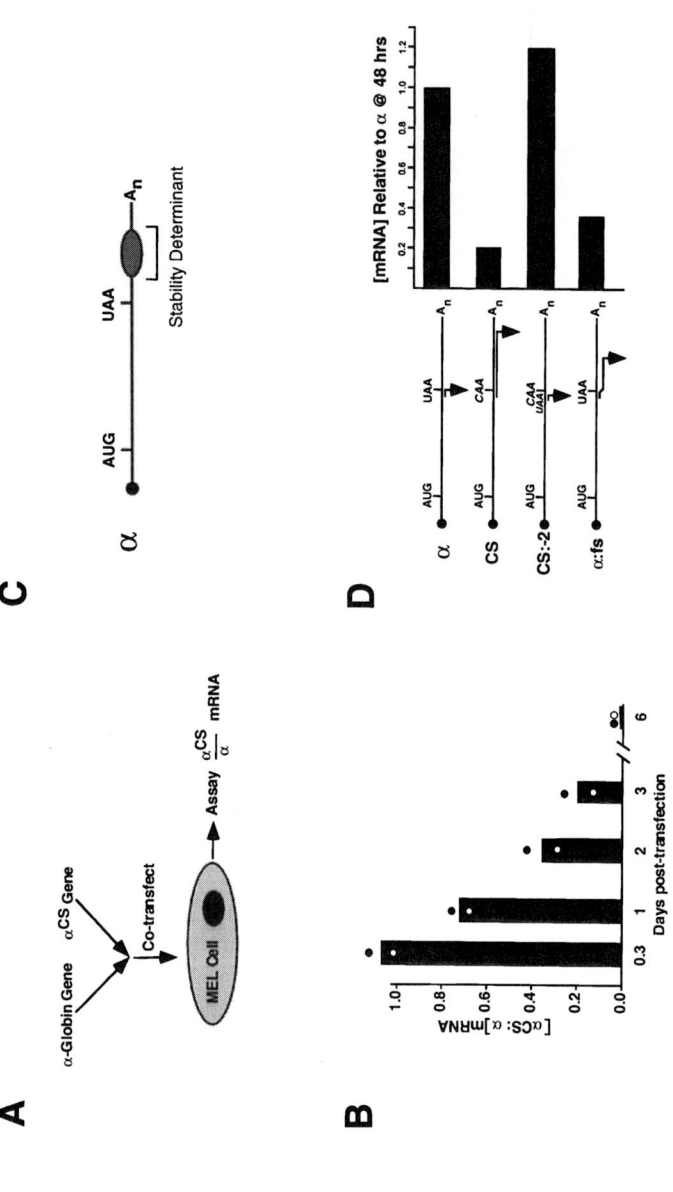

to 0.25 at 48 hr posttranscription; by 6 days posttranscription the α^{CS} mRNA is undetectable despite the continued presence of α^{wt} (Fig. 6B). This fall in the α^{CS}:α^{wt} mRNA ratio parallels the fall in this ratio observed in individuals heterozygous for the α^{CS} mutation (23). This parallel validates the MEL cell system for use in the detailed analysis of the mechanisms of CS-mediated α-globin mRNA destabilization.

B. Evidence for a Stability Determinant within the 3′ UTR

Although the CS mutation triggers accelerated α-globin mRNA decay, it was initially unclear whether this instability was a consequence of the antitermination effect of the CS mutation or was due to the the single nucleotide substitution *per se*. To discriminate between these two possibilities, site-specific mutations were introduced into the α-globin gene that reproduced three additional naturally occurring α-globin antitermination mutations (73). Each mutation destabilized α-globin mRNA to the same extent as the α^{CS} mutation. Thus, α-globin mRNA destabilization does not depend on the specific

---

FIG. 6. A transiently transfected erythroid cell line models the translation-dependent destabilizaton of α-globin mRNA by the Constant Spring mutation. (A) Schematic overview of an assay for α-globin mRNA stability. Equimolar concentrations of otherwise identical plasmids containing the α^{CS} and wild-type α genes are transiently coexpressed in mouse erythroleukemia (MEL) cells. Cells are harvested at defined intervals posttransfection and the level of α^{CS} mRNA is determined relative to the level of α mRNA. Pilot experiments indicated that the transfected genes were expressed early, but underwent transcriptional silencing between 18 and 24 hr; therefore, a fall in the α^{CS}:α level over subsequent time intervals indicates the relative instability of the α^{CS} mRNA. (B) α^{CS} mRNA is destabilized in transiently transfected MEL cells. The α^{CS}:α mRNA levels were determined in MEL cells at defined times postelectroporation as described in A. The fall in the level of α^{CS} mRNA relative to α mRNA (which can be detected up to 6 days posttransfection) confirms the known instability of α^{CS} mRNA *in vivo* and validates MEL cells as an appropriate system in which to study α^{CS} mRNA destabilization. (C) Hypothetical α-globin stability determinant located in the 3′ UTR of the α-globin mRNA. The stability determinant is postulated to contain both *cis*- and *trans*-acting components (shaded oval). The function of this determinant could be disrupted by the α^{CS} UAA→CAA mutation either through the change in primary sequence *per se*, or through the permissive effect that the antitermination mutation has upon ribosomal transit into the 3′ UTR. In the latter case, the translating ribosome would be expected to disrupt physically the stability determinant. (D) Ribosomal readthrough into the 3′ UTR destabilizes the α mRNA. The stability of three α-globin mRNA variants was determined relative to wild-type a-globin mRNA as described in A. The CS mRNA and CS:-2 mRNAs both contain a UAA→CAA antitermination mutation (italics); the CS:-2 also contains a second site, an in-frame UAA nonsense mutation immediately 5′ to the position of the native UAA termination codon. The α:fs mRNA contains a UCC→UC deletion 10 nt upstream of the native UAA termination codon that permits ribosomes to frameshift past the native UAA termination codon and read into the 3′ UTR for an additional 16 nt. Horizontal arrows with downward deflections indicate the 3′ extent of ribosomal transit (translational termination); the α:fs frameshift is indicated by a deflection in that line.

identity of the antitermination base substitution, but rather reflects their effect on translating ribosomes that continue translation into the 3′ UTR.

The above-mentioned studies suggest that the α-globin mRNA contains a stability-determining element within its 3′ UTR (Fig. 6C). This hypothesis was directly tested in two ways (Fig. 6D). In the first approach, the α^{CS} mutation was left intact and a mutation creating an in-frame termination codon was introduced two codons 5′ to the normal (destroyed) terminator. This second-site mutation stabilized the encoded α^{CS} mRNA. In a second approach, a mutation was introduced into the normal α-globin gene that permitted ribosomes to frameshift past the normal (intact) termination codon. This frameshifted α-globin mRNA was unstable. These two experiments confirmed that α^{CS} mRNA destabilization is mediated by ribosomal entry into the 3′ UTR. This mechanism suggests that the 3′ UTR contains a determinant(s) important for mRNA stability, and that physical disruption of this determinant by the translating ribosome destabilizes α^{CS} mRNA.

C. Stability Element Function

To determine whether the α-globin mRNA 3′ UTR stability element was functional in non-erythroid cells, the relative stabilities of the α^{CS} and α^{wt} mRNAs were determined in mouse fibroblasts (C127) and HeLa cells (73). In contrast to the time-dependent fall in the α^{CS}: α^{wt} mRNA ratio observed in erythroid MEL cells, the ratio remained constant over a prolonged posttransfection interval. Thus the stabilizing function of the α-globin mRNA 3′ UTR differs for eythroid and nonerythroid cells. Two mechanisms could account for this observed difference. First, the α-globin mRNA 3′ UTR stability determinant might function in an erythroid-specific manner. In this case the α^{wt} and α^{CS} mRNAs would be equally unstable in nonerythroid cells. Second, if α-globin mRNAs were not translated in nonerythroid cells, there would be no translation-dependent destabilization of α^{CS} mRNA, and hence the two mRNAs would both be stable. These two possibilities were explored by assessing the absolute stability of α-globin mRNA and its translational activity in nonerythroid cells (73). Direct measurement of α-globin mRNA $t_{1/2}$ in nonerythroid C127 cells was carried out by introducing the α^{wt}-globin gene under the control of a serum-inducible *fos* promoter (75). Serum induction of the stably integrated chimeric genes resulted in a limited burst of α-globin mRNA transcription. The α-globin mRNA was shown to be translationally active in nonerythorid COS-1 cells by demonstrating its presence on polysomes. Based on its slow rate of decay, α-globin mRNA appeared to be highly stable in these cells, with a $t_{1/2}$ in excess of 24 hr. Thus the α-globin mRNA appears to be stable in nonerythroid cells and, in distinction to the situation in erythroid cells, this stability is not dependent on the 3′ UTR stability determinant. A possible model to explain these observations is that the stability determinant protects α-globin mRNA against an erythroid cell-specific destabilizing factor (such as an endo- and exonuclease). In this way a functional 3' UTR stability element would not be required to maintain α-globin mRNA stability in nonerythroid cells where such factors were not present. Such erythroid cell-specific destabilizing factors have yet to be identified.

D. The Stability Element Maps to a Series of C-Rich Pyrimidine Tracks in the 3' UTR

The precise location of the 3' UTR stability element was mapped using two functionally independent approaches (73, 75). Based on the observation that a translating 80S ribosome could physically disrupt the stability element, a "ribosome interference" assay was devised in which mutations encoding in-frame translation-termination codons were introduced at different sites within the α^{CS} gene. The destabilizing effects of readthrough ribosomes that terminate at each of these new termination sites could be used to map the 5' border of the stability element. Entry of the ribosome 1 codon into the 3' UTR had no adverse effect, whereas entry of four or more codons into this region destabilized the mRNA. Because the leading edge of the 80S ribosome extends approximately six codons from its center (76), the 5' border of the *cis* component of the stability element maps between codons 7 and 10 of the α-globin mRNA 3' UTR.

The internal anatomy and 3' border of the stability element were mapped by "linker scanning" (75). Thirteen sets of base substitutions were introduced into the α^{wt} gene between the normal termination codon and the poly(A) addition site. Each set of substitutions created a *Hind*III site that could be used for detection and quantitation of the specific mRNA. The normal translation termination codon was left intact to prevent ribosomal readthrough into the 3' UTR. The mutant genes were each cotransfected into MEL cells with an equimolar amount of the α^{wt} gene, and the relative levels of the two mRNAs were determined 48 hours posttransfection (Fig. 7). Most mutations had minimal or no effect on mRNA levels; however, mutations at 3'UTR codon positions 9, 10, 13, 19, and 21 destabilized α-globin mRNA to levels observed with the α^{CS} mutation. These mutations clustered in three discontinuous segments, encompassing codons 9 through 23 of the α-globin mRNA 3' UTR. Inspection of the three identified segments disclosed that they are pyrimidine rich and contain one or more repeats of the sequence CCUCC. The most 5' destabilizing mutations, beginning at codon 9, overlap with the 5' border of the stability element predicted by the ribosome interference assay. Hence, data from the ribosome interference and linker scanning assays both support

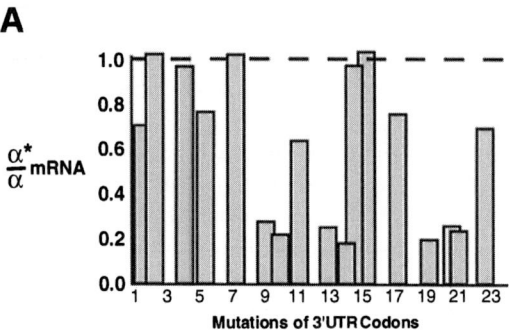

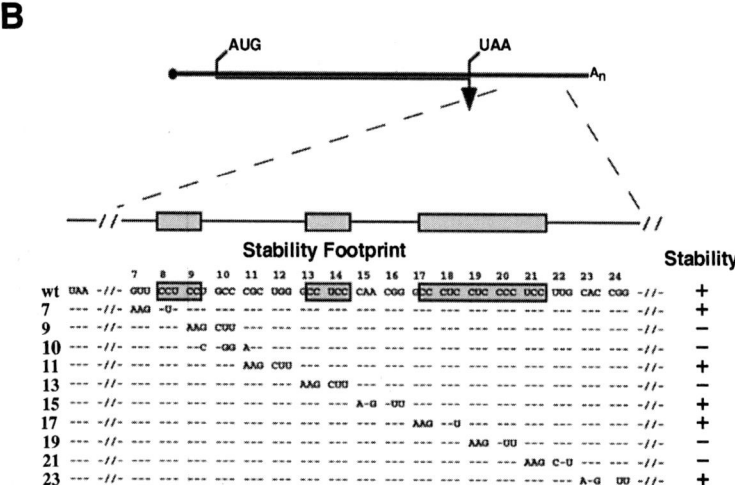

Fig. 7. Linker scanning mutagenesis identifies three noncontiguous segments of the α-globin 3′ UTR important to its stability. (A) Effects of clustered base substitutions on α-globin mRNA stability. Globin genes containing defined clusters of base substitutions were transiently cotransfected into MEL cells with an equimolar concentration of the wild-type α-globin gene. Levels of the mutant (α*) mRNA were measured relative to levels of α mRNA at 48 hr and plotted as a histogram. The dashed horizontal line represents the stability of the wild-type α-globin mRNA. The α* mRNAs are identified by the position of their nucleotide substitutions (in codons) relative to the native UAA termination codon. For example, mRNA 13 contains substitutions that cluster around codon 13 in the α-globin mRNA 3′ UTR. (B) Partial 3′ UTR sequences of informative α* variants. The sequence of the wild-type α-globin mRNA 3′ UTR is shown on the first line, and the sequence of informative clusters of base substitutions are on subsequent lines. These mRNAs are identified in A. The numbers at the top indicate the position of residues (in codons) relative to the native translation termination site. The normal (+) or decreased (−) stability of α* mRNAs determined in A is reiterated to the right of the corresponding 3′ UTR sequence. A stability footprint constructed from the data in A indicates three pyrimidine-rich tracks of the α-globin mRNA 3′ UTR that are likely to be important to RNA stability (shaded boxes). An expanded diagram of the α-globin mRNA 3′ UTR is included above the sequences to position the three pyrimidine-rich tracks relative to the full-length mRNA.

the existence of a *cis*-acting stability determinant within the α-globin mRNA 3' UTR and agree upon its 5' border.

E. Role of RNP Formation in Stabilization of Human α-Globin mRNA

Although useful in defining cis components of a stability-determining element, the ribosome interference and linker scanning assays do not address whether the polypyrimidine segments function directly or whether they mediate their effect via interaction with trans-acting factors. The association of trans-acting factors with this mRNA might result in a functional ribonucleoprotein (RNP) complex. To detect such a complex, a gel mobility-shift assay was performed. The α-globin 3' UTR was internally ^{32}P labeled by *in vitro* transcription, incubated with MEL cell cytosolic S100 extract, digested with a combination of RNases A and T1, and electrophoresed on a nondenaturing polyacrylamide gel (34). As seen in Fig. 8A, a segment of the α-3' UTR was protected from RNase digestion by protein factors present in the S100 extract. The sequence specificity of the RNase-resistant "α complex" was demonstrated by the ability of unlabeled excess α-globin 3' UTR to compete for cytosolic factors making up the RNP complex, and by demonstrating that the α complex did not assemble on a variety of other nonrelated RNA probes.

The functional importance of the α complex was established by determining whether there was a parallel between the effects of mutations on the stability of α-globin mRNA *in vivo* and their effects on complex assembly *in vitro* (Fig. 8B). The correlation between α-complex assembly and mRNA stability is exact: mutations within the 3' UTR that do not affect α-globin mRNA stability permit assembly of a normal α complex. In contrast, the five linker mutations that destabilize the mRNA in transfected cells (mutations at 3' UTR codons 9, 10, 13, 19, and 21) block assembly of a normal α complex. This correlation between structure and function suggests that the α complex is a major determinant of α-globin mRNA stability.

The three discontinuous pyrimidine-rich segments comprising the *cis* stability determinant contain a total of five repeats of the CCUCC motif. A single copy of this motif is present in each of the first two segments, and three overlapping motifs are found in the third and most extensive track. This suggested that these C-rich elements might comprise recognition sites for one or more RNA-binding proteins, and was tested by repeating the gel mobility-shift assay using each of four homoribopolymers as unlabeled competitor (Fig. 8C). Only poly(C) was effective in inhibiting assembly of the α complex, suggesting that one (or more) of the protein components in the α complex exhibits significant poly(C) binding activity.

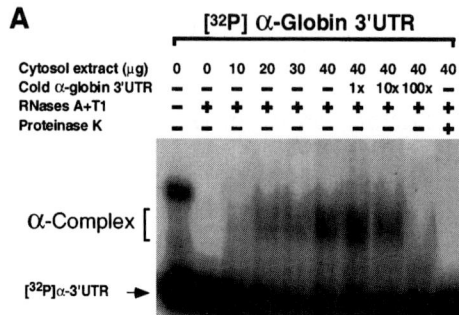
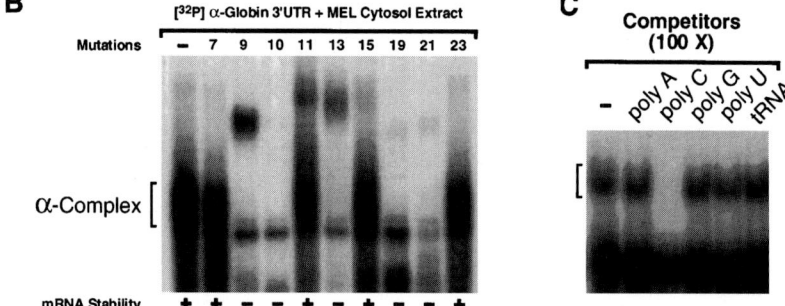

FIG. 8. A sequence-specific RNP complex (α complex) assembles on the 3′ UTR of the human α-globin mRNA. (A) Detection of the α complex. Internally ^{32}P-labeled α-globin mRNA 3′ UTR was incubated with increasing amounts of MEL cell (S100) cytosolic extract, and subsequently digested with RNases A and T1. RNase-resistant RNP complexes were resolved on a nondenaturing polyacrylamide gel and autoradiographs exposed. The position of a complex (α complex), which is inhibited by excess unlabeled α-mRNA 3′ UTR and is fully sensitive to proteinase K digestion, is indicated. (B) The α complex formation *in vitro* correlates with the stabilities of α-globin mRNA variants *in vivo*. The ^{32}P-labeled 3′ UTRs of α-globin mRNAs containing clusters of nucleotide substitutions (see Fig. 7) were tested for their ability to assemble the α complex. The position of the α-complex is indicated to the left. The stability of the mRNAs containing each of these mutations, as determined by the linker scanning assay (Fig. 7A), is referenced below the appropriate lane. (C) The α complex is highly sequence specific. Assembly of an α complex on the 3′ UTR of wild-type α-globin mRNA is resistant to competition by high levels of poly(A), poly(G), poly(U), and tRNA, as well as the 3′ UTRs from β-globin, ζ-globin, human growth hormone, and other mRNAs (not shown). However, a 100× excess of unlabeled poly(C) effectively competes for the α complex. The position of the α complex is indicated by a bracket.

F. Identification of Trans-Acting Factors Involved in Human α-Globin mRNA Stability

The protein constituents of the α complex were more fully defined by a series of RNA affinity chromatography studies (34). The 3' UTR of the α-globin mRNA was bound to poly(U)-linked Sepharose beads through base pairing to its A_{30} tail. The beads were incubated in S100 extracts prepared from MEL cells metabolically labeled with [^{35}S]methionine (Fig. 9). The beads were extensively washed, and the bound proteins were eluted and visualized by SDS-PAGE. Three proteins that bound specifically to the poly(U)/α-3' UTR beads were identified (39, 44, and 46 kDa). The chromatography was performed in parallel with poly(C)-linked Sepharose beads, and these bound only the 39-kDa protein. The poly(C) binding activity of the identified proteins was assessed by transferring them to nitrocellulose and probing the membrane with [^{32}P]poly(C) ("Northwestern analysis"). A single 39-kDa band was visualized in both the poly(C) and the poly(U)/α-3' UTR eluates. This result suggested that the 39-kDa constituent of the α complex exhibited specificity for C-rich sequences, subsequently termed the *polyC binding* activity (PCB).

Functional assessment of the 39-kDa protein demonstrated that although it was a necessary component of the α complex, additional factors were required for complex assembly (34). To assess its importance to α-complex assembly, the 39-kDa protein was purified (77). Incubation of the purified protein with the α-globin mRNA 3' UTR did not result in assembly of an α complex. MEL cell extracts depleted of the 39-kDa protein by repeated adsorptions on poly(C)-linked Sepharose beads were also unable to assemble an α complex on the α-globin mRNA 3' UTR. However, when biochemically purified 39-kDa protein was added to the depleted MEL cell extract, α-complex assembly was restored. Thus the 39-kDa poly(C) binding protein appears to be essential to, but not sufficient for, α-complex assembly.

G. Identification of a 39-kDa Poly(C) Binding Protein, αCP

The 39-kDa protein was digested with trypsin and three peptide fragments were harvested by high-performance liquid chromatography (HPLC) and microsequenced (77). A homology search of protein sequence databanks revealed that the first and third peptide fragments were each identical to translated segments of two anonymous cDNA clones. The second peptide fragment comprised a mixture of two distinct fragments. The sequence of each of these fragments matched one or the other of the original anonymous proteins. The two proteins, illustrated in Fig. 10, were named αCP-1 and -2 (α-complex protein; the genetic loci were named *αCP-1* and *αCP-2*, respectively) (77). The protein sequences of αCP-1 and -2 are 80% identical; each

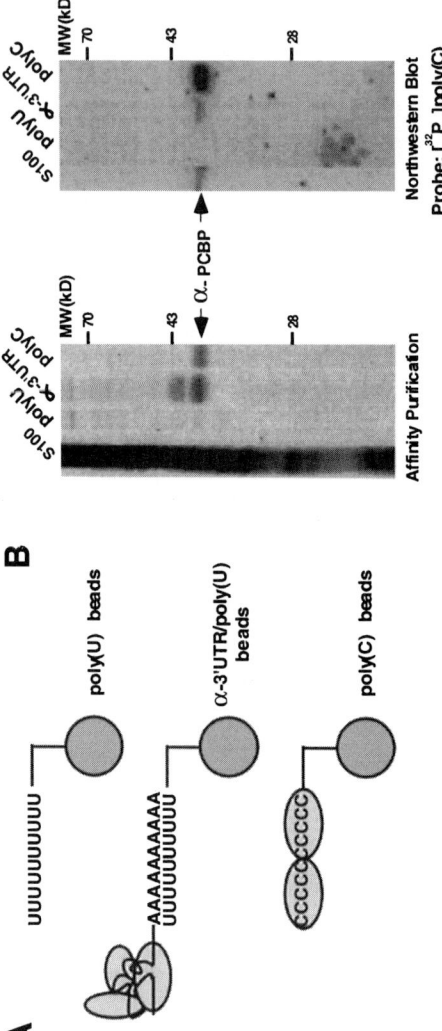

FIG. 9. Protein components of the α complex can be identified by RNA affinity chromatography. (A) Schematic of 3' UTR affinity chromatography. Synthetic poly(U) was linked to Sepharose beads (shaded circles). The α-globin 3' UTR sequences transcribed *in vitro* from a polymerase chain reaction-generated cDNA template were hybridized via the poly(A) tail to the poly(U)–Sepharose beads. The beads were subsequently incubated with a ^{35}S-labeled MEL cell S100 extract. Components of the α complex (shaded ovals) that bound to this matrix were eluted and resolved by SDS-PAGE and visualized by autoradiography (see B). The extract was also incubated with poly(C)–Sepharose beads to directly isolate poly(C) binding activity (horizontal shaded oval). The purification was carried out in parallel with unhybridized poly(U)–Sepharose to control for nonspecific protein binding. (B) Analysis of proteins enriched by affinity chromatography. Left panel: SDS-PAGE and autoradiography of proteins eluted from an unpurified metabolically ^{35}S-labeled MEL cell S100 extract and from each of the three affinity resins in A. Right panel: Northwestern analysis of a parallel set of unlabeled affinity purified proteins and S100 extract probed with [^{32}P]poly(C). Size markers and the position of a 39-kDa protein with poly(C) binding activity (PCBP; αCP; are noted between the two autoradiographs.

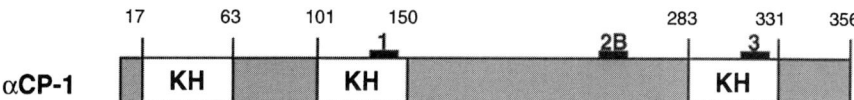

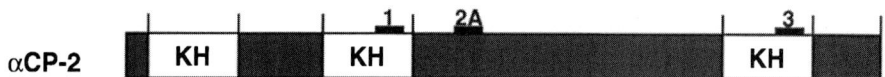

FIG. 10. PCB activity is encoded by two related proteins which containing multiple RNA binding domains. PCB activity purified by affinity chromatography was found to reside in two proteins that share 80% sequence identity: αCP-1 and -2. Each protein contains three highly conserved repeats of the KH-type RNA binding motif first described in hnRNP K (K-homology, or KH, domain; open boxes). The four peptide fragments, purified from a tryptic digest of the PCB protein and microsequenced, were used in a homology search of protein sequence databases and are illustrated as heavy lines (1, 2A, 2B, and 3). The sizes of αCP-1 and -2 are indicated (in amino acid residues) above each protein.

contains three KH domain repeats. The KH (K-homology) domain is a 45-residue sequence first identified in the hnRNP K protein (78) and later found in a number of proteins with RNA binding activity, including FMR1, which is encoded by the fragile-X mental retardation locus (79, 80). Both αCP-1 and αCP-2 contain a second conserved RNA binding motif, the RGG box. Thus, two closely related proteins that contain at least two structural motifs characteristic of a subfamily of RNA binding proteins appear to account for poly(C) binding activity characteristic of the α complex. We next asked whether both of these proteins could be assembled in the α complex.

The incorporation of αCP-1 and αCP -2 proteins into the α complex *in vivo* was directly demonstrated using coimmunoprecipitation studies (77). αCP-1 and αCP-2 cDNAs were individually fused to *myc*-epitope tags, cloned into expression vectors, and transfected into U293 cells. S100 extracts prepared from the transfected cells were incubated with *in vitro*-transcribed [^{32}P]α-3′ UTR. The components of the incubation were then immunoprecipitated with an anti-*myc* monoclonal antibody (MoAb). Labeled RNA was coprecipitated by anti-*myc* MoAb, suggesting that it was bound by the recombinant αCP-1 or αCP-2 in an α complex. As a control for the specificity of this approach, a parallel experiment was carried out using a [^{32}P]α-3′ UTR probe that contained a mutation from the linker scanning assay that destabilized α-globin mRNA and inhibited α-complex formation. The mutant mRNA was not precipitated by the anti-*myc* MoAb, confirming the specificity of the assay. These data demonstrated that αCP-1 and αCP -2 are both constituents of the α complex. Whether the two proteins are functionally unique or must be present simultaneously for assembly of an α complex is not yet known.

H. Deadenylation of Stable and Unstable Human α-Globin mRNAs

Although *cis* and *trans* components of the stability-determining α complex have been identified, the mechanism(s) through which the complex mediates α-globin mRNA stability remains unclear. In yeast, deadenylation of the poly(A) tail appears to precede the degradation of certain mRNAs (*9–12*). A link between mRNA degradation and its deadenylation is generally believed to exist in higher organisms as well, although specific mechanisms have been difficult to delineate. The generation of mice transgenic for the human $α^{wt}$ and $α^{CS}$ genes permitted evaluation of the linkage between poly(A) tail length and degradation of α-globin mRNA during erythroid differentiation. This mouse model system accurately recapitulates the destabilization of human α-globin mRNA by the CS mutation (*65a*). Total RNA was harvested from the bone marrow erythroid cells of mice carrying the wild-type α-globin gene and mice carrying the CS gene. The lengths of the poly(A) tails of both mRNAs clustered at discrete intervals of 20–25 adenylate residues. This periodicity corresponds to the length of the poly(A) tail thought to be protected by binding of a single molecule of poly(A) binding protein (PABP) (*81*). However, the poly(A) tails of $α^{CS}$ mRNA were notably shorter that those of the wild-type α-globin mRNA. These results suggest a linkage between destabilization of α-globin mRNA and phased shortening of its poly(A) tail (*65a*). The phased pattern of poly(A) tail lengths might be expected if destabilization of the $α^{CS}$ mRNA proceeded *via* accelerated removal of individual PABP molecules from its poly(A) tail with subsequent hydrolysis of the exposed adenylate residues. The mechanism by which the α complex establishes or preserves poly(A) tail length, perhaps through physical interaction with PABP, remains undefined.

I. Structural Basis of α-Globin mRNA Stability: A Model

Based on the above data, we have constructed a model for the regulation of α-globin mRNA stability (Fig. 11). The stability of the human α-globin mRNA is determined by the assembly of an RNP complex at the sites(s) of polypyrimidine-rich tracks within the α-globin mRNA 3′ UTR. This α complex comprises at least three distinct proteins, one of which (αCP) recognizes and binds cytidine-rich RNA sequences. The α-complex protects the α-

Stabilized α-Globin mRNA

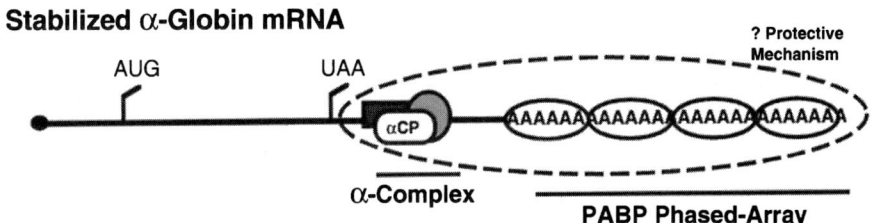

Destabilized α-Globin mRNA

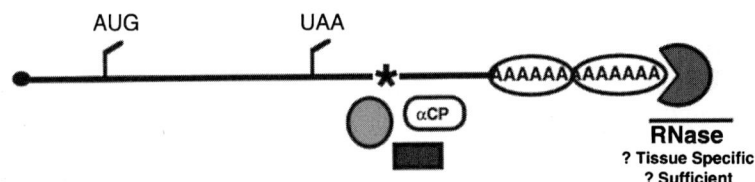

FIG. 11. Model of mRNA stabilization by the α complex. Under normal conditions the α complex, composed of αCP and other trans-acting factors, assembles on conserved pyrimidine-rich tracts within the 3' UTR of the α-globin mRNA. The complex confers stability to the α-globin mRNA through an undetermined mechanism. The α-globin mRNA can be destabilized either by ribosomal readthrough with subsequent physical disruption of the α complex, or through the introduction of site-specific mutations (indicated by an asterisk) that directly interfere with α-complex function. The failure to assemble a functional α complex is believed to target the α-globin mRNA for rapid degradation. The link between destabilization of the $α^{CS}$ mRNA and its shortened poly(A) tail suggests that deadenylation may be an initiating or rate-limiting step in the decay pathway, although a mechanistic link between destabilization and deadenylation has not been firmly established. At least one of the RNases that initiates decay is likely to be erythroid cell specific. The role played by PABP, which appear to bind the poly(A) tail in a phased array (open ovals), is also unclear. The cap (•), translation initiation and termination sites (angled line), and poly(A) tail (AAAAA) are indicated.

globin mRNA from degradation through an unknown mechanism; its function can be disrupted through ribosomal interference or by specific point mutations within the polypyrimidine tracks. In the case of $α^{CS}$ mRNA, destabilization by ribosomal interference appears to accelerate shortening of its poly(A) tail, suggesting a link between destabilization and deadenylation. Although the specific RNases that mediate α-globin mRNA decay have not been identified, it is likely that they are restricted to cells of the erythroid lineage, because disruption of α-complex assembly in nonerythroid cells does not adversely affect mRNA stability.

IV. Stability of Other Globin mRNAs

A. Determinants of α-Globin mRNA Stability Are Highly Conserved

The general mechanisms that control globin gene expression in a variety of vertebrates have been remarkably conserved during evolution. In all vertebrates studied to date, these mechanisms ensure balanced expression of the α-like and β-like globin genes. The organization of the globin gene clusters, structures of individual globin genes, and general mechanisms of transcriptional control maintain an overall similarity from species to species. The identification of *cis* determinants crucial to maintaining human α-globin mRNA stability permits the investigation of whether these elements are also conserved. The 3' UTRs from representative vertebrate α-globin mRNAs are aligned in Fig. 12. The overall conservation of polypyrimidine tracks in this region suggests that they mediate an important facet of globin gene expression. Our studies of the human α-globin mRNA have determined that this function is likely related to mRNA stabilization. All three of the polypyrimidine tracks identified in the human α-globin mRNA are maintained in other primates and in rodents. In more distantly related mammals, the first two polypyrimidine tracks are lost as recognizable entities whereas the third segment, which is also the most extensive, is maintained. The loss of the two shorter tracks during mammalian evolution suggests that they may be only marginally important to mRNA stability; in contrast, conservation of the third polypyrimidine track predicts its importance to this process. The loss of all three polypyrimidine tracks in avians is not surprising when one considers that avian erythroid cells remain transcriptionally active to a late stage of differentiation and retain their nucleus as mature circulating erythrocytes. Persistent globin gene transcription in the erythrocytes of these animals may obviate the need for selective stabilization of globin mRNAs to achieve high-level globin gene expression. Such a possibility is speculative, but would be of some interest to prove.

B. Parallel Evolution of Cis- and Trans-Acting Components of the Stability-Determining Complex from Mouse to Humans

We tested whether the evolutionary conservation of the polypyrimidine tracks reflects conservation of stability complex function by determining the importance of this region to the stability of murine (m) α-globin mRNA (82). The ability of the murine 3' UTR to form an RNP complex was determined by incubating ^{32}P-labeled mα-globin 3' UTR with cytosolic extract from

HUMAN	...TAAGCTGGAGCCTCGGTAGCCGTT	CTTCTGCCG	GCTGGC	CTTCCC	ACGGG..CCCTTCTGTCTTTG	AATAAAGTCTGAGTGGCGGC
ORANGUTAN	...TAAGCTGGAGACTCGGTAGCCGTT	CTTCTGCCG	GCTGGG	CTCCC	AAAGG..CCCTTCCGTCTTTG	AATAAAGTCTGAGCGGGCG
RHESUS	...TAAGCTGGAGCCTCGGTGGCCATG	GTT TTGCG	TTGGG	CTT.CC	CCAGG..CCCTTCCGTGTCTTTG	AATAAAGTCTGAGTGGCGGC
CHIMP	...TAAGCTGGAGCCTCGGTGGCCATG	GTT TTGCG	CTTGGG	CTTC	CCAGG..CCTCTCCCTT.TCC	CCACCTGTACCCCCGTGTCTTTG	AATAAAGTCTGAGTGGCGGC
MOUSE	....	TAAGCTG	GGGCTTG	GCCATG	GCACCTGTACC..TCTTGGTCTTTG	AATAAAGCCTGAGTAGGAAGAG
RAT	....	TAAGCCG	CT CTCACAG	GGGCTTT	GGCATC AACCAGG	GCACCTATACC..TCTTGGTCTTTG	AATAAAGCCTGAGTAGGAAGC
GOAT	....	TAAGCTGGAGCCTCGGCCACCCTACCCTGGCTGGAG	GCCCTTT	GCCCTC	GGGGACTCTACC..TCCTGATCTTTG	AATAAAGTCTGAGTGGGCTGCA
SHEEP	....	TAAGCTGGAGCCTCGGGGACCCCTACCCTGGCCTGGAG	GCCCCTTT	GGGCTC	GGGGGCTCTACC..TCCTGATCTTTG	AATAAAGTCTGAGTGGGCTGC
HORSE	....	TAAGCTGGAGCCACGGGGATCCCTGCCCCGCCGGG	......	....	GGTGCCCGCACT.TCCCTATCTTTG	AATAAAGTCTGA.TGGGCTGCATG
RABBIT	....	TAAGCTGGAGCCTGGGAG.........	CCGGCCTGCCC	......	CCCCCA	GCCGTCTCTCTGCAT.TCCCCCA	GCCCCACCTGTCTCTTTT	AATAAAGTCTGAGTGAGTGGCA
CHICKEN	TAA..GACGGCACGTGGCTAGAGCT	GGGGCCAACCCATGC	......	GCCGCC	GACAGGGAGCAGCC.AAATGAGATGA	AATAAATCTGTTGCATTTGTGCTCC
DUCK	TAG...ACGGCACCCTGGCTAGAGCT	GGACCCACCCTGTTG	......	CAGCTT	AACTGGGAGCAGCC.AAATGATCTGA	AATAAATCTGTTGCATTTGTGCTCC
PIGEON	....	TAAGATGCGGCCACCATGCCTAGAGCT.GGACACAACCTGCTCC.AGCCTG	AGCGAGCAACC.AAATGATCTGA	AATAAATCTTTCATTTGTGCTCC
SALMON	TAA......GACCATCATGAA...GTCCAAAATTGGACTC	AGTTCC	GCTCTGTTATC..ACA........	AATAAACAGGCAATGAATG
XENOPUS	TAAGGCTCAGCAACACAGCAGCCAGAAGTCTCAACA....TCAGACATCGTTAATTATATGC........	AATCAAACTGACAAGCTT

FIG. 12. Cis-elements important to α-globin stability are conserved in evolution. The α-globin mRNA 3' UTRs from diverse species are presented using gaps (···) to permit optimal alignment. Highly conserved pyrimidine-rich tracks, which were initially identified in human α-globin mRNA using the linker scanning assay (Fig. 7), are highlighted in gray. The poly(A) hexanucleotide signal (AAUAAA) is boxed.

MEL cells. As is the case for the human (h) α-3′ UTR, the mα-3′ UTR assembles a sequence-specific RNP complex. However, the human and mouse RNP complexes display different mobilities (Fig. 13A). This distinction suggests that there may be one or more differences in the proteins that comprise the hα and mα complexes. Although the overall structures of the polypyrimidine tracks of the human and murine 3′ UTRs are similar, the human sequences have a higher cytosine:uridine ratio than do the mouse sequences (Fig. 13B). The repeated human CCUCC motifs appear as CUUCC motifs in the mouse. The demonstration that the mα complex was resistant to competition by excess unlabeled poly(C) is a functional consequence of this difference. Thus, poly(C) binding activity, a defining characteristic of the human α complex, is not necessary for mα-complex assembly. The prediction that the mα complex might be successfully competed by excess unlabeled poly(CU) was subsequently tested and confirmed.

To determine whether the mouse α-3′ UTR polypyrimidine tracks, like their human counterparts, are important to maintaining mRNA stability, the destabilizing effects of mutations in these segments of the mouse mRNA were assessed (Fig. 13C) (82). Although mutations within the second polypyrimidine track did not adversely affect mRNA stability, clustered base substitutions in the third track resulted in major destabilization of the mα-globin mRNA. As in humans, the destabilizing effects of these base substitutions paralleled their ability to block the assembly of the mα complex. Thus, assembly of a 3′ UTR mRNP complex in the mouse, as in humans, appears to be linked to stabilization of α-globin mRNA.

Further study of the mα complex by Northwestern analysis indicated that poly(CU) binding activity associates with a 48-kDa cytosolic protein [poly(CU) binding protein, or CUBP] (Fig 14). Unlike the hα complex, the mα complex does not appear to contain αCP-1 or αCP-2. Thus αCP and CUBP appear to be polypyrimidine-binding components that are specific to the hα and mα complexes, respectively. The identity of the additional components necessary for hα- and mα-complex assembly are still not known, although the hα complex is likely to require the 44- and 46-kDa protein components that were identified by affinity chromatography (34). Whether the mα and hα complexes share one or more of these or other still uncharacterized factors is unknown. Thus, the mode of evolution of the α-RNP complex from mouse to human is quite unusual; a minor shift in the base composition of the polypyrimidine track from CU rich to C rich has resulted in the parallel substitution of one polypyrimidine binding protein (αCP) for another (CUBP) (82);. The two related yet physically distinct α-complexes appear to serve an identical function in mRNA stabilization, perhaps through the same mechanism.

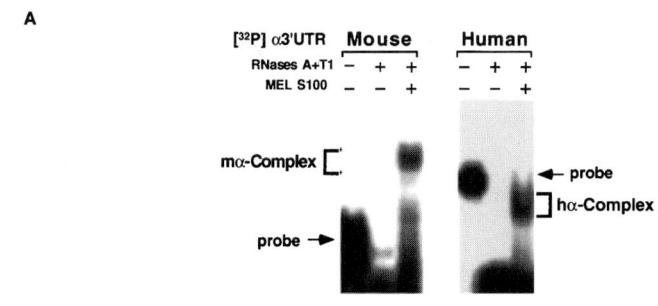

FIG. 13. Humans and mice have evolved differences in the cis- and trans-acting components of the α-globin mRNP stability complex. (A) The mouse and human α-globin mRNA 3′ UTRs assemble different mRNP complexes. Gel mobility-shift assays were carried out using either ^{32}P-labeled mouse or human α-globin mRNA 3′ UTR in MEL cell S100 extract. The positions of the mouse and human α complexes, which migrate differently, are bracketed; free probe is indicated by an arrow. (B) Comparison of the primary sequence of human and mouse α-globin mRNA 3′ UTRs. The two sequences are aligned and gaps introduced to maximize local homology. Three conserved pyrimidine-rich tracks are highlighted in gray. The five human CCUCC motifs are underlined; the corresponding three mouse CCUUC sequences are double underlined. The UAA termination codon is boxed. The 3′-terminal 38 and 40 nucleotides of the human and mouse sequences, respectively, have been omitted for clarity. (C) Parallel effects of mutations in the mα-globin 3′ UTR on complex formation and mRNA stability. Five sets of base substitutions were introduced into the mouse α-globin mRNA 3′ UTR (heavy lines). The effects of these substitutions on mα complex formation and mRNA stability were determined by gel mobility-shift analysis and in vivo analysis, respectively. Positive (+) and negative effects (−) of the mutations are indicated.

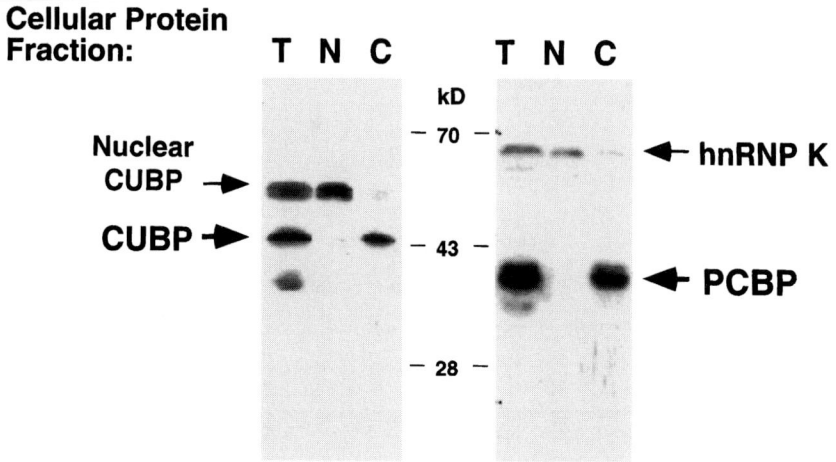

FIG. 14. CUBP and PCBP (αCP) are distinct polypyrimidine-binding proteins (82). Fractionated MEL cell extracts were resolved in duplicate by SDS-PAGE, transferred to nylon, and probed with [^{32}P]oligo(CT) (left) or [^{32}P]oligo(C) (right). The positions of the CUBP, αCP, and hnRNP K proteins are indicated. The molecular weight standards are positional between the two autoradiographs. Fractions: T, = total; N = nuclear; C,= cytoplasmic.

C. Human β-Globin mRNAs Are Stabilized by One or More Elements within the 3′ UTR

Normal erythroid differentiation and function depends on balanced synthesis of α- and β-globin proteins. The α- and β-globin genes derive from a common ancestral gene (83), and share both general and specific structural features. Based upon these similarities, it would be reasonable to predict that the α- and β-globin mRNAs also share functional and structural determinants of stability. In addition, it is well established that the α- and β-globin mRNAs are equally stable in human erythroid cells. To screen the β-globin mRNA 3′ UTR for the presence of a stability element comparable to the α-3′ UTR element, the destabilizing effect of ribosomal entry into the β-globin mRNA was assessed (84). Unlike the α^{CS} mRNA, simple antitermination mutations in the β-globin mRNA permit ribosomes to read only 21 codons into the 3′ UTR before they encounter an in-frame termination codon. To permit ribosomes to translate as far as the poly(A) hexanucleotide signal, as they do in the destabilized α^{CS} mRNA, a β-globin gene was constructed containing two mutations: a -1 frameshift mutation 2 codons 5′ to the normal termination codon and an in-frame antitermination mutation 10/11 codons into the 3′ UTR (Fig. 15A). The combination of these two mutations permits ribosomes to translate 37 codons into the β-globin mRNA 3′ UTR where, as in

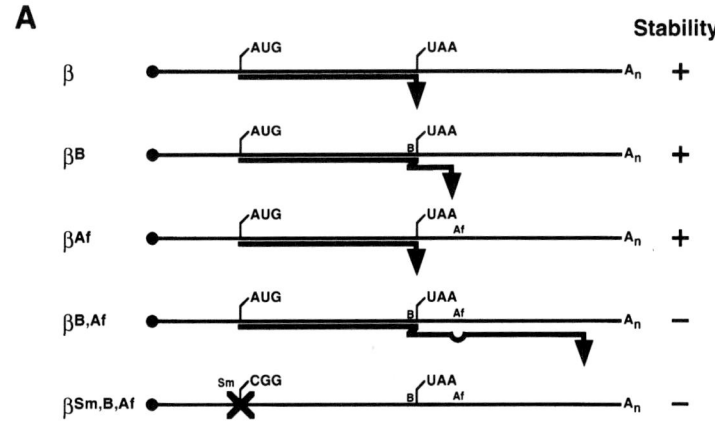

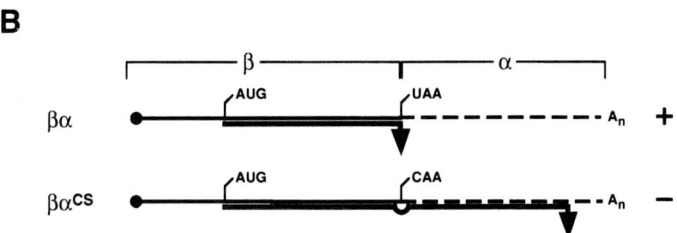

FIG. 15. The stability of the human β-globin mRNA is dependent on structural determinants positioned within its 3' UTR (84). (A) Human β-globin mRNA is destabilized by two tandem mutations. Plasmids containing each of four human β-globin gene variants were transiently cotransfected into MEL cells with a control plasmid containing the wild-type β-globin gene (β). The stability of the variant β-globin mRNAs was subsequently determined relative to the stability of the β control. Variant mRNAs contained one or more of the following mutations: mutation B is a 2-bp deletion located two codons 5' to the termination codon, which results in a −1 frameshift; mutation Af is an antitermination mutation in the −1 frame located 10/11 codons into the 3' UTR; mutation Sm is an AUG→CCG mutation at the initiation codon. Horizontal arrows beneath each mRNA indicate the open translational reading frame (vertical inflection, translation termination; backslash, frameshift; depression, antitermination mutation; X, block in translation initiation). The stability of each mRNA is indicated (+, stable; −, unstable). The cap (•), translation initiation and termination sites (tick marks), and poly(A) tail (A_n) are indicated. (B) The α-complex function can be transferred to the β-globin mRNA. Chimeric globin genes were transiently expressed in MEL cells and the stability of their mRNAs determined. The genes in both cases encoded the human β-globin 5' UTR and coding region (solid line) linked to the α-globin 3' UTR (dashed line). The $βα^{CS}$ gene contained an $α^{CS}$-like antitermination codon which permits ribosomal readthrough into the 3' UTR in a manner identical to that of authentic $α^{CS}$ mRNA. The stability of the two mRNAs is indicated by a +, stable; −, unstable). Details of the mRNA structures and of translational open reading frames (bold lines) are as detailed in A.

the α^{CS} mRNA, translation is terminated by an in-frame nonsense codon within the 5' end of the poly(A) hexanucleotide signal (AAUAAA). The mutant β-globin mRNA was substantially less stable than β^{wt} mRNA in transfected MEL cells. Thus, as with α^{CS}-globin mRNA, readthrough into the β-globin mRNA 3' UTR destabilizes the mRNA. This suggested the presence of one or more stability elements within the β-globin 3' UTR.

The functional equivalence of the α- and β-globin 3' UTRs to maintaining mRNA stability was tested by assessing whether they provided the same level of stability to a linked β-globin coding region (Fig. 15B). The data suggest that the α- and β-globin 3' UTRs are equally important in determining the stability of their cognate coding regions, and that both the α and β stability elements are sensitive to disruption by readthrough ribosomes.

D. Human α- and β-Globin mRNAs Are Stabilized by Structurally and Mechanistically Distinct Elements

Although they appear to function in a similar manner, the structures of the α- and β-globin 3' UTR stability elements differ in several important respects. The β-globin 3' UTR does not contain extensive polypyrimidine tracks, nor does it support assembly of an RNP complex comparable to the α complex (our unpublished data). The extent of readthrough necessary for mRNA destabilization also differs for the two mRNAs (84). Whereas the α-globin mRNA is destabilized by ribosomal readthrough for as few as 4 codons into the 3' UTR, destabilization of the β-globin mRNA requires ribosomal readthrough for greater than 10 codons. These observations suggest that the corresponding α- and β-globin mRNA stability elements are structurally distinct and are positioned differently in their respective 3' UTRs.

The *cis* and *trans* components that determine β-globin mRNA stability are less well defined compared to the corresponding elements of the α-globin mRNA. To destabilize β-globin mRNA it is necessary to introduce two tandem mutations; neither mutation destabilizes the mRNA when introduced alone. The fact that mRNA stability is destroyed only when two mutations separated by 35 nucleotides are present in *cis* suggests that the β-globin mRNA stability motif might be composed of two independent and redundant elements. This redundancy may explain why there are no naturally occurring mutations presently described that have as their primary defect destabilization of the human β-globin mRNA.

The apparently dissimilar natures of the determinants important to α- and β-globin mRNA stability is surprising. However, the difference in their stabilizing mechanisms, although seemingly wasteful, parallels distinct differences in the developmental stage-specific expression of the α- and β-globin genes (reviewed in 63), the composition and fine details of their promoters

(85, 86), the differences in the chromatin structure of the α- and β-globin gene clusters (54, 87, 88), and the mechanisms of α- and β-globin cluster activation (89–92).

E. Possible Function of RNP Complexes Related to the α Complex in Stabilization of Nonglobin Erythroid mRNAs

The capacity of the hα-mRNA 3' UTR to assemble α complexes in S100 extracts from a variety of cells suggests that the protein components of this complex may serve additional functions in nonerythroid cells. A search of available databases has identified several additional very stable mRNAs that contain pyrimidine-rich 3' UTR segments that may participate in assembly of an α-like complex (92a). It is possible that the α complex is widely distributed among different cell types, or else represents one of a number of related complexes that comprise partially overlapping combinations of protein subunits. The later model would parallel combinatorial mechanisms of gene transcription control that are utilized by most eukaryotic promoters. It is also possible that these RNA binding proteins may mediate functions other than mRNA stability. Data from one study suggests that αCP or a closely related protein may mediate translational repression (93; also see 82 for discussion). By investigating the spectrum of related RNP complexes that assemble on a variety of mRNAs in a host of unrelated cell types, these themes can be tested and developed.

V. Potential Therapeutic Applications

The realization that mRNA stability can have a profound impact on gene expression and, ultimately, on cellular phenotype leads to speculation on potential novel therapeutic applications. A more thorough understanding of the general and specific mechanisms through which mRNA stability is modulated may permit design of specific approaches to common genetic diseases that work through the alteration of the stability of crucial target mRNAs. For example, the severity of disease in individuals with β-thalassemia is directly related to the relative excess in α-globin synthesis. Individuals who coinherit an α-thalassemic and a β-thalassemic determinant exhibit an unexpectedly mild phenotype because the balance between α- and β-globin gene expression is normalized (94–97). For this reason, therapies that decrease α-globin gene expression are predicted to ameliorate the clinical severity of β-thalassemia (95). If the stability of α-globin mRNA could be manipulated through regulation of the expression or the function of factors that comprise the α complex (such as αCP), α-globin production might be partially blocked. This therapeutic approach might also be applied to other diseases where alteration in the stability of a particular mRNA would benefit the patient.

The specificities of RNA binding proteins or subsets of these proteins involved in mRNA stabilization suggest a general approach that might be employed to modulate gene expression. It is clear from *in vitro* complex assembly assays that the addition of specific competitor nucleic acids can inhibit complex formation. As an extension, introduction of such sequences into living cells might have a similar effect. Depending on the function of the RNP complex, the disruption of the complex might result in a major change in the stability of the mRNA and, therefore, its steady-state level of expression. The expression of high levels of an RNA that mimics a cognate RNA binding site (e.g., iron response element, Rev response element, or α-globin 3′ UTR stability element) might serve as a "molecular decoy" by competing the binding protein(s) away from its primary target. Consequently, target mRNAs lacking their normal complement of stabilizing or destabilizing RNP complexes would be disregulated. The success of such an approach would depend on identification of the critical RNA binding site(s), determination of involved *trans*-factors, and design of decoy RNAs with high affinity and appropriate specificity to one or more protein factors. Success would also depend on the level of decoy expression, the concentration of the target *trans*-factors, and the affinity of the decoy for these factors relative to that of the native target RNA. The specificity of decoy action would reflect the extent to which the targeted *trans*-factor is selective for a particular mRNA or class of mRNAs. Shared specificities for RNP formation (as might exist among related mRNAs) might prove either problematic or fortuitous, depending on the desired spectrum of effects. The success of any approach would also depend on the importance of other functions mediated by the target factor and the degree to which they are affected by the intervention. Additional theoretical and practical therapies are likely to follow as the complex series of events that determine mRNA stability for a variety of specific mRNAs is established.

The potential of therapeutic interventions based on alteration of globin gene expression has became a possibility as the importance of mRNA stability to globin gene expression has become established and its basis delineated. Future investigations that more clearly define the additional molecular species involved in this process, their structures, and the mechanisms through which they contribute to the determination of globin mRNA stability are likely to expand and refine the repetoire of potentially beneficial therapeutic approaches. Parallel advances in other gene systems may yield a wealth of additional possibilities and significantly widen the scope of molecular therapeutics.

References

1. R. A. Shivdasani and S. H. Orkin, *Blood* **87,** 4025 (1997).
2. J. Ross, *Mol. Biol. Med.* **5,** 1 (1988).
3. J. Ross, in "Microbiological Reviews" (W. K. Joklik, ed.), Vol. 59, pp. 423–450. 1995.
4. A. B. Sachs, *Cell* **74,** 413 (1993).
5. P. Surdej, A. Riedl, and M. Jacobs-Lorena, *Annu. Rev. Genet.* **28,** 263 (1994).
6. C. A. Beelman and R. Parker, *Cell* **81,** 179 (1995).
7. D. Muhlrad and R. Parker, *Nature (London)* **340,** 578 (1994).
8. D. Muhlrad, C. J. Decker, and R. Parker, *Genes Dev.* **8,** 855 (1994).
9. A. B. Shyu, J. G. Belasco and M. E. Greenberg, *Genes Dev.* **5,** 222 (1991).
10. S. W. Peltz, G. Brewer, P. Bernstein, P. Hart and J. Ross, *CRC Crit. Rev. Eucar. Gene Exp.* **1,** 99 (1991).
11. C. J. Decker and R. Parker, *Genes Dev.* **7,** 1632 (1993).
12. C. J. Decker and R. Parker, *Trends Biol. Sci* **19,** 336 (1994).
13. R. Binder, J. A. Horowitz, J. P. Basilion, D. M. Koeller, R. D. Klausner, and J. B. Harford, *EMBO J.* **13,** 1969 (1994).
14. A. Jacobson and S. W. Peltz, *Annu. Rev. Biochem.* **65,** 693 (1996).
15. G. Shaw and R. Kamen, *Cell* **46,** 659 (1986).
16. C. Y. Chen and A. B. Shuy, *Trends Biol. Sci.* **20,** 465 (1995).
17. H. Aviv, Z. Volloch, R. Bastos and S. Levy, *Cell* **8,** 495 (1976).
18. J. Ross and T. D. Sullivan, *Blood* **66,** 1149 (1985).
19. L. Hamalainen, J. Oikarinen and K. Kivirikko, *J. Biol. Chem.* **260,** 720 (1985).
20. A. Maata, E. Elkholm and R. P. Penttinen, *Biochim. Biophys. Acta* **1260,** 294 (1995).
21. X. A. Li and D. C. Beebe, *Dev. Biol.* **146,** 239 (1991).
22. J. B. Clegg, D. J. Weatherall and P. F. Milner, *Nature (London)* **234,** 337 (1971).
23. S. A. Liebhaber and Y. W. Kan, *J. Clin. Invest.* **68,** 439 (1981).
24. D. M. Hunt, D. R. Higg, P. Winichagoon, J. B. Clegg and D. J. Weatherall, *Br. J. Haematol.* **51,** 405 (1982).
25. R. D. Klausner, T. A. Rouault, and J. B. Harford, *Cell* **72,** 19 (1993).
26. M. W. Hentze, S. W. Caughman, T. A. Rouault, J. G. Barriocanal, A. Dancis, J. B. Harford, and R.D. Klausner, *Science* **238,** 1570 (1987).
27. T. A. Rouault, M. W. Hentze, D. J. Haile, J. B. Harford, and R. D. Klausner, *Proc. Natl. Acad. Sci. U.S.A.* **86,** 5768 (1989).
28. S. Kaptain, W. E. Downey, C. Tang, C. Philpott, D. Haile, D. G. Orloff, J. B. Harford, T. A. Rouault, and R. D. Klausner, *Proc. Natl. Acad. Sci. U.S.A.* **88,** 10109 (1991).
29. J. A. Atwater, R. Wisdom, and I. M. Verma, *Annu. Rev. Genet.* **24,** 519 (1990).
30. C. G. Burd and G. Dreyfuss, *Science* **265,** 615 (1994).
31. P. F. Predki, L. M. Nayak, M. B. Gottlieb, and L. Regan, *Cell* **80,** 41 (1995).
32. M. L. Zapp and M. R. Green, *Nature (London)* **342,** 714 (1989).
33. H. S. Olsen, P. Nelbock, A. W. Cochrane, and C. A. Rosen, *Science* **247,** 845 (1990).
34. X. Wang, M. Kiledjian, I. M. Weiss, and S. A. Liebhaber, *Mol. Cell. Biol.* **15,** 1769 (1995).
35. L. Jia, C. Bonaventura, J. Bonaventura, and J. S. Stamler, *Nature (London)* **380,** 221 (1996).
36. H. W. Dickerman, T. C. Cheng, H. Kazazian, and J. L. Spivak, *Arch. Biochem. Biophys.* **177,** 1 (1976).
37. M. J. Evans and J. B. Lingrel, *Biochemistry* **8,** 3000 (1969).
38. J. Hunt, *Biochem. J.* **138,** 487 (1974).
39. A. G. Stewart, P. M. Clissold, and H. R. Arnstein, *Eur. J. Biochrm.* **65,** 349 (1976).
40. R. N. Bastos, Z. Volloch and H. Aviv, *J. Mol. Biol.* **110,** 191 (1977).
41. K. S. Kabnick and D. E. Houseman, *Mol. Cell. Biol.* **8,** 3244 (1988).

42. V. Volloch and D. Housman, *Cell* **23**, 509 (1981).
43. V. Volloch and D. Housman, *J. Cell Biol.* **93**, 390 (1982).
44. R. N. Bastos and H. Aviv, *J. Mol. Biol.* **110**, 205 (1977).
45. A. Krowczynska, R. Yenofsky and G. Brawerman, *J. Mol. Biol.* **181**, 231 (1985).
46. K. Lowenhaupt and J. Lingrel, *Cell* **14**, 337 (1978).
47. S. A. Liebhaber, Z. Wang, F. Cash, B. Monks, and J. E. Russell, *Mol. Cell. Biol.* **6**, 2637 (1996).
48. M. Albitar, M. Katsumata, and S. A. Liebhaber, *Mol. Cell. Biol.* **11**, 3786 (1991).
49. G. Gourdon, J. A. Jacqueline, D. Wells, W. G. Wood, and D. R. Higgs, *Nucleic Acids Res.* **22**, 4139 (1994).
50. D. E. Sabath, E. A. Spangler, E. M. Rubin, and G. Stamatoyannopoulos, *Blood* **82**, 2899 (1993).
51. J. A. Sharpe, D. J. Wells, E. Whitelaw, P. Vyas, D. R. Higgs, and W. G. Wood, *Proc. Natl. Acad. Sci. U.S.A.* **90**, 11262 (1993).
52. E. A. Spangler, K. A. Andrews, and E. M. Rubin, *Nucleic Acids Res.* **18**, 7093 (1991).
53. J. E. Russell, Z. Wang, B. R. Monks, F. E. Cash, and S. A. Liebhaber, *Blood* **84**, (abstract) 506a (1994).
54. M. Yagi, R. Gelinas, J. T. Elder, M. Peretz, T. Papayannopoulou, G. Stamatoyannopoulos, and M. Groudine, *Mol. Cell. Biol.* **6**, 1108 (1986).
55. B. E. Serjeant, K. P. Mason, and G. R. Serjeant, *Brit. J. Haematol.* **39**, 259 (1978).
56. R. A. Spritz, J. K. DeRiel, B. G. Forget, and S. W. Weissman, *Cell* **21**, 639 (1980).
57. R. M. Winslow and V. M. Ingram, *J. Biol. Chem.* **241**, 1144 (1966).
58. J. Ross and A. Pizarro, *J. Mol. Biol.* **167**, 607 (1983).
59. N. Lumelsky and B. G. Forget, *Mol. Cell. Biol.* **11**, 3528 (1991).
60. N. Lumelsky, *Biochim. Biophys. Acta* **1261**, 265 (1995).
61. A. Albrecht, A. Krowczynska and G. Brawerman, *J. Mol. Biol.* **178**, 88 (1984).
62. R. Bandyopadhyay, M. Coutts, A. Krowczynska, and G. Brawerman, *Mol. Cell. Biol.* **10**, 2060 (1990).
63. J. E. Russell and S. A. Liebhaber, in "Advances in Genome Biology" (R. S. Verma, ed.), Vol. 2, p. **283**. JAI Press, Greenwich, Connecticut, 1993.
64. P. F. Milner, J. B. Clegg, and D. J. Weatherall, *Lancet* **702**, 729 (1971).
65. Y. W. Kan, D. Todd and A. Dozy, *Br. J. Haematol.* **28**, 103 (1974).
65a. J. Morales, J. E. Russell, and S. A. Liebhaber, *J. Biol. Chem.* (1997), in press.
66. T. B. Bradley, R. C. Wohl, and E. J. Smith, *Clin. Res.* **23**, 131A (1975).
67. W. W. De Jong, P. Meera Khan, and L. F. Bernini, *Am. J. Hum. Genet.* **27**, 81 (1975).
68. J. B. Clegg, D. J. Weatherall, I. Contopolou-Griva, K. Caroutsos, P. Poungouroas, and H. Tsrerenvis, *Nature (London)* **251**, 245 (1974)
69. G. Flatz, J. L. Kinderlerer, J. V. Kilmantin, and H. Lehmann, *Lancet* **1**, 732 (1971).
70. H. F. Bunn, G. J. Schmidt, D. N. Haney, and R. G. Dluhy, *Proc. Natl. Acad. Sci. U.S.A.* **72**, 3609 (1975).
71. J. Delanoe-Garin, Y. Blouquit, N. Arous, J. Kister, C. Poyart, M.L. North, J. Bardakdjian, C. Lacombe, J. Rosa, and F. Galacteros, *Hemoglobin* **12**, 337 (1988).
72. G. Brewer and J. Ross, *Methods Enzymol.* **181**, 202 (1990).
73. I. M. Weiss and S. A. Liebhaber, *Mol. Cell. Biol.* **14**, 8123 (1994).
74. P. A. Marks, Z. Chen, J. Banks, and R. A. Rifkind, *Proc. Natl. Acad. Sci. U.S.A.* **80**, 2281 (1983).
75. I. M. Weiss and S. A. Liebhaber, *Mol. Cell. Biol.* **15**, 2457 (1995).
76. S. L. Wolin and P. Walter, *EMBO J.* **7**, 3559 (1988).
77. M. Kiledjian, X. Wang, and S. A. Liebhaber, *EMBO J.* **14**, 4357 (1995).
78. M. J. Matunis, W. M. Michael, and G. Dreyfuss, *Mol. Cell. Biol.* **12**, 164 (1992).

79. C. T. Ashley, K. D. Wilkinson, D. Reines, and S. T. Warren, *Science* **262**, 563 (1993).
80. H. Siomi, M. C. Siomi, R. L. Nussbaum, and G. Dreyfuss, *Cell* **74**, 291 (1993).
81. B. W. Baer and R. D. Kornberg, *Proc. Natl. Acad. Sci. U.S.A.* **77**, 1890 (1980).
82. X. Wang and S. A. Liebhaber, *EMBO J.* **15**, 5040. (1996).
83. M. Goodman, M. L. Weiss, and J. Czelusniak, *Systemat. Zool.* **31**, 376 (1982).
84. J. E. Russell and S. A. Liebhaber, *Blood* **87**, 5314 (1996).
85. S. A. Liebhaber, M. Goosens, and Y. W. Kan, *Proc. Natl. Acad. Sci. U.S.A.* **77**, 7054 (1980).
86. R. M. Lawn, A. Efstratiadis, C. O'Connell, and T. Maniatis, *Cell* **21**, 647 (1980).
87. M. Groudine, T. Kohwi-Shigematsu, R. Gelinas, G. Stamatoyannopoulos, and T. Papayannopou, *Proc. Natl. Acad. Sci. U.S.A.* **80**, 7551 (1983).
88. A. P. Jarman and D. R. Higgs, *EMBO J.* **7**, 3337 (1988).
89. D. Tuan, D. Solomon, Q. Li, and I. M. London, *Proc. Natl. Acad. Sci. U.S.A.* **82**, 6384 (1985).
90. W. C. Forrester, C. Thompson, J. T. Elder, and M. Groudine, *Proc. Natl. Acad. Sci. U.S.A.* **83**, 1359 (1986).
91. D. R. Higgs, W. G. Wood, A. P. Jarman, J. Sharpe, J. Lida, I. M. Pretorius, and H. Ayyub, *Genes Dev.* **4**, 1588 (1990).
92. A. P. Jarman, W. G. Wood, J. A. Sharpe, G. Gourdon, H. Ayyub, and D. R. Higgs, *Mol. Cell. Biol.* **11**, 4679 (1991).
92a. M. Holcik and S. A. Liebhaber, *Proc. Natl. Acad. Sci* (1997), in press.
93. A. Ostareck-Lederer, D. H. Ostareck, N. Standart, and B.J. Thiele, *EMBO J.* **13**, 1476 (1994).
94. A. E. Kulozik, E. Kohne, and E. Kleihauer, *Ann. Hematol.* **66**, 51 (1993).
95. X. Ponnazhagan, M. Nallari, and A. Srivastava, *J. Exp. Med.* **179**, 733 (1994).
96. E. Kanavakis, J. S. Wainscoat, W. G. Wood, D. J. Weatherall, A. Cao, M. Furbetta, R. Galanello, D. Georgiou, and T. Sophocleous, *Br. J. Haematol.* **52**, 465 (1982).
97. C. Camashella, U. Mazza, A. Roetto, and M. Cappellini, *Am. J. Hematol.* **48**, 82 (1995).

Self-Glucosylating Initiator Proteins and Their Role in Glycogen Biosynthesis

PETER J. ROACH AND
ALEXANDER V. SKURAT

*Department of Biochemistry and
Molecular Biology
Indiana University School of Medicine
Indianapolis, Indiana 46202*

I. Glycogenin and the Pathway of Glycogen Biogenesis	290
A. Discovery of the Initiation Step	290
B. Elongation and the Proglycogen Hypothesis	291
II. Biochemistry of Glycogenin and the Glg Proteins	293
A. Primary Structure	293
B. Subunit Structure	294
C. Enzymatic Functions	298
III. Molecular Biology of Glycogenin and Glycogenin-like Proteins	301
A. Preface	301
B. Mammalian Glycogenins	302
C. Glycogenin-like Genes in Other Species	304
D. Conserved Domains in Glycogenin and Glycogenin-like Proteins	305
IV. Physiological Function of Glycogenin and the Glg Proteins	309
A. Role of Glycogenin *in Vivo*	309
B. Is Glycogenin Rate Determining for Glycogen Synthesis?	311
C. Regulation of Glycogenin and the Glg Proteins	311
D. Other Possible Roles of Glycogenin	313
E. Glycogenin and Diabetes	313
V. Conclusion	314
References	315

Glycogen, a branched polymer of glucose found in cell types ranging from bacteria to humans, is believed to serve as a glucose reserve, synthesized during favorable nutritional conditions for utilization in times of deprivation. The main polymerization of glucose is via α-1,4-glycosidic linkages with branch points introduced by α-1,6-glycosidic linkages. Research on glycogen metabolism over many years has led to several important conceptual advances in understanding the molecular basis of cellular regulation. A novel process discovered recently during the study of glycogen biogenesis is that of an initiation stage in polymer biosynthesis mediated by an enzymatically active self-priming[1] protein. In this case, the protein self-glucosylates to form an oligosaccharide primer for the elongation reaction. Though clearly definable as enzymes, the ultimate fate of such self-glucosylating proteins is to become covalently attached to the polysaccharide product, not a usual characteristic of enzymes. The cloning of cDNAs or genes encoding the mammalian protein, glycogenin, and two homologs, Glg1p and Glg2p from the yeast *Saccharomyces cerevisiae*, ushers in a new phase in the study of self-glucosylating proteins. In addition, several related sequences have appeared in the sequence databases, suggesting the existence of multiple isoforms and perhaps a wider distribution for proteins of this class. Other reviews have described the earlier stages of this story (*1–4*). Though seeking to be reasonably self-contained, this article will concentrate on some of the more recent developments in the study of self-glucosylating proteins and the biogenesis of glycogen in eukaryotic cells.

I. Glycogenin and the Pathway of Glycogen Biogenesis

A. Discovery of the Initiation Step

As is true for most biopolymers, the biosynthesis of glycogen involves both a specific initiation phase as well as a phase of elongation (Fig. 1). Until quite recently, much less emphasis had been given to the initiation stage. Part of the reason is that most studies on the regulation of eukaryotic glycogen metabolism had focused on the allosteric and covalent control of the enzymes glycogen synthase and phosphorylase, which mediate the formation and lysis, respectively, of the α-1,4-glycosidic linkages. In addition, the biochemical definition and characterization of the initiator proteins has lagged behind work on glycogen synthase and phosphorylase.

There was initially debate as to whether a specific primer for the biosynthesis of glycogen even existed. Purified glycogen synthase can use glucose as an acceptor for the transfer of glucose from UDP-glucose to form maltose, potentially allowing for the synthesis of glycogen without any requirement for a separate initiation reaction (*5*). The K_m for glucose is highly unfavorable, close to molar (*5*), but one could make the argument that the reaction did not need to be very efficient if glycogen molecules were never completely degraded. This situation would be especially relevant in terminally differentiated cells such as muscle and liver, which are the two major repositories of

---

[1] The term "priming" is used to describe the process that establishes competency of the initiator protein for elongation. For glycogenin and the Glg proteins, this involves a self-glucosylation reaction; protein that has undergone this reaction is described as being "primed."

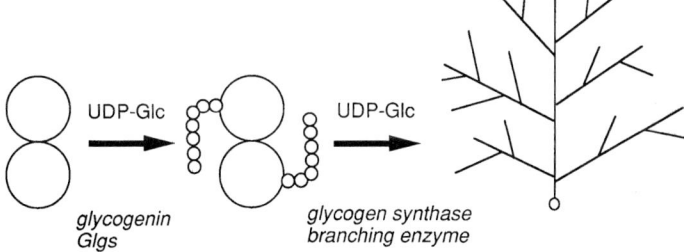

FIG. 1. Model for glycogen synthesis. In this scheme, a self-glucosylating protein is responsible for the autocatalytic generation of an oligosaccharide primer that can be elongated by glycogen synthase. The action of this latter enzyme, together with the branching enzyme, would result in the formation of a mature glycogen molecule. The self-glucosylation is depicted as an intramolecular, intersubunit reaction.

glycogen. Advancing a qualitatively different model, Krisman and Barengo (6) postulated the existence of a protein backbone to which glycogen would be attached. Attachment of carbohydrate would require also a "glycogen initiator synthase." Whelan (1) and co-workers went on to identify a protein, designated "glycogenin," that was covalently attached to glycogen via tyrosine residues. Glycogenin was also found to be a well-known "contaminant" of rabbit muscle glycogen synthase preparations, suggesting additionally that it interacted directly with glycogen synthase (7). The laboratories of Whelan (8) and Cohen (9) then showed that glycogenin had self-glucosylating activity, giving rise to the type of model shown in Fig. 1. Glycogenin, via self-glucosylation, would form an oligosaccharide chain, up to about ten residues long (10), that would then serve as an effective substrate for glycogen synthase (section II,C). A role for glycogenin, or a similar self-glucosylating protein, in the biosynthesis of glycogen is now fairly well established and, in the yeast S. cerevisiae, genetic experiments have demonstrated that a self-glucosylating protein is required for glycogen accumulation (Section IV).

B. Elongation and the Proglycogen Hypothesis

In vitro, it is possible to show that glycogenin and the yeast Glg proteins, once primed, can serve as substrates for glycogen synthase and can be converted to high M_r species (see Section II, C). Thus, one could propose that glycogenin and glycogen synthase, together with the participation of the branching enzyme, are sufficient to account for glycogen synthesis. Recently, Lomako et al. (11, 12) have proposed the existence of an additional discrete intermediate form of glycogen, termed "proglycogen" (see Ref. 3 for a detailed account of this topic). Proglycogen is defined as being insoluble in trichloroacetic acid and with a mass of ~400,000 Da. Proglycogen scarcely enters the gel in discontinuous SDS–PAGE (*11*). It was also proposed that a separate "proglycogen synthase" exists to synthesize proglycogen from primed glycogenin (*11, 12*). Proglycogen synthase activity would either be due to a novel gene product or else mediated by some specialized phosphorylation state of conventional glycogen synthase. A proteoglycan from liver described by Ercan *et al.* (*13*) had some of the characteristics ascribed to proglycogen. If there were a major stable intermediate like proglycogen, then it would constitute an additional step in the glycogen synthetic pathway and it might necessitate a revision of how we view glycogen metabolism. For example, it could mean new sites of regulation relevant to understanding the control of glycogen metabolism. Furthermore, it would require a specific interaction of glycogenin with proglycogen synthase rather than with glycogen synthase.

Several results do not support the existence of a discrete intermediate like proglycogen. When Rat1 fibroblasts labeled with [^{14}C]-glucose were analyzed by discontinuous SDS–PAGE, most radioactivity was observed either in the stacking gel or as an apparently discrete band at the entry to the separating gel (*14*), similar to what was described by Lomako *et al.* (*11*). The pattern was not significantly changed by the use of gels with a lower percentage of acrylamide. However, when the continuous gel electrophoretic system of Weber and Osborn (*15*) was employed, it was no longer clear that there was a discrete polysaccharide species but rather a continuum of species spanning a wide range of molecular weights (*14*). Application of two-dimensional gel electrophoresis similarly revealed a continuum of glucosylated species ranging from low molecular weight to the top of the separating gel (*14*). From these, and related experiments, no evidence was found for a discrete intermediate species like proglycogen, but instead a smooth distribution of species from glycogenin to polysaccharide of high molecular mass.

Another issue is that of the putative proglycogen synthase, which in mammalian systems has yet to be well defined in molecular terms. One hope might be that the simpler yeast model could help clarify whether such an activity exists. In *S. cerevisiae*, there are two genes encoding glycogen synthase, *GSY1* and *GSY2* (*16*). *GSY1* is expressed at a low consitutive level whereas *GSY2* encodes the major nutritionally regulated form of the enzyme. The two proteins share significant sequence homology, and deletion of both genes is needed to eliminate completely glycogen synthesis (*16*). Yeast also contains two-self glucosylating proteins, Glg1p and Glg2p, with sequence similarity to that of glycogenin (*17*), thus seeming biochemically (Section II) and functionally (Section III) related to glycogenin. In wild-type yeast, the self-glucosylating protein Glg2p is attached to glycogen and cannot be detected as a free form by SDS–PAGE (*18*). In the *gsy1 gsy2* mutant cells, which cannot synthesize glycogen, free Glg2p accumulates around its predicted molecular mass as judged by SDS–PAGE and Western blotting (18). Loss of conventional glycogen synthase activity generated free Glg2p and not a larger form. Thus, any proglycogen synthase activity in yeast would have to be associated with Gsy1p or Gsy2p. Completion of the sequencing of the S. cerevisiae genome also reveals that there are no other proteins with any significant sequence homology to Gsy1p or Gsy2p.

The existence of proglycogen would appear, in our judgment, to be controversial at this time. There is, of course, no doubt that intermediates do exist between glycogenin and the mature glycogen molecule. What is in question is whether there is a discrete intermediate, of size ~400,000 Da, that significantly accumulates *and* whose concentration is defined by the action of specific enzymes or enzyme activities. In our hands, the apparent discreteness of the SDS–PAGE polysaccharide species just entering the separating gel seems to be due to the use of discontinuous gel electrophoresis, all molecules above a certain size collecting at this point, and may not represent a significant accumulation of polysaccharide of discrete size.

II. Biochemistry of Glycogenin and the Glg Proteins

A. Primary Structure

Mammalian glycogenin[2] is a protein of 332 residues with a predicted subunit M_r of ~37,000. It is an acidic protein, composed of 12% Asp plus Glu residues, to give a predicted pI of 4.84. This protein has been purified from rabbit skeletal muscle (7, 8) and liver (19). Immunoreactive species were also detected in rabbit heart, retina, and liver (20). A self-glucosylating kidney protein has also been reported (21, 22). The apparent M_r from SDS–PAGE reported in the literature spans a suprisingly wide range, from 32,000 to 44,000, as is discussed in some detail by Alonso *et al.* (3). The reason for this variability is not entirely clear but in part it undoubtedly stems from the fact that, depending on experimental conditions, the protein can have variable amounts of carbohydrate attached, which can affect electrophoretic migration. Additionally, the protein may undergo proteolysis to generate a 32-kDa active fragment (23). Perhaps it would be wise to reevaluate these results, as regards both the tissue distribution and the apparent molecular weights, in

---

[2] We refer here to the form of the enzyme as initially purified from rabbit muscle. Much of the biochemical analysis of glycogenin has addressed this protein. All work to date on the recombinant protein, from rabbit muscle (in our laboratory) or human muscle (in the laboratory of W. Whelan), has dealt with this form. However, the existence of isoforms of glycogenin (Section III) means that we may have to reevaluate some of the experiments with whole cells or tissues with regard to which isoform(s) of glycogenin may have been present.

light of the realization that at least four isoforms of human glycogenin exist (Section III).

The yeast Glg proteins (Fig. 2) have 55% sequence identity to each other over an NH_2-terminal segment of about 260 residues and ~33% identity to mammalian glycogenin in this region, which likely contains the catalytic domain. Database searches and multiple sequence alignments allow us to define a domain structure in glycogenin and glycogenin-like proteins, as discussed in detail in Section IV. The mammalian and S. *cerevisiae* proteins share domains II, III, and IV but align poorly in the region of domain I. One might predict that domains II and III have special importance for catalysis. Glg1p, with a predicted M_r of 69,836, has a large insert compared to the other two proteins. Within this region is a segment, domain VI, that has similarity to a COOH-terminal sequence in glycogenin. Glg1p and Glg2p also share two other segments, domains V and VII. Domain V aligns with an alternatively spliced isoform of glycogenin (Section III) and domain VII may interact with glycogen synthase. Recombinant Glg1p, produced as a poly(His)-tagged protein in *Escherichia coli*, runs aberrantly on SDS–PAGE, with an apparent M_r of 117,000 (predicted M_r of 72,000 for the fusion protein). The reason is not known though it is notable that Glg1p is, like glycogenin, very acidic with predicted pI values of 3.95; the predicted pI of Glg2p is 5.60.

B. Subunit Structure

Glycogenin in solution most likely exists as an oligomer. Evidence comes from gel filtration experiments (7, 8, 24) as well as X-ray crystallography (25).

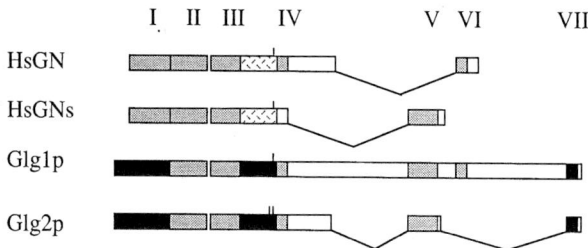

FIG. 2. Domain structure of glycogenin and the Glg proteins. The yeast Glg proteins are aligned with mammalian glycogenin, exemplified here by the human form (HsGN). Also shown is an alternate form (HsGNs) that is identical up to residue 202 but then has a distinct COOH terminus (section III,B). Regions of the Glg proteins and mammalian glycogenin that are similar and that may define conserved domains (Roman numerals) are shaded. Regions of similarity in Glg1p and Glg2p only are indicated by filled boxes. The hatched regions in HsGN and HsGNs are identical. The open boxes indicate no particular similarity. The vertical ticks indicate sites of self-glucosylation.

Small crystals of recombinant glycogenin were obtained and, though not suitable to solve the complete structure of the protein, the diffraction data allowed conclusions to be drawn regarding the unit cell. There was a noncrystallographic twofold axis of symmetry consistent with the polypeptide being arranged in dimers or possibly even higher oligomers such as a tetramer. This is relevant to considerations of enzyme mechanism as discussed below.

Glycogenin also interacts with glycogen synthase. Mammalian glycogenin copurified with muscle glycogen synthase and was identified as a "contaminant" of the enzyme preparation by Pitcher et al. (7). In addition, glycogenin and glycogen synthase have been shown to coimmunoprecipitate from extracts of cultured cells using antiglycogenin antibodies (A. S. Skurat and P. J. Roach, unpublished results). Pitcher et al. (7) proposed that glycogenin is a "subunit" of glycogen synthase. In liver, however, the ratio of glycogenin to glycogen synthase is much lower, making a stoichiometric association similar to that seen in muscle improbable (19). Yeast Glg proteins are also likely to interact with glycogen synthase (17). Glg2p was first identified in a yeast two-hybrid screen using Gsy2p as the bait, and the initial clone contained only the COOH-terminal 47 residues of Glg2p, which contains domain VII (Fig. 2). Domain VII is likely involved, therefore, in interacting with Gsy2p (Section II, C, 4). The interaction does not require the COOH-terminal 81 residues of Gsy2p, the region responsible for covalent regulation of the synthase (17).

From the preceding discussion, it is probable that, at least under some conditions, cells contain complexes composed of multiple copies of both glycogen synthase and glycogenin. The exact role of such complexes in the process of glycogen synthesis is not yet clear. Are the complexes permanent? Must there be a stoichiometric association between glycogen synthase and glycogenin? Much evidence supports the idea that primed glycogenin is an effective substrate for glycogen synthase and, because glycogenin and glycogen synthase interact, one might expect that, at least initially, this would be an intramolecular reaction (i.e., glycogen synthase acting on glycogenin in the same complex), even though there is actually no in vitro evidence on this point. However, one might predict serious constraints on the ability of the glycogen synthase to elongate the branched carbohydrate chains on a bound glycogenin partner in such an intramolecular reaction once the polysaccharide reaches a certain size. Several hypotheses can be formally advanced to avoid this conceptual problem (Fig. 3).

First, the growing glycogen molecule could be cut loose from its tether to glycogenin at some early stage in its maturation (model I). In this model, glycogenin would be recycled prior to completion of the glycogen molecule that it initiated, and so there could be a pool of glycogenin not covalently at-

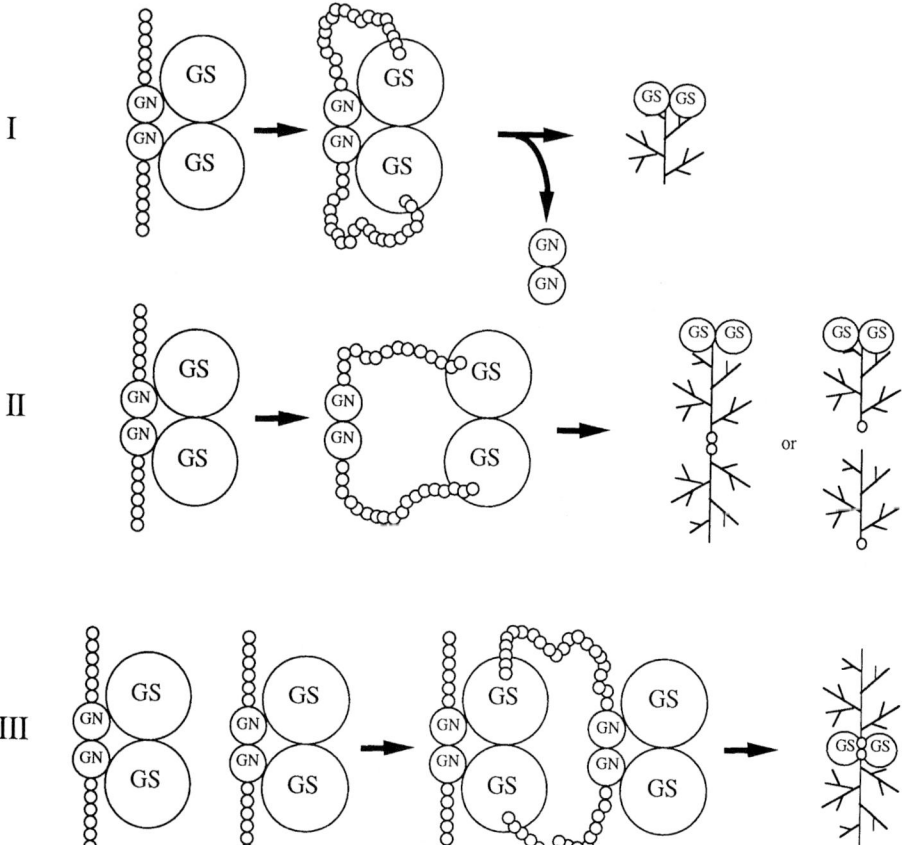

FIG. 3. Models for glycogenin (GN) and glycogen synthase (GS) interactions during glycogen synthesis. Glycogenin and glycogen synthase can interact, and one interesting question is the fate of this interaction through the course of the synthesis of a glycogen molecule. Another issue is whether the synthesis involves glycogen synthase acting on carbohydrate attached to a bound glycogenin in an intramolecular reaction. For the sake of simplicity in this figure, glycogenin and glycogen synthase start out as dimers that interact to form a heterotetramer. However, the existence of larger complexes is not excluded. In model I, glycogenin is cleaved from the growing glycogen molecule prior to its completion. In model II, growth of the glycogen molecule in an intramolecular reaction at a certain stage forces a dissociation of glycogen synthase and glycogenin, and the latter remains covalently attached to the glycogen. Whether glycogenin remains as an oligomer is not specified. In model III, glycogen synthase and glycogenin remain associated and the elongation reaction is achieved through intermolecular reactions.

tached to glycogen (and which may or may not be noncovalently bound to synthase). Under several circumstances, such free glycogenin is not detectable in cells. For example, most of the glycogenin in resting rabbit muscle is associated with glycogen, and electrical or hormonal stimulation, with concomitant glycogen breakdown, is needed to release glycogenin (24). In another example, essentially all of the rabbit glycogenin ectopically expressed in Rat1 fibroblasts is covalently linked to glycogen at glucose concentrations greater than 5 mM; prolonged exposure with low or no glucose is needed to generate low-M_r glycogenin (14). In addition, this model would require a mechanism, presumably a novel enzyme activity, to cleave the glycogenin from the growing glycogen molecule. This model has little experimental support. However, if reinitiation of glycogenin were very efficient, the pool of free glycogenin could be small and not easily detected.

Model II is essentially that proposed by Smythe and Cohen (2) in which glycogen synthase and glycogenin would become dissociated in the course of the synthesis of the glycogen particle. One would predict on this model that, after dissociation, the glycogen synthase would be free to act on any glycogen molecule, whether it had initiated it or not. It is not known whether glycogenin subunits remain associated after synthesis of a glycogen molecule. The main support for this model comes from one study in which electrical stimulation or epinephrine treatment of rabbit skeletal muscle generated a significant proportion of low-M_r glycogenin that could be separated from glycogen synthase by rapid gel filtration of muscle extracts (24). Only after incubation for a period of hours did the glycogen synthase and the glycogenin reassociate. Purified glycogenin and glycogen synthase appear not to need any such prolonged incubation to interact and so some factor, physiological or artifactual, must presumably impede their association in the tissue extracts.

In model III, glycogen synthase and glycogenin would remain associated, whether in the presence of glycogen or not. Steric hindrance to chain elongation by glycogen synthase in an intramolecular reaction is avoided by allowing the glycogen synthase to act intermolecularly, on other glycogenin–glycogen synthase complexes. Models II and III predict association of both glycogenin and glycogen synthase with glycogen. However, in model III, glycogen breakdown, as in the experiments of Smythe *et al.* (24), would have predicted the generation of a glycogen synthase–glycogenin complex, unless there were an excess of glycogenin.

Pitcher *et al.* (7) proposed a strict 1:1 stoichiometry of glycogen synthase and glycogenin. However, none of the models above actually fails if such a stoichiometry does not exist. If there is excess glycogen synthase, we only require that the glycogenin present interacts with the glycogen synthase. Such may be the case in liver, at least with respect to the known isoform of glycogenin. Indeed, one can ask why muscle needs so much glycogenin compared with liver. Were glycogenin in molar excess over glycogen synthase, all of the models would likewise remain viable.

Probably, most current evidence supports some form of model II. However, discussion of different formal possibilities seems appropriate as novel isoforms of glycogenin are identified. These new proteins may or may not behave in every detail like the glycogenin that has been studied to date. Note also that the models are not necessarily exclusive. Thus, with model II or III, there could still be cleavage of the linkage of glycogenin to carbohydrate, as suggested in model I.

The concept of a glycogen particle, with an associated collection of relevant metabolic enzymes, goes back to the early work of Fischer and colleagues (26, 27). Much of their emphasis was on the degradative enzymes, such as phosphorylase and phosphorylase kinase. Interestingly, a recent study of Polishchuk et al. (28) identified glycogenin-like species in protein–glycogen complexes, as might be expected, but also in sarcoplasmic reticulum and in phosphorylase kinase peparations. Nor should we exclude the possible involvement of other molecules in such complexes. For example, the glycogen-binding type 1 phosphatase, PP1G, is composed of a catalytic subunit and a large regulatory subunit R_{Gl} that is thought to interact with glycogen (29) and with membranes (30). In S. cerevisiae, there are two genes, GAC1 (31) and PIG1 (32), that encode proteins with some sequence homology to R_{Gl}. Gac1p is functionally similar to R_{Gl} and seems to act as a phosphatase subunit in yeast (31, 33). There is weaker evidence for a similar role of Pig1p (32). Interestingly, a portion of the Pig1p protein was identified in the same yeast two-hybrid screen that found Glg2p (17). Also, there is evidence from yeast two-hybrid assays that Gac1p can interact with the yeast glycogen synthase Gsy2p (K. Tatchell, personal communication). Therefore, these phosphatase subunits may not only interact with the phosphatase catalytic subunit but also with glycogen synthase, maybe phosphorylase kinase, and somehow be involved in organzational aspects of glycogen synthesis. A rather fanciful arrangement of some of these proteins is shown in Fig. 4. How this organization would relate to the process of glycogen metabolism is not known. Note that the composition of the complex might not be constant. For example, during the early stages of synthesis, binding of degradative enzymes like phosphorylase might not be favored, thus allowing the formation of small glycogen molecules even in the face of active phosphorylase.

C. Enzymatic Functions

1. Self-Glucosylation

The discovery by the groups of Cohen (8) and Whelan (9) that glycogenin had enzymatic activity, catalyzing a self-glucosylation reaction with UDP-glucose as a donor, was a landmark in understanding glycogenin function. The

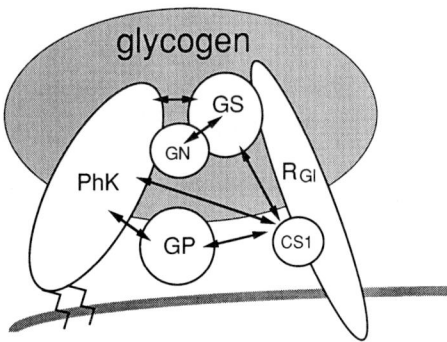

FIG. 4. Arrangement of proteins involved in glycogen metabolism, indicating some possible interactions among proteins and glycogen. Overlaps indicate known or proposed protein–protein interactions. Likewise, overlap with glycogen indicates interaction (or covalent attachment in the case of glycogenin). Double-headed arrows indicate known catalytic interactions (e.g., phosphorylase is a substrate for phosphorylase kinase). R_{Gl} and phosphorylase kinase are shown interacting with a membrane. GN, Glycogenin; CS1, catalytic subunit of type 1 phosphatase; GP, glycogen phosphorylase; PhK, phosphorylase kinase; GS, glycogen synthase; R_{Gl}, type 1 phosphatase regulatory subunit.

K_m for UDP-glucose is low, ~5 μM, and the reaction is favored by the presence of Mn^{2+} ions (8, 35). (More detailed discussion of the reaction and its kinetics can be found in Refs. 2–4.) The glucose residues are attached to each other by α-1,4-glycosidic linkages, as in glycogen, and to the protein by a single glucose-1-O-tyrosyl linkage. The Tyr was identified by protein chemical methods as Tyr-194 (34), as was later confirmed by site-directed mutagenesis (35, 36). Because of the different chemical nature of these linkages, there was initially debate as to whether a separate glucosyltransferase was needed to mediate the initial step. When recombinant glycogenin was expressed in *E. coli*, the protein was active for self-glucosylation (37) and, from analysis of the glucosylated peptide by mass spectrometry, contained up to seven attached glucose residues (35). This result is consistent with glycogenin catalyzing the formation of both linkages. However, formally, one could not exclude the possibility that a bacterial enzyme existed to mediate the initial priming step. Alonso *et al.* (10) showed that recombinant glycogenin produced in a mutant *E. coli*, defective in UDP-glucose production, was still active for self-glucosylation, providing evidence that there is no additional glycogenin-modifying enzyme.

Recombinant Glg1p and Glg2p proteins are capable of self-glucosylation with kinetic characteristics not unlike those observed for their mammalian counterpart (17, 18). From sequence alignment, Glg2p has two Tyr residues and Glg1p has one Tyr in the vicinity of Tyr-194. Interestingly, the sequence similarity to the mammalian protein is not very strong in this region. Mutation of Tyr-232 of Glg1p to Phe virtually eliminated self-glucosylation, whereas in Glg2p it was necessary to mutate both Tyr-230 and Tyr-232 to Phe in order to achieve the same result (18). Thus Glg2p appears to have two sites of glucosylation. Matters are more complicated, however, because elimination of self-glucosylation by such mutations was not sufficient to confer a fully glycogen-defective phenotype in yeast cells (Section IV,A).

2. ALTERNATE GLUCOSYL ACCEPTORS

Glycogenin can also transfer glucose to other acceptor molecules, such as maltose (25), p-nitrophenyl maltosides (38), and n-dodecyl-β-D-maltoside (39). The Glg proteins have similar properties (18). Interestingly, glycogenin mutated at Tyr-194, and hence incapable of self-glucosylation, is still active in such reactions indicating that Tyr-194 is required only as an acceptor of the transferred glucose (35, 36). In fact, the mutants are slightly more active, perhaps because there is no competition with the entropically favored self-glucosylation. In a similar way, Glg proteins no longer competent for self-glucosylation are active in transglucosylation, although with these proteins the rate is significantly diminished (18). Both glycogenin (23) and the Glg proteins (18) catalyze a slow hydrolysis of UDP-glucose. There is so far no evidence for any physiological relevance to this reaction, which may simply be a side reaction due to the catalytic mechanism. More intriguing is to ask whether transglucosylation has any relevance physiologically and whether there might be more interesting acceptors than maltose or other oligosaccharides. Teleologically, it is hard to see why, after having evolved a complex self-glucosylating system to initiate glycogen synthesis, there would be a need for other sites of initiation. Perhaps transglucosylation would not be related to glycogen synthesis. We should keep an open mind, especially as new forms of glycogenin or glycogenin-like genes are analyzed.

3. MECHANISM OF SELF-GLUCOSYLATION

In two studies (9, 35), the rate of glycogenin self-glucosylation was found to be first order with respect to enzyme concentration, i.e., constant specific activity, prompting the proposal that the process was intramolecular. Because glycogenin is an oligomer, the intramolecular reaction could be either intrasubunit or intersubunit, as discussed by Cao et al. (25). Though definitive evidence is lacking, we favor the intersubunit model as being esthetically more appealing in that it gives a teleological rationale for the existence of glycogenin in a multimeric form. Alonso et al. (23) recently reported that the reaction followed higher order kinetics at the lowest protein concentrations examined. This result could be accommodated if the oligomeric glycogenin complex dissociated at high dilution.

4. ELONGATION BY GLYCOGEN SYNTHASE

In models of glycogen biogenesis, such as in Fig. 1, the key properties of the initiator protein are, first, the ability to mediate formation of an oligosaccharide primer and, second, the ability to act as a substrate for glycogen synthase or, in Whelan's model, proglycogen synthase. Purified recombinant mammalian glycogenin (35, 37), Glg1p, or Glg2p (17) can all, after an initial self-glucosylation reaction, be converted to species of high molecular weight by incubation with purified glycogen synthase. This ability to serve as a substrate for glycogen synthase is directly related to the amount of self-glucosylation of the glycogenin (35). Because the polysaccharide formed is linear, not branched, it is hard to compare its migration on SDS–PAGE with natural glycogen, but it is of such a size that it barely enters the gel in discontinuous SDS–PAGE.

Although the presence of attached carbohydrate is clearly important to elongation, protein–protein interactions also play a role. For example, there is species specificity in the efficacy with which glycogen synthase elongates the self-glucosylating proteins. *In vitro*, yeast Gsy2p works poorly to elongate rabbit muscle glycogenin, and mammalian glycogen synthase is less effective with the Glg proteins than with glycogenin (17). Because glycogen synthase and the self-glucosylating proteins interact physically (Section II,B), it is likely that this interaction underlies the efficiency of the elongation step. In the case of the Glg proteins, domain VII in the extreme COOH terminus is important for interaction with glycogen synthase. Truncation in this COOH-terminal domain severely impairs elongation of Glg2p by Gsy2p. Mutation of Tyr-367 has a similar, though less strong, effect (18). Sequences in mammalian glycogenin important for glycogen synthase interaction have not been identified to date (see also Section III,D).

III. Molecular Biology of Glycogenin and Glycogenin-like Proteins

A. Preface

With the current explosion in gene discovery, some of what is presented in this section will almost certainly be out of date prior to publication. The initial cloning of cDNAs for mammalian glycogenin (37) and of the yeast *GLG* genes (17) were directed ventures. Now, related sequences are entering the databases either as part of random cDNA sequencing projects, as ESTs, or as sequences from systematic genome sequencing projects.

B. Mammalian Glycogenins

Full-length cDNAs encoding glycogenin have been obtained from rabbit (*37*) and human (*40*) muscle (Table I). The rabbit sequence was identical in all but two positions to that obtained by sequencing the protein that copurified with glycogen synthase (*41*). In addition, polymerase chain reaction (PCR) fragments corresponding to bovine and rat glycogenin were also sequenced (*37*). There is >90% identity among these sequences. The human gene has been assigned to chromosome 3q25.1 (*40*) or 3q24 (*40a*). In addition, two glycogenin-related sequences were mapped on chromosomes 12 and 13, but these are most likely intronless pseudogenes (*40*). A pseudogene was also found in rabbit (E. Viskupic and P. J. Roach, unpublished data; *37*).

Three human cDNA sequences are currently in GenBank. Two sequences from skeletal muscle (accession numbers U31525 and U44131) are identical in their coding regions and exhibit high identity over their entire lengths with the rabbit muscle glycogenin. We designate the corresponding protein HsGN. The third is a keratinocyte cDNA (accession number X79537) that has an identical sequence to the other cDNAs except for a 220-bp deletion within the coding region. Thus, the protein sequence predicted by the keratinocyte cDNA is identical to HsGN through 202 residues, but then changes abruptly to a different reading frame to give a 279-residue protein. Sequences from two other independent EST clones (EST99320 from thyroid and EST60802 from white blood cells) span this fusion point, suggesting that the keratinocyte cDNA did not result from a cloning artifact. More likely, the cDNA derives from an incompletely processed message or, as a more interesting possibility, an alternatively spliced form. However, the insert region was not flanked by conventional splice junction sequences. Information on the gene structure will help clarify these issues. We refer to the protein predicted from the keratinocyte cDNA as HsGNs (for "short"). At this time, it is not known whether HsGNs is restricted to keratinocytes nor is anything known about its function. The start of the alternate COOH terminus is beyond the equivalent of Tyr-194, making it likely that HsGNs would self-glucosylate.

Northern analysis using glycogenin cDNA as a probe indicated a dominant mRNA of 2.4 kb in humans (*40*) and 1.8 kb in rabbits (*37*). The mRNA distribution was similar and widespread in both species, with strong signals from muscle and heart. The signal from liver was very weak, consistent with the observation that liver contains low levels of glycogenin protein compared to muscle. Analysis of purified liver glycogenin by protein sequencing indicated identity to the muscle glycogenin, suggesting that the same gene is ex-

TABLE I
SEQUENCES OF GLYCOGENIN AND GLYCOGENIN-RELATED GENES[1]

Species	Designation	Type	Accession Number(s)	Source	Description	Ref.
Human	HsGN	cDNA	U44131	Skeletal muscle	Glycogenin isoform	40
	HsGN	cDNA	U31525	Skeletal muscle	Characterized biochemically, self-glucosylates *in vitro*	—
	HsGNs	cDNA	X79537	Keratinocyte	Alternate COOH terminus	—
	Hs-H04891	EST	H04891	Brain	Isoform?	—
	Hs-R71874	EST	R71874	Breast	Isoform?	—
Rabbit	RabGN	cDNA	L01791	Skeletal muscle	Equivalent to HsGN	37
Mouse	MmGN	EST	W34042, W29358		—	—
Saccharomyces cerevisiae	Glg1p	Genomic	U25546		Self-glucosylates *in vitro*	17
	Glg2p	Genomic	U25436		Self-glucosylates *in vitro*	17
Caenorhabditis elegans	CeGlg-a	Genomic	Z72514		T10.B10.8	—
	CeGlg-b	EST	D74982	Embryo	—	—
Caenorhabditis briggsae	CbGlg-a	EST	R03811		Merged sequences	—
	CbGlg-a	EST	R05077		—	—
	CbGlg-b	EST	R04793	Mixed stage	—	—
Arabidopsis thaliana	AtGlg-a	EST	R64916	Mixed tissues	—	—
	AtGlg-b	EST	R30162	Mixed tissues	—	—
	AtGlg-c	EST	T22877	Mixed tissues	—	—
Oryza sativa	WS176	cDNA	D26537	Callus	Protein induced by water stress	—
Schizosaccharomyces pombe	SpGlg	Genomic	Z49811		SPAC5H10.12c	—
Rhodobacter sphaeroides	RsGlg	Genomic	M89780		ORF2 of sulfolipid operon	43

[1] Glgs; as of June, 1996.

pressed in both tissues (19). Examination of the source of ESTs matching glycogenin in sequence databases does not, of course, give reliable quantitative information about tissue distribution, but it is informative that sequences from numerous human tissues are present.

Mammals express at least two other forms of glycogenin. Two human EST sequences present in GenBank would encode proteins similar but not identical to the glycogenin previously discussed. One clone from brain (accession number HO4891) encodes 79 amino acids, with a stop codon in correspondance with that in glycogenin. The COOH-terminal 52 residues have over 40% identity with rabbit and human glycogenin. Northern blotting indicates a very different tissue distribution from glycogenin. The 5'-end of the second clone (accession number R71874), which is from breast, starts at residue 76 of glycogenin and the sequence, over 136 residues, predicts a protein with about 68% identity to human glycogenin. Restriction mapping suggests that the two cDNAs are not identical. Work is ongoing to clone full-length cDNAs and to study the function of these novel glycogenin-like molecules.

C. Glycogenin-like Genes in Other Species

Searches of the sequence databases reveal a number of entries with some degree of similarity to glycogenin (Table I). It must be stressed, though, that so far the only instance in which there is genetic and/or biochemical evidence to support functional homology to glycogenin is for the yeast Glg proteins, and the following attempts at generalization should be taken as provisional. For the other entries in Table I, it remains to be established whether the proteins have glucosyl transferase activity. However, the sequence similarity between the Glg proteins and glycogenin is only ~33% (over domains I–III), and so some entries in Table I have a good chance to be functional glycogenin homologs. The nematodes *Caenorhabditis briggsae* and *Caenorhabditis elegans* each have at least two genes encoding proteins with clear similarity to glycogenin. An open reading frame (ORF) from the fission yeast *Schizosaccharomyces pombe* encodes a protein with low identity to both Glgs and glycogenin. No true bacterial homolog of glycogenin is known so far, but there is a match to a *Rhodobacter sphaeroides* protein encoded by a gene involved in sulfolipid biosynthesis, a pathway that involves UDP sugars (43). Where names had not already been given to the corresponding sequences, we temporarily designate them Glg, for glycogenin-like gene, preceded by a species abbreviation.

Especially interesting is the fact that there are plant-derived sequences that have similarity to glycogenin. Studies of the initiation of starch synthesis in plants have suggested that principles might operate similar to those involved in glycogen biosynthesis in animals. Indeed, recent work has identified a sweet potato self-glucosylating protein called amylogenin, to which attachment of the carbohydrate moiety is through an arginine (44). Protein sequences derived from amylogenin, though matching some rice and *Arabidopsis thaliana* sequences in GenBank, did not match glycogenin. However, there are also three separate EST sequences from *A. thaliana* that encode proteins with some similarity to glycogenin. Potentially, then, *A. thaliana* has self-glucosylating proteins of both the amylogenin and the glycogenin families. In response to water stress, rice expresses a protein, WS176, that has unknown function and that has some sequence similarity to glycogenin.

D. Conserved Domains in Glycogenin and Glycogenin-like Proteins

Multiple sequence alignment of the proteins listed in the previous section is hampered by the fact that several sequences are partial and several are from raw EST data. The latter may contain mistakes and only limited efforts have been made to edit sequences. The absence of a sequence in an alignment is sometimes because the sequence is missing and sometimes because it does not align. Additionally, not all of the proteins may turn out to be functionally similar to glycogenin and the Glg proteins. They may be related more distantly, and sequence similarity may be based on some more restricted shared property such as, say, nucleoside or phosphate binding. The following is a preliminary attempt to find identifiable domains by comparisons of the available sequences of glycogenin and glycogenin-like proteins. The mammalian sequences included are limited to those available from rabbit muscle and from humans; in the latter case, the sequences potentially defining four isoforms were included. Despite the limitations noted above, some regions of strong conservation emerge and some residues are conserved even in the most remote members, the rice WS176, *S. pombe*, and *R. sphaeroides* proteins.

1. Domain I

The NH_2-terminus of glycogenin is conserved in mammalian and some worm sequences (Fig. 5). In this region, there is little sequence similarity with the Glg proteins, which, however, display strong identity to one another in the same segment. The NH_2-terminal portion of domain I is present in SpGlg and RsGlg, but is scarcely seen in the rice WS176 protein.

2. Domain II

This is a region of very strong conservation (Fig. 6). Though the NH_2-terminal part is absent from some sequences (CeGlg-a, WS176, SpGlg, and RsGlg), it is otherwise well conserved. It is characterized by the motif R-P-D/E-L. The COOH-terminal portion, present in all the sequences available, contains an invariant Lys residue and several other residues that are highly

```
RabGN     9   T T N D A Y A K G A L V L G S S L K Q H R T S R R L A V L T T P Q
HsGN     10   T T N D A Y A K G A L V L G S S L K Q H R T T R R L V V L A T P Q
HsGNs    10   T T N D A Y A K G A L V L G S S L K Q H R T T R R L V V L A T P Q
CeGlg-b  13   A T N D R Y A Q G A L T L L N S L H A S G T T R R I H C L I T N E
CbGlg-a   9   A T N D R Y A Q G X - T L L N S L H S S G T N R K I H C L I T N E
CbGlg-b   2   - - - - - - - - - - - - - - - - - - - - - - - - - - - - - - -
WS176    26   A G D G D Y W K G V V G L A K G L R R V R S A Y P L V V A V L P D
Sp-Glg   61   T E E D Y Y F N A T R V L I H R L K Y H P T T K S K Y P I H I L A
Rs-Glg    1   - T N A D Y A L G A R A L L R S L A L S G T T A D R V V L H T D V

RabGN    42   V S D T M R K A L E I V F D E V I T V D I L D S G
HsGN     43   V S D S M R K V L E T V F D E V I M V D V L D S G
HsGNs    43   V S D S M R K V L E T V F D E V I M V D V L D S G
CeGlg-b  46   I S N S V R E K L V N K F D E V T V V D I F N S N
CbGlg-a  41   I S S S V R Q E L E D K F D E V T V V D V F N S N
CbGlg-b   2   - - - - - - - - - - - - - - - I V D V F N S N
WS176    59   V P G E H R R K L V E Q G C V V R E I Q P V Y P P
Sp-Glg   94   L R G V D E W K I E R F R K D G A S V I V I D P I
Rs-Glg   33   P E E A L A P L R A L G A R L V R V E L L P T S P
```

FIG. 5. Domain I. The sequences of Table I were aligned using the MACAW program (44), with subjective decisions over which alignments to accept. Several regions of significant conservation emerged, as shown in this and in Figs. 6–11. Residues identical in more than half of the aligned sequences are boxed.

conserved. Notable is the D-A-D motif where the Asp residues are invariant and the Ala is changed in only three cases.

3. Domain III

Also a region of strong identity, this domain is most conserved at the NH_2 and COOH termini and less in the middle (Fig. 7). Notable motifs in domain III include F-N-X-G and L-N-X-F-F.

4. Domain IV

This small segment follows the poorly conserved region that is known to contain sites of glucosylation in the HsGN, RabGN and the Glg proteins (Fig. 8). It is characterized by the K-P-W motif.

5. Domain V

This is a region in which the human glycogenin with the alternate COOH terminus, HsGNs, aligns with the Glg1p and the Glg2p sequences (Fig. 9).

6. Domain VI

Mammalian glycogenins, HsGN and RabGN, have a region of modest similarity to Glg1p in domain VI (Fig. 10). Thus, the COOH termini of HsGN

```
RabGN      66   D S A H L T L M K R P E L G V T L T K L H C W S L T Q Y S K C V F
HsGN       67   D S A H L T L M K R P E L G V T L T K L H C W S L T Q Y S K C V F
HsGNS      67   D S A H L T L M K R P E L G V T L T K L H C W S L T Q Y S K C V F
Hs-R71874   1   - - - - - - - - K R P E L G L T L T K L H C W T L T H Y S K C V F
Glg1p      88   N S E N L A L L E R P E L S F A L I K A R L W E L T Q F E Q V L Y
Glg2p      86   N K A N L E L L K R P E L S H T L L K A R L W E L V Q F D Q V L F
CeGlg-a    64   - - - - - - - - - - - - - L Y T K I R L W A M T E F D V I V H
CeGlg-b    71   D S E N L S L I G R P D L G V T F T K F H C W R L T Q Y S K A V F
CbGlg-a    65   D S D N L A L I G R P D L G V T F T K I H C W R L T Q Y T K A V F
CbGlg-b    11   D S D N L K L I E R P D L G V T F T K L H C W R L T Q Y T K C V F
AtGlg-b    34   - - - - - - - - K R S Y N X W N Y S K X R V W Q V T D Y D K L V F
AtGlg-c     1   - - - - - - - - - - - - - - - - - - - - - - - - - - - - - - -
WS176      97   - - - - - - - - - - - - - - - N Y S K L R I W E F V E Y E R M V Y
SpGlg     138   - - - - - - - - - - - - - - - M F S K L R I F E Q I Q F D K I C V
RsGlg      59   - - - - - - - - - - - - - - - N F A K L R L W Q L V D Y R S V V F

RabGN      99   M D A D T L V L A N I D D L F E
HsGN      100   M D A D T L V L A N I D D L F D
HsGNS     100   M D A D T L V L A N I D D L F D
Hs-R71874  26   L D A D T L V L S N V D E L F D
Glg1p     121   L D S D T L P L N K E F L K L F
Glg2p     119   L D A D T L P L N K E F F E I L
CeGlg-a    82   L D L D I L P T R D I S T L F E
CeGlg-b   104   L D A D T M I I R N S D E L F E
CbGlg-a    98   L D A D T M V I R N S D E L F E
CbGlg-b    44   L D A D T L V I R N A D E L F T
AtGlg-b    59   I D A D F I I V K N I D Y L X S
AtGlg-c     1   - - - D I Q V Y E N I D H L F D
WS176     115   L D A D I Q V F D N I D H L F D
SpGlg     156   I D S D I L I M K N I D D I F D
RsGlg      77   I D A D A L V L R N V D R L F D
```

FIG. 6. Domain II. See legend to Fig. 5.

and HsGNs have regions of similarity to different segments of the large insert in Glg1p. The significance of this intriguing observation is not clear.

7. Domain VII

This 19-residue segment is thought to be involved in interaction of Glg proteins with glycogen synthase, as discussed previously (Fig. 11). Mutation of the Tyr residue in Glg2p impairs elongation by Gsy2p. Note that this Tyr is in the motif W-E-X-X-D-Y-L. A similar motif, W-E-X-X-X-D-Y-M/L, is present in the unaligned region of several other sequences, including some mammalian glycogenins. Whether this motif is related to a common function, such as glycogen synthase interaction, and defines a functional domain is not known but merits investigation.

The most highly conserved regions are domains II and III and these are

```
RabGN      127   - W P D C F N S G V F V Y Q P S V E T Y N Q L L H V A S E Q G S F
HsGN       128   - W P D C F N S G V F V Y Q P S V E T Y N Q L L H L A S E Q G S F
HsGNs      128   - W P D C F N S G V F V Y Q P S V E T Y N Q L L H L A S E Q G S F
Hs-R71874   53   - W P D C F N S G V F V F Q P S L H T H K L L L Q H A M E H G S F
Glg1p      155   - W P D M F N S G V M M L I P D A D T A S V L Q N Y I F E N T S I
Glg2p      153   - W P D M F N T G V L L L I P D L D M A T S L Q D F L I K T V S I
CeGlg-a    109   - - - D M F N S G V F V L K T N E T V F H D M E Q H V A S A E S Y
CbGlg-b     70   - W P D S F N S G V F V F V P N H E T Y R Q L V G L C R D T R I L
AtGlg-a     32   - P P X Y F N A G M F V Y E P N L S T Y H N L L E T V K I X P P T
AtGlg-b     86   - - - V L F N S G V M V L E P S A C L F E X X M L K S F K X G S Y
AtGlg-c     59   - P A L Y F N A G M F L Y E P N L E T Y E V L L R T L K I T P P T
WS176      177   - P P L Y F N A G M F V H E P G L G T A K D L I D A L V V T P P T
SpGlg      250   - D T P Y F N A G L M L I R P S E L H F N R I L K I G R F P Y M Y
RsGlg      138   - D F H R M N S G V F T A R P S T D T Y A R M L E A L D V P G A F

RabGN      159   D G - G D Q G L L N T F F
HsGN       160   D G - G D Q G I L N T F F
HsGNs      160   D G - G D Q G I L N T F F
Hs-R71874   85   X G - A D Q G L L N S F F
Glg1p      187   D G - S D Q G I L N Q F F
Glg2p      185   D G - A D Q G I F N Q F F
CeGlg-a    139   D G - G D Q G F L N T Y F
CbGlg-b    102   R W E G - - - - - - - -
AtGlg-a     64   L F - A E Q D F L N M Y F
AtGlg-b    116   N G - G A Q G F L N G Y F
AtGlg-c     91   P F - A E Q X F L X M Y F
WS176      209   P F - A E Q D F L N M F F
SpGlg      290   E N A K M M E Q S L L N L
RsGlg      170   W R R T D Q S F L Q Q F F
```

FIG. 7. Domain III. See legend to Fig. 5.

likely involved in catalysis. Domain I is not well conserved between mammals and yeast even though proteins from both are known to self-glucosylate, and it is not yet clear whether this NH_2-terminal region is involved in self-glucosylation. The fact that the human isoforms HsGN and HsGNs are identical through 202 residues could also be used to argue that their distinct COOH termini are not involved in basic catalytic function. Interestingly, the region containing the sites of self-glucosylation in the mammalian and yeast

```
RabGN     206   N A K V V H F - L G Q T K P W N Y
HsGN      207   S A K V V H F - L G R V K P W N Y
Glg1p     244   S I K L I H F - I G K H K P W S L
Glg2p     244   H I R L I H F - I G T F K P W S R
CeGlg-a   196   D P A I F H Y T L G P T K P W L W
WS176     249   Q V K V V H Y C A A G S K P W R F
Rs-Glg    209   Q I R I L H F - - Q Y E K P W Q A
```

FIG. 8. Domain IV. See legend to Fig. 5.

```
HsGNs   223   - - S R L Y P R K N G R N D G N R A R L I I W E Q I P L T T S R G
Glg1p   378   N E S R E Y S K E N D N N I I N S S S N R D Q E S P P N S T Q E L
Glg2p   315   H G K R E N Q K H V D L D I T S V D R N A S Q K S T A E K H D I E

HsGNs   254   N L T L T S S R N T A F F C E H I H
Glg1p   411   N S S Y S V V S T Q A D S D E H Q N
Glg2p   348   K P T S K P Q S A - - - - - - - - -
```

FIG. 9. Domain V. See legend to Fig. 5.

proteins is not very well conserved and does not yield any consensus domain. More structure–function analysis, and ultimately a three-dimensional structure, will be needed to understand fully the significance of the conserved sequences.

IV. Physiological Function of Glycogenin and the Glg Proteins

A. Role of Glycogenin *in Vivo*

The role proposed for glycogenin, in models like that of Fig. 1, is based in large part on biochemical analyses of glycogen and glycogenin; indeed, many properties of glycogenin fit with this postulated function. However, formal proof that glycogenin is required for glycogen accumulation *in vivo* may have to await the production of animals in which the gene has been disrupted by homologous recombination. The existence of multiple isoforms of mammalian glycogenin may make even this strategy more complicated. If other isoforms have functions overlapping with glycogenin, then a single gene knock-out might not elicit a significant phenotype (see below for the Glg proteins). Alternatively, the other isoforms may have cell-specific functions or functions that do not involve glycogen biosynthesis. In this section, it should be recalled that the studies of mammalian systems have addressed the familiar glycogenin and any discussion of the functions or regulation of the other forms of the protein is purely speculative at this time.

In *vitro*, the properties of yeast Glg1p and Glg2p proteins are quite similar to each other and to those of glycogenin, in terms of enzyme activity and elongation by glycogen synthase (*17*). Also, Glg2p is associated covalently

```
HsGN    280   S G A I S H L S L G E I P A M A Q P F V S
RabGN   279   S G A V S H L S L G E T P A T T Q P F V S
Glg1p   442   S G E E S H L D D I S T A A S S N N N V S
```

FIG. 10. Domain VI. See legend to Fig. 5.

```
Glg1p  591  F K F DWE D S D Y L S K V E R C F P
Glg2p  359  F K F DWE S T D Y L D R V Q R A F P
```

FIG. 11. Domain VII. See legend to Fig. 5.

with glycogen extracted from yeast cells and is released by treatment with α-amylase. In mutants lacking glycogen synthase activity, Glg2p is found as a free polypeptide of expected $M_r \sim 43{,}000$ (18). The ease of genetic manipulation of yeast has allowed the physiological role of the self-glucosylating Glg proteins to be probed more deeply than is so far possible with mammalian systems. Deletion of either GLG1 or GLG2 causes no significant change in the extent of glycogen accumulation, suggesting that the encoded proteins have overlapping functions, consistent with their similar properties in vitro. However, cells deleted for both genes cannot accumulate glycogen, providing the best evidence so far that a self-glucosylating protein is involved in and is required for glycogen biosynthesis. Expression of mammalian glycogenin partially restores glycogen accumulation, providing evidence that the mammalian protein can fulfill a similar function. Whether Glg1p and Glg2p differ functionally in some more subtle way has yet to be resolved.

There remains one puzzling result, however. When either GLG1 or GLG2 was mutated at the Tyr residues thought to be sites of glucosylation (Tyr-232 for Glg1p, Tyr-230 and Tyr-232 for Glg2p), self-glucosylating activity in vitro by recombinant proteins produced in E. coli was essentially abolished. When expressed in yeast cells lacking Glg1p and Glg2p, however, these same mutants supported some glycogen accumulation even though the extent was reduced as compared to wild-type. Additional mutations in the COOH-terminal domain VII, the region implicated in glycogen synthase interaction, were necessary for the complete abolition of glycogen storage (18). Either truncation of domain VII or mutation of the conserved Tyr (Tyr-600 in Glg1p or Tyr-367 in Glg2p) was required. Two main possibilities exist to explain these results. First, the COOH-terminal Tyr residue could be a minor site of self-glucosylation, difficult to detect in vitro but sufficient to support some glycogen synthesis in vivo. Alternatively, mutant Glg proteins, no longer capable of self-glucosylation, may allow glycogen synthase to synthesize glucan chains from alternate primers. These substitute primers could be spurious and due to the presence of undirected glycogen synthase activity. However, we cannot formally exclude the possibility that the transglucosylase activity of the Glg proteins is utilized physiologically for initiation from other legitimate, but as yet unidentified, proteins. Mutations of the Glg proteins in domain VII would then disrupt interaction with glycogen synthase and so suppress this mode of synthesis.

B. Is Glycogenin Rate Determining for Glycogen Synthesis?

The initiation stage is an attractive site at which to regulate the synthesis of any polymer and this idea holds true also for glycogen accumulation. One could reason that the activity or level of glycogenin would determine the number of glycogen molecules in the cell, and hence the amount of glycogen stored. The genetic experiments described above with yeast prove that Glg proteins are necessary for physiological glycogen accumulation and impaired activity can reduce glycogen storage. However, these are powerful mutations and any enzyme in a pathway becomes rate determining as its activity approaches zero. Thus, a different test is to overexpress the protein and look for increased flux through the relevant pathway. This has been done in yeast (*18*) and animal (*45, 46*) cells. Overexpression, by severalfold, of Glg2p in yeast had no effect on glycogen storage, suggesting that its level is not normally limiting for glycogen synthesis. A similar result was obtained from transient overexpression of glycogenin in COS cells (*45, 46*). A more detailed study was made of a stably transformed Rat1 fibroblast cell line that overexpresses rabbit muscle glycogenin (*14*). Again, the overexpression did not alter the total amount of glycogen accumulated but did affect its distribution, promoting the presence of soluble glycogen, of smaller molecular weight.

C. Regulation of Glycogenin and the Glg Proteins

Even if overexpression of glycogenin or the Glg proteins does not increase glycogen deposition, this does not mean that these proteins are not subject to regulation. If the activity were negatively regulated, for example, then overexpression of the wild-type protein need not affect the pathway. A good example of a similar phenomenon is provided by glycogen synthase, where substantial overexpression of the wild-type enzyme does not affect glycogen storage unless there are activating mutations that overcome negative control by phosphorylation (*45, 46*).

So far, information on possible modes of regulation of glycogenin function is scant. There has been some controversy as to whether glycogenin is a phosphoprotein. Lomako and Whelan (*47*) reported that Ser-43 is phosphorylated *in vitro* by cAMP-dependent protein kinase, although we and others have been unable to substantiate this result. Lomako *et al.* (*48*) reported ^{32}P-labeling *in vivo* of a species purported to be glycogenin. We saw no phosphorylation of rabbit glycogenin expressed in COS cells under conditions where we could easily detect labeling of glycogen synthase (*45*). However, similar experiments with a stable Rat1 fibroblast line overexpressing glycogenin revealed weak labeling of glycogenin but to a level much lower than that of glycogen synthase (A.V. Skurat and P. J. Roach, unpublished). In extracts from yeast cells metabolically labeled with ^{32}P we could not detect phosphorylation of endogenous Glg2p, but when purified recombinant Glg2p was added to a yeast extract together with ATP/Mg^{2+}, a trace phosphorylation of purified Glg2p was detected (J. Mu and P. J. Roach, unpublished). Although a role for phosphorylation of glycogenin or Glg proteins cannot be totally excluded, the evidence to date must be considered weak.

The presence of glycogenin affects the activity state of glycogen synthase. When glycogenin is stably expressed in Rat1 cells, it increases the activation state of glycogen synthase (A.V. Skurat and P. J. Roach, unpublished). Conversely, mutant yeast cells that do not express Glg proteins have glycogen synthase that is inactivated compared to controls (17). It is not clear if this is an evolved regulatory strategy or a consequence of interfering with the assemblage of proteins involved in glycogen biosynthesis. Recall that among the proteins that interact with yeast Glg2p are phosphatase regulatory subunits (see Fig. 4), and so phosphatase recognition or accessibility might be somehow be altered.

Glucose concentration alone can have an important influence on the metabolism of mammalian cells, especially cells such as hepatocytes, whose glucose transport system is not dependent on the strictly regulated Glut4 receptors. In COS cells transiently expressing glycogenin, the protein was detected at its free molecular weight and it was not converted, at least in major amounts, into glycogen (45, 46). Increasing the glucose concentration caused a shift to a slightly higher apparent M_r that was attributed to a higher degree of priming of the protein with oligosaccharide. It was proposed that the glucose concentration might thus regulate the initiation step by controlling the glucosylation state of glycogenin (49), which *in vitro* determines its efficacy as a glycogen synthase substrate (35, 37). A possible mechanism to mediate this control was provided by the observation that glycogen phosphorylase can utilize primed glycogenin as a substrate for phosphorylsis (50). Glucose inhibition of phosphorylase could then favor the self-glucosylation of glycogenin. However, increased glucose still could not overcome negative controls of glycogen synthase in COS cells, and significant increases in the total level of glycogen accumulation were observed only after expression of a hyperactive glycogen synthase mutant (45, 46).

In terms of glycogen synthesis, the response of Rat 1 fibroblasts to glucose concentration is very different from that of COS cells. Increased glucose concentration promotes glycogen accumulation and in cells stably expressing rabbit muscle glycogenin, all of the glycogenin can be converted to glycogen (14). The exact locus of glucose control in this system is unknown. It could be at the initiation stage, as postulated above, or through control of glycogen synthase or both. Glycogen synthase is allosterically activated by the glucose metabolite glucose-6-P (4), which may also influence the phosphorylation of glycogen synthase (51–53). In liver cells, it is thought that glycogen phosphorylase is a key locus for glucose interaction (54). Understanding this difference in behavior between COS and Rat1 cells will be interesting, but whether it relates more to the elongation or the initiation step is uncertain.

D. Other Possible Roles of Glycogenin

A role for glycogenin in generating an oligosaccharide primer for glycogen synthesis would seem well established, but this does not preclude its having other functions. One such function might be in intracellular targeting. Glycogen and glycogen synthase in cultured cells or in freshly prepared hepatocytes are distributed between the soluble and pellet fractions after low-speed centrifugation (49, 55). Ectopically expressed glycogen synthase is recovered, as an active enzyme, mainly in the pellet fraction (45). In contrast, glycogenin expressed in cultured cells is found predominantly in the soluble fraction, whether attached to glycogen or not. In COS and Rat1 cells, overexpression of glycogenin caused a redistribution of glycogen synthase to the soluble fraction. Concomitantly, it increased the soluble glycogen at the expense of the pellet glycogen and may favor the formation of smaller glycogen molecules (14, 45). One possible explanation is that the pellet glycogen fraction contains the largest glycogen molecules. The glycogenin level might simply dictate how many glycogen molecules are made and not how much glucose is polymerized. Thus, increased amounts of glycogenin would correlate with smaller glycogen molecules. An alternative hypothesis, which is perhaps inherently more interesting, is that the glycogenins have targeting functions. The glycogenin studied to date would target glycogen synthase to make soluble glycogenin. Perhaps other forms of glycogenin might be involved in targeting to other cellular locations, such as structures that precipitate in the low-speed pellet. Perhaps some of the different characteristics of glycogen in different tissues, liver and muscle being the notable ones, are conferred by the complement of glycogenins present.

E. Glycogenin and Diabetes

Non-insulin-dependent diabetes mellitus (NIDDM), the most prevalent form of diabetes, is a collection of disorders that are believed to have a genetic basis (56). Only for a few rare syndromes has that genetic basis been elucidated. It is believed that the majority of cases of NIDDM will involve multiple genetic loci that combine with environmental factors and aging in the development of the disease (57). Identification of such genetic defects will be difficult. One approach is to consider candidate genes for the disease based on current understanding of physiology and to look for any link between affected individuals and mutations in the gene.

Impaired disposal of blood glucose is the critical defect of diabetes mellitus. In the normal individual, most glucose postprandially is converted into glycogen and therefore a deficiency in glycogen storage could promote hyperglycemia. Numerous factors can affect glycogen accumulation and included would be mutations affecting enzymes of the biosynthetic pathway. Glycogen synthase (muscle isoform) (57) and the type 1 phosphatase regulatory subunit, R_{Gl} (58), have both been considered candidate genes for NIDDM. So far, there is no clear indication that mutations in these genes are linked to the disease in the populations studied, but this does not exclude their relevance in other groups. One study has reported that a widespread polymorphism in the R_{Gl} subunit is associated with insulin resistance (59). The glycogenin genes are obvious candidates for predisposition to NIDDM because their defect could impair glucose disposal.

V. Conclusion

Many key features of the model of Fig. 1, which was built on the studies of C. Krisman, W. Whelan, and P. Cohen, now seem quite well established from the work of the past few years. Some interesting details remain to be worked out but the overall scheme seems to hold. Glycogenin-like genes and proteins are being discovered in a variety of different species, generating satisfaction that the model might be of widespread applicability. As often happens in science, simplification may be fleeting and is to be enjoyed quickly before the picture again becomes more complex. Lurking in the databases is evidence for novel isoforms of glycogenin. Will their study simply confirm the model with minor variations in different cell types? Or will these new forms force us to revise the model completely? Either way, the next few years promise to be exciting in the study of glycogenin and self-glucosylating proteins.

Acknowledgments

Work from the authors' laboratories was supported by grants from the National Institute of Diabetes and Digestive and Kidney Disease, Grants R23 DK27221 and R01 DK42576 (P.J.R.), and an American Diabetes Association grant (A.V.S.).

REFERENCES

1. W. J. Whelan, *BioEssays* **5**, 136 (1986).
2. C. Smythe and P. Cohen, *Eur. J. Biochem.* **200**, 625 (1991).
3. M. D. Alonso, J. Lomako, W. M. Lomako, and W. J. Whelan *FASEB J.* **9**, 1126 (1995).
4. A. V. Skurat and P. J. Roach, in "Diabetes Mellitus: A Fundamental and Clinical Text" (D. LeRoith, J. M. Olefsky, and S. I. Taylor, eds.), pp. 213. Lippincott-Raven Pub., Philadelphia, 1996.
5. E. Salsas and J. Larner, *J. Biol. Chem.* **250**, 1833 (1975).
6. C. R. Krisman and R. Barrengo, *Eur. J. Biochem.* **52**, 117 (1975).
7. J. Pitcher, C. Smythe, D. G. Campbell, and P. Cohen, *Eur. J. Biochem.* **169**, (1987).
8. J. Lomako, W. M. Lomako, and W. J. Whelan *FASEB J.* **2**, 3097 (1988).
9. J. Pitcher, C. Smythe, and P. Cohen *Eur. J. Biochem.* **176**, 391 (1988).
10. M. D. Alonso, J. Lomako, W. M. Lomako, W. J. Whelan, and J. Preiss, *FEBS Lett.* **352**, 222 (1994).
11. J. Lomako, W. M. Lomako, and W. J. Whelan, *FEBS Lett.* **279**, 223 (1991).
12. J. Lomako, W. M. Lomako, W. J. Whelan, R. S. Dombro, J. T. Neary, and M. D. Norenberg, *FASEB J.* **7**, 1386 (1993).
13. N. Ercan, M. C. Gannon, and F. Q. Nuttall, *J. Biol. Chem.* **269**, 22328 (1994).
14. A. V. Skurat, S.-S. Lim, and P. J. Roach, submitted (1996).
15. K. Weber and M. Osborn, *J. Biol. Chem.* **244**, 4406 (1969).
16. I. Farkas, T. A. Hardy, M. G. Goebl, and P. J. Roach *J. Biol. Chem.* **266**, 15602 (1991).
17. C. Cheng, J. Mu, I. Farkas, D. Huang, M. G. Goebl, and P. J. Roach, *Mol. Cell. Biol.* **15**, 6632 (1995).
18. J. Mu, C. Cheng, and P. J. Roach, *J. Biol. Chem.* **271**, 26554 (1996).
19. C. Smythe, C. Villar-Palasi, and P. Cohen, *Eur. J. Biochem.* **183**, 205 (1989).
20. I. R. Rodrigues and S. J. Fliesler, *Arch. Biochem. Biophys.* **260**, 628 (1988).
21. E. Meezan, S. Ananth, S. Manzella, P. Campbell, S. Siegal, D. J. Dillion, and L. Roden, *J. Biol. Chem.* **269**, 11503 (1994).
22. L. Roden, S. Ananth, P. Campbell, S. Manzella, and E. Meezan, *J. Biol. Chem.* **269**, 11509 (1994).
23. M. Alonso, J. Lomako, W. M. Lomako, and W. J. Whelan, *J. Biol. Chem.* **270**, 15315 (1994).
24. C. Smythe, P. Watt, and P. Cohen, *Eur. J. Biochem.* **189**, 199 (1990).
25. Y. Cao, L. K. Steinrauf, and P. J. Roach, *Arch. Biochem. Biophys.* **319**, 293 (1995).
26. F. Meyer, L. M. G. Heilmeyer, R. H. Haschke, and E. H. Fischer, *J. Biol. Chem.* **245**, 6642 (1970).
27. L. M. G. Heilmeyer, F. Meyer, R. H. Haschke, and E. H. Fischer, *J. Biol. Chem.* **245**, 6649 (1970).
28. S. V. Polishchuk, N. R. Brandt, H. E. Heilmeyer, M. Varsanyi, and L. M. G. Heilmeyer, Jr., *FEBS Lett.* **263**, 271 (1995).
29. P. Stralfors, A. Hiraga, and P. Cohen, *Eur. J. Biochem.* **149**, 295 (1985).
30. P. M. Tang, J. A. Bondor, K. M. Swiderek, and A. A. DePaoli-Roach, *J. Biol. Chem.* **266**, 15782 (1991).
31. J. François, S. Thompson-Jaeger, J. Skroch, W. Zellenka, W. Spevak, and K. Tatchell, *EMBO J.* **87 (1992).**
32. C. Cheng, D. Huang, and P. J. Roach, *Yeast* **13**, 1 (1997).
33. J. Skrotch Stuart, D. L. Frederick, C. M. Varner, and K. Tatchell, *Mol. Cell. Biol.* **14**, 896 (1994).
34. C. Smythe, F. B. Caudwell, M. Ferguson, and P. Cohen, *EMBO J.* **7**, 2681 (1988).

35. Y. Cao, A. M. Mahrenholz, A. A. DePaoli-Roach, and P. J. Roach, *J. Biol. Chem.* **268,** 14687 (1993).
36. M. D. Alonso , J. Lomako, W. M. Lomako, and W. J. Whelan, *FEBS Lett.* **342,** 38 (1994).
37. E. Viskupic, Y. Cao, W. Zhang, C. Cheng, A. A. DePaoli-Roach, and P. J. Roach, *J. Biol. Chem.* **267,** 25759 (1992).
38. J. Lomako, W. M. Lomako, and W. J. Whelan, *FEBS Lett.* **264,** 13 (1990).
39. S. M. Manzella, L. Roden, and E. Meezan, *Anal. Biochem.* **216,** 383 (1994).
40. F. Barbetti, M. Rocchi, M. Bossolasco, R. Cordera, P. Sbraccia, P. Finelli, and G. G. Gonsalez, *Biochem. Biophys. Res. Commun.* **220,** 72 (1996).
40a. J. Lomako, K. Mazurak, W. M. Lomako, M. D. Alonso, W. J. Whelan and I. R. Rodriguez, *Genomics* **33,** 519 (1996).
41. D. G. Campbell and P. Cohen, *Eur. J. Biochem.* **185,** 119 (1989).
42. C. Benning and C. R. Somerville, *J. Bacteriol.* **174,** 6479 (1992).
43. D. G. Signh, J. Lomako, W. M. Lomako, W. J. Whelan, H. E. Meyer, M. Serwe, and J. W. Metzger, *FEBS Lett.* **376,** 61 (1995).
44. G. D. Schuler, S. F. Altschul, and D. J. Lipman, *Proteins* **9,** 180 (1991).
45. A. V. Skurat, Y. Wang, and P. J. Roach, *J. Biol. Chem.* **269,** 25534 (1994).
46. A. V. Skurat, H.-L. Peng, H.-Y. Chang, J. F. Cannon, and P. J. Roach, *Arch. Biochem. Biophys.* **328,** 283 (1996).
47. J. Lomako and W. J. Whelan, *Biofactors* **1,** 261 (1988).
48. J. Lomako, W. M. Lomako, and W. J. Whelan, *Eur. J. Biochem.* **234,** 343 (1995).
49. A. V. Skurat, Y. Cao, and P. J. Roach, *J. Biol. Chem.* **268,** 14701 (1993).
50. Y. Cao, A. V. Skurat, A. A. DePaoli-Roach, and P. J. Roach, *J. Biol. Chem.* **268,** 21717 (1993).
51. J. C. Lawrence Jr., and J. Larner, *J. Biol. Chem.* **253,** 2104 (1978).
52. J. Francois and H. G. Hers, *Eur. J. Biochem.* **174,** 561 (1988).
53. C. Villar-Palas, *Biochim. Biophys. Acta* **1244,** 203 (1995).
54. H. G. Hers, *Annu. Rev. Biochem.* **42,** 167 (1976).
55. J. M. Fernandez-Novell, J. Arino, S. Vilaro, and J. J. Guinovart, *Biochem. J.* **281,** 443 (1992).
56. S. S. Fajans, in "Diabetes Mellitus: A Fundamental and Clinical Text" (D. LeRoith, J. M. Olefsky, and S. I. Taylor, eds.), p. 251. Lippincott-Raven Pub., Philadelphia, 1996.
57. A. R. Shuldiner and K. D. Silver, in "Diabetes Mellitus: A Fundamental and Clinical Text" (D. LeRoith, J. M. Olefsky, and S. I. Taylor, eds.), p. 565. Lippincott-Raven Pub., Philadelphia, 1996.
58. C. Bjorbaek, T. A. Vik, S. M. Echwald, P. Y. Yang, H. Vestergaard, J. P. Wang, G. C. Webb, K. Richmond, T. Hansen, R. L. Erikson *et al.*, *Diabetes* **44,** 90 (1995).
59. L. Hansen, T. Hansen, H. Vestergaard, C. Bjorbaek, S. M. Echwald, J. O. Clausen, Y. H. Chen, M. X. Chen, P. T. W. Cohen, and O. Pedersen, *Hum. Mol. Genet.* **4,** 1313 (1995).

Molecular Genetics of Yeast TCA Cycle Isozymes

LEE MCALISTER-HENN[1]
AND W. CURTIS SMALL

Department of Biochemistry
University of Texas Health Science Center
San Antonio, Texas 78284

I. Malate Dehydrogenases	319
A. Genes and Mutant Phenotypes	319
B. Structures and Expression	321
C. Cellular Localization and Compartmental Function	325
D. Remaining Questions about Function	326
II. Isocitrate Dehydrogenases	326
A. Genes and Mutant Phenotypes	327
B. Structures and Expression	329
C. Cellular Compartmentation	330
III. Citrate Synthases	331
A. Genes and Mutant Phenotypes	331
B. Structures and Regulation of Expression	332
IV. Aconitase and Fumarase Isozymes	333
A. Aconitase	333
B. Fumarase	334
V. Conclusions and Perspective	335
References	337

The tricarboxylic acid (TCA) cycle is ubiquitous in organisms with oxidative metabolism. Associated with mitochondrial compartmentation of the cycle in eukaryotic cells is replication of some enzymatic activities in other compartments. Because communication between pathways separated by membrane barriers depends on selective transport of a limited number of common metabolites, the TCA cycle isozyme families are postulated to be critical points for control of metabolic flux. A classic example is the pair of mitochondrial and cytosolic malate dehydrogenases that direct the flux of carbon and of reducing equivalents between the TCA cycle and cytosolic pathways.

In *Saccharomyces cerevisiae*, the existence of redundant TCA cycle enzymatic activites (Fig. 1) has been apparent since early purification and cel-

[1] To whom correspondence may be addressed.

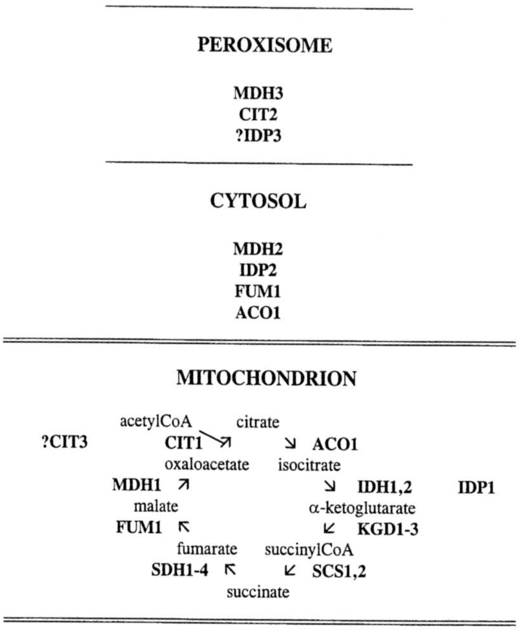

```
                      PEROXISOME

                        MDH3
                        CIT2
                        ?IDP3


                       CYTOSOL

                        MDH2
                        IDP2
                        FUM1
                        ACO1


                     MITOCHONDRION

            acetylCoA     citrate
   ?CIT3      CIT1 ↗       ↘ ACO1
            oxaloacetate    isocitrate
          MDH1 ↗             ↘  IDH1,2    IDP1
           malate              α-ketoglutarate
          FUM1 ↖              ↙ KGD1-3
            fumarate    succinylCoA
          SDH1-4 ↖       ↙ SCS1,2
                       succinate
```

FIG. 1. Compartmental localization of yeast TCA cycle isozymes. MDH, Malate dehydrogenase; CIT, citrate synthase; IDP, $NADP^+$-specific isocitrate dehydrogenase; FUM, fumarase; ACO, aconitase; IDH, NAD^+-specific isocitrate dehydrogenase; KGD, α-ketoglutarate dehydrogenase; SCS, succinyl CoA synthetase; SDH, succinate dehydrogenase.

lular fractionation studies. In comprehensive analyses by Duntze et al. (1) and Perlman and Mahler (2), nonmitochondrial activities were identified for malate and isocitrate dehydrogenases, citrate synthase, fumarase, and aconitase. Among metabolic functions proposed for several of these enzymes is catalysis of reactions in the glyoxylate cycle (Fig. 2), which permits growth on the two-carbon sources, ethanol and acetate. Catalytic roles of other enzymes were unclear.

This review focuses on progress from this and other laboratories toward characterization of structural, metabolic, and genetic relationships within the TCA cycle isozyme families in yeast. This simple eukaryote provides obvious advantages for molecular and genetic manipulation. In particular, analysis of metabolic function is facilitated by straightforward techniques for gene disruption, and, in several instances, sequential disruption of genes within a family has been essential for identification and purification of residual isozymes. Also, recent completion of genome sequence analysis has helped clarify some relationships and has revealed several homologs of unknown function. The availability of complete collections of related genes for an organism clearly

MOLECULAR GENETICS OF YEAST TCA CYCLE ISOYZMES   319

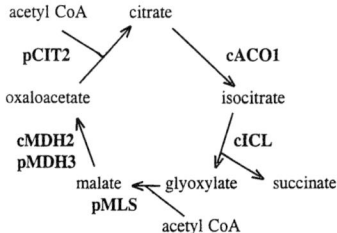

FIG. 2. Compartmental localization of glyoxylate cycle enzymes. ICL, Isocitrate lyase; MLS, malate synthase; p, for peroxisomal, c, cytsolic; other abbreviations are as for Fig. 1.

provides an enormous advantage for defining structural and compartmental determinants. This review also describes some of the unexpected outcomes of research in this area, including discovery of novel mechanisms for compartmentalization and for regulation of gene expression.

I. Malate Dehydrogenases

There are at least two forms of malate dehydrogenase in most eukaryotic cells, a mitochondrial enzyme that functions in the TCA cycle and a cytosolic enzyme that catalyzes the first step in gluconeogenesis from pyruvate. These isozymes also participate in the malate/aspartate shuttle cycle for the indirect exchange of reducing equivalents between cellular compartments. Because of the equilibrium of the malate dehydrogenase reaction, supply of oxaloacetate, which is compartmentally restricted by transport barriers, is rate limiting for these metabolic processes. Also because of the unfavorable equilibrium for oxaloacetate production, it has been proposed that malate dehydrogenase may physically interact with enzymes within specific metabolic pathways to ensure efficient delivery of that product (3). Thus, questions motivating research on this isozyme family include not only structural relatedness and mechanisms for compartmental localization, but investigation of potential structural determinants for function within different pathways.

A. Genes and Mutant Phenotypes

Nuclear genes (*MDH1* and *CIT1*) encoding mitochondrial malate dehydrogenase and citrate synthase were among the first TCA cycle enzyme genes to be cloned and disrupted in yeast (4–6). These disruptions produced a similar and, at that time, an unexpected growth phenotype, the inability to grow with acetate as a carbon source. The ability of strains containing these disruptions to grow on other nonfermentable carbon sources, including glycerol, clearly distinguishes these mutants from nuclear *pet* or mitochondrial *petite* mutants that characterize numerous other mitochondrial dysfunctions in S. *cerevisiae*.

Definition of this distinct phenotype led to coordinated efforts to examine defects in a collection of yeast acetate nonutilizing, glycerol$^+$ (ACN) mutants isolated by Dr. Mark T. McCammon (7). Mutants in two complementation groups were found to have defects in mitochondrial citrate synthase and malate dehydrogenase. Further preliminary analyses of phenotypes of ACN and related yeast mutants led to formulation of some useful working generalities. First, mutants with defects in TCA cycle enzymes but with other isozyme counterparts in the cell exhibit an energetic acetate$^-$ growth phenotype, i.e., an inability to grow with acetate as a carbon source on either rich or minimal medium; the ability of these mutants to grow on glycerol is attributed to bypass reactions catalyzed by residual isozymes. Second, several mutants with defects in gluconeogenic or glyoxylate cycle enzymes can be distinguished from the TCA cycle mutants by a more narrow auxotrophic acetate$^-$ growth phenotype, i.e., the ability to grow on rich but not minimal medium with acetate as a carbon source, reflecting the role of these pathways in carbon assimilation. Some additional features of these phenotypes are discussed below.

Immunochemical screening led to identification of an ACN complementation group with defects in the cytosolic isozyme of malate dehydrogenase (MDH2) (8). Mutants in this group and an *MDH2* gene disruption mutant exhibit the auxotrophic acetate$^-$ growth phenotype; unfortunately, the specific nutritional requirement(s), present in rich but absent in minimal medium, has eluded definition. Haploid mutant strains with both *MDH1* and *MDH2* gene disruptions were found to be acetate$^-$ but still capable of growth on glycerol, and extracts from these mutants were found to have residual malate dehydrogenase activity. The *MDH3* gene was cloned following purification of the third isozyme from such extracts (9). Isozyme-specific antisera and cellular fractionation techniques were used to demonstrate localization of MDH3 (9) and of low levels of MDH2 to peroxisome fractions (10); however, MDH2 is primarily found in soluble, nonorganellar fractions.

Disruption of *MDH3* produces no dramatic growth phenotype in some strains (9) but an inability to utilize oleate as a carbon source in other strains (11; L. McAlister-Henn and W. C. Small, unpublished observations). The latter phenotype has been interpreted as indicating an important role for MDH3 in reoxidizing NADH generated during β-oxidation of fatty acids in peroxisomes. Because disruption of *MDH2* affects acetate or ethanol utilization as a carbon source, whereas disruption of *MDH3* does not, MDH2 may be the critical enzyme for glyoxylate metabolism. Peroxisomal localization does not appear to be essential for this function because other glyoxylate pathway enzymes, including aconitase and isocitrate lyase, appear to be soluble cytosolic activities (1, 10). This suggests significant intercompartmental flux of intermediates (Fig. 2).

B. Structures and Expression

The three yeast malate dehydrogenases are all homodimers, and they exhibit similar kinetic properties (9). The aligned amino acid sequences of the similarly sized subunits (Table I) have residue identities ranging from 43 to 55%. Among strictly conserved residues are those with catalytic functions and those that participate in cofactor binding based on comparisons with crystallographic structures determined for homologous mammalian and *Escherichia coli* enzymes (12, 13). Among salient differences are regions pertinent to organellar targeting (Fig. 3). The mitochondrial MDH1 enzyme has a 17-residue amino-terminal extension not present on the other isozymes; this presequence is removed on mitochondrial import (14). The peroxisomal MDH3 enzyme has a unique carboxy-terminal Ser-Lys-Leu tripeptide, a canonical type I peroxisomal localization sequence (15); this tripeptide has been shown to be essential for organellar localization of MDH3 (16).

Cellular levels of all three malate dehydrogenases are reduced in yeast cells cultivated with glucose as a carbon source. For MDH1 and MDH3, this appears to be the result of catabolite repression of gene expression, a common effect of glucose on many oxidative functions in yeast. Levels of MDH2 are subject to more dramatic regulation by catabolite inactivation. This phenomenon involves rapid and specific inactivation and degradation of several gluconeogenic and glyoxylate pathway enzymes when glucose is added to an actively respiring yeast culture (17). At least one unique structural feature of MDH2, the amino-terminal 12-residue extension not present on the other isozymes (Fig. 3), is a determinant for catabolite inactivation. Removal of the extension makes MDH2 refractory to catabolite inactivation (Fig. 4) (18) and also eliminates phosphorylation of MDH2, an early event in inactivation and turnover (19). Expression of the amino-terminal truncated form of MDH2 was found to have the unexpected effect of also retarding rates of catabolite inactivation of other enzymes, including fructose-1,6-bisphosphate and isocitrate lyase. These molecular changes were found to result in a significant lag in the time required for yeast cultures to adapt from slow acetate growth rates to rapid growth rates on glucose (Fig. 5), the first experimental demonstration of a selective advantage for the catabolite inactivation phenomenon.

Maximum cellular levels of expression of the malate dehydrogenase isozymes are attained with growth on C_2 sources, ethanol and acetate. Under such conditions, MDH2 contributes approximately 65% and MDH3 approximately 10% of the total cellular activity. With growth on glucose, over 90% of the much lower total cellular activity is attributed to MDH1 (9).

TABLE I
SUMMARY—YEAST TCA CYCLE ISOZYMES

Isozyme[a]	Function[b]	Size[c]	Residues[d]	Gene	Chromosome	Expression
mMDH	TCA cycle	2 × 33,800	317 (17)	MDH1	XI	glucose repression
cMDH	gluconeogenesis, (?)glyoxylate cycle	2 × 40,700	376[e]	MDH2	XV	glucose repression, catabolite inactivation
pMDH	NAD$^+$ regeneration, (?)glyoxylate cycle	2 × 37,100	343	MDH3	IV	glucose repression
mNAD-IDH	TCA cycle, glutamate synthesis	4 × 38,000 / 4 × 37,800	349 (11) / 354 (15)	IDH1 / IDH2	XIV / XV	glucose repression, glutamate repression
mNADP-IDH	glutamate synthesis	2 × 46,400	412 (16)	IDP1	IV	constitutive
cNADP-IDH	glutamate synthesis	2 × 46,600	412	IDP2	XII	glucose repressed
p(?)NADP-IDH	?	(2 ×?) 47,000	420	IDP3	XIV	?
mCS	TCA cycle, glutamate synthesis	2 × 49,000	479 (?)	CIT1	XIV	glucose repression, glutamate repression
pCS	glutamate synthesis, (?)glyoxylate cycle	2 × 51,400	460	CIT2	III	retrograde regulation
m(?)CS	?	(2 ×?) 51,200	486 (?)	CIT3	XVI	?
m/cACON	TCA cycle, glutamate synthesis	82,800	778 (?)	ACO1	XII	glucose repression, glutamate repression
(?)ACON	?	86,600	789	ACO2	X	?
m/cFUM	TCA cycle	4 × 50,200	488 (?)	FUM1	XVI	?

[a] CS, Citrate synthase; ACON, aconitase; m, mitochondrial; c, cytosolic; p, peroxisomal; other abbreviations are as in Fig. 1.
[b] Functions are based on mutant phenotypes.
[c] Molecular weights are for the mature polypeptide.
[d] Numbers of amino acid residues are given for mature polypeptides, with precursor sizes in parentheses or for the entire open reading frame if the amino terminus of the mature polypeptide is unknown.
[e] The open reading frame at the MDH2 locus contains 422 amino acid residues. The number shown is based on amino acid sequence analysis of the purified protein (8, 89).

AMINO TERMINI

```
                                                 (Y)            (A)           (N)
                                                  ↑              ↑             ↑
MDH1: [MLSRVAKRAFSSTVANP]YKVTVLGAGGGIGQPLSLLLK------LNHKVTD--LRLYDLK--GAKGV (40)
                         ||||||||||||||||||         ::.  |||  ||::.  :  |||
MDH3:                    MVKVAILGASGGVGQSLSLLLK------LRSYVSE--LALYDIR--AAEGI (41)
                         .:.|||||||||:|||||||         :. .| |||:|  .:   :.
MDH2: PHSVTPSIEQDSLKIAILGAAGGIGQPLSLLLKAQLQYQLKESLRSSVTHIHLALYDVNQEAINGV (66)
```

CARBOXY TERMINI

```
MDH1:  LKKNIEKGVNFVASK        (317)
       ||||:||.||.:||.
MDH2:  LKKNIDKGLEFVASRSASS    (376)
       .:.|||.|:.|||
MDH3:  LRKNIEKGKSFILDSSKL     (343)
```

FIG. 3. Aligned amino- and carboxy-terminal sequences of yeast malate dehydrogenase isozymes. Identical (|) and similar (:) residues are indicated, with dashes representing gaps necessary to maximize alignment. The mitochondrial targeting presequence of MDH1 is shown in brackets. Residue changes in MDH1 described in the text are shown above the sequence.

323

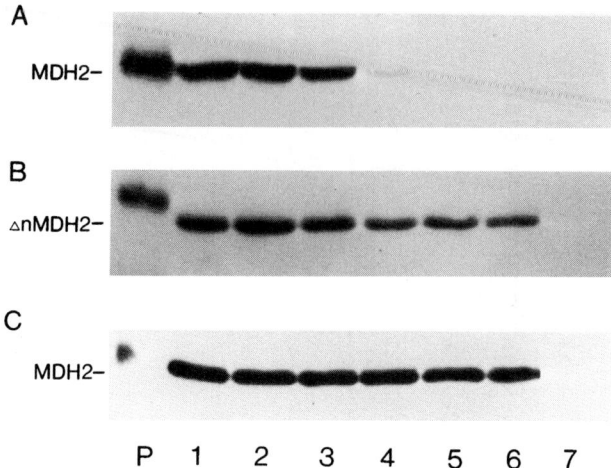

Fig. 4. Glucose-induced changes in levels of MDH2. Yeast strains expressing MDH2 (A), a truncated form of MDH2 lacking amino acid residues 1–12 (B), or elevated levels of MDH2 using a multicopy expression vector (C) were shifted from acetate to glucose growth medium. Protein extracts (100 μg) were prepared from cells harvested at 0 (lanes 1), 15 (lanes 2), 30 (lanes 3), 60 (lanes 4), 90 (lanes 5), and 120 minutes (lanes 6) following the shift and were used for immunoblot analysis with anti-MDH2 antiserum. Lanes 7 contained protein extracts from cells harvested during steady-state growth with glucose. Lanes P contained 0.5 μg of purified MDH2. (Modified from Ref. 18, with permission.)

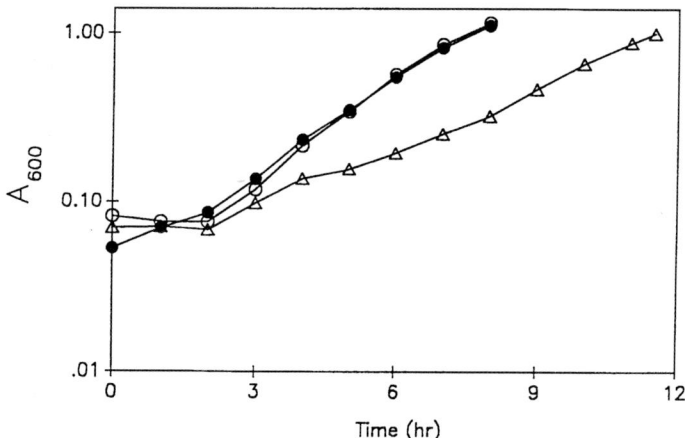

Fig. 5. Growth rates of yeast strains expressing various forms of MDH2 following a shift to glucose medium. Yeast strains expressing MDH2 (○), the amino-terminal truncated form of MDH2 (△), or multicopy levels of MDH2 (●) were pregrown in rich medium with acetate then shifted to rich medium with glucose as the carbon source. Growth rates were measured spectrophotometrically. Normal levels of the truncated enzyme, but not elevated levels of the authentic enzyme, retard the transition to rapid glucose growth rates. (Modified from Ref. 18, with permission.)

C. Cellular Localization and Compartmental Function

One goal of recent research has been to alter cellular localization of the malate dehydrogenase isozymes to assess structural features important for function in specific compartmental or metabolic pathways. Because of the presence of canonical targeting sequences on the amino terminus of MDH1 and the carboxy terminus of MDH3, efforts have focused on obtaining cytosolic localization of these organellar isozymes to determine if they can functionally compensate for loss of cytosolic MDH2.

For peroxisomal MDH3, altered localization is easily achieved. Expression of an *MDH3* gene lacking the carboxy-terminal codons for the Ser-Lys-Leu tripeptide eliminates peroxisomal localization *in vivo* and results in a catalytically active, stable cytosolic form of the enzyme (16). The cytosolic form of MDH3 was determined to be competent for replacement of MDH2 in that expression in an *MDH2* disruption mutant restores parental growth rates with acetate or ethanol on minimal medium. Interestingly, addition of the unique amino-terminal 12-residue extension from MDH2 to the cytosolic form of MDH3 does not alter patterns of expression, demonstrating that the extension is necessary but not sufficient for the phenomenon of catabolite inactivation.

Altering localization of mitochondrial MDH1 has proved to be more complicated. Removal of codons for the amino-terminal presequence and expression of the coding region for the mature polypeptide was determined to have little or no effect *in vivo*, i.e., the presequence appears to be unnecessary for mitochondrial localization (20). A putative cryptic targeting sequence near the amino terminus of the mature form of MDH1 was identified by comparison of yeast polypeptide sequences (Fig. 3); in particular, several positively charged residues (Lys-25, Arg-30, and Lys-38) spaced at intervals reminiscent of mitochondrial targeting sequences are not present in the other isozyme sequences. Simultaneous alteration of these MDH1 residues to residues present in the other isozymes, as indicated in Fig. 3, produces an active enzyme and, in combination with removal of the presequence, eliminates mitochondrial localization (W. C. Small and L. McAlister-Henn, unpublished observations). Multicopy expression was found to be necessary to obtain relatively normal cellular levels of the cytosolic form of MDH1. In contrast with results obtained with cytosolic MDH3, this form of MDH1 only partially compensates for loss of authentic MDH2. Thus, characteristics other than catalysis appear to be important for compartmental function in this case. Interestingly, expression of the cytosolic form of MDH1 was found to compensate for loss of the mitochondrial MDH1 isozyme, i.e., expression restores growth of an *MDH1* disruption mutant on acetate media. Similar results obtained for isocitrate dehydrogenase are described below along with speculation about the basis for these observations.

D. Remaining Questions about Function

Disruption of all three malate dehydrogenase genes in a haploid yeast strain eliminates measurable cellular activity. However, such a mutant retains some capacity for growth on nonfermentable carbon sources relative to a *petite* mutant of the same parental strain. This suggests the existence of alternative cellular sources for essential products of the malate dehydrogenase reaction. It seems unlikely that other isozymes exist because no close homologs of the cloned yeast genes have emerged in the completed genome sequence nor, importantly, have homologs of the mammalian cytosolic enzyme (21). The latter is significantly different from the mammalian mitochondrial enzyme and the closely related yeast and *E. coli* enzymes. The absence of a homolog of the mammalian cytosolic enzyme and undetectable levels of MDH2 in cells grown on glucose suggest that yeast may lack a functional malate/aspartate shuttle for reoxidizing NADH for glycolysis. This notion is supported by the slow glucose growth rates of strains lacking alcohol dehydrogenase isozyme I (22), the enzyme responsible for a major fermentation route.

Above, an argument based on growth phenotypes is presented for function of MDH2 and not of MDH3 in the glyoxylate cycle. However, the growth deficiencies on ethanol and acetate observed for *MDH2* mutants are also observed for mutants with defects in phospho*enol*pyruvate (PEP) carboxykinase (7), suggesting they may reflect dysfunctions in gluconeogenesis. That these phenotypes are not displayed by *MDH3* mutants formally means that the peroxisomal enzyme is not essential for growth on two-carbon sources. However, by analogy with growth of TCA cycle mutants on glycerol, the other cellular isozymes may compensate for functions normally provided by MDH3.

II. Isocitrate Dehydrogenases

The oxidative decarboxylation of isocitrate to form α-ketoglutarate is catalyzed by mitochondrial nicotinamide adenine dinucleotide (NAD^+)-specific and differentially compartmentalized $NADP^+$-specific enzymes. This reaction in the tricarboxylic acid cycle is considered a committed step because it is essentially irreversible under physiological conditions and, as expected for a regulatory enzyme, the multisubunit NAD^+-specific enzyme exhibits complex allosteric modulation of activity. In particular, activation by adenosine monophosphate (AMP) in yeast and adenosine diphosphate (ADP) in mammals, and inhibition by adenosine triphosphate (ATP), are properties associated with sensitive regulation of respiratory rates by cellular energy charge (23). More recently, an additional possible role in direct regulation of expression has emerged with the intriguing observation that yeast NAD^+-

specific isocitrate dehydrogenase specifically binds to 5' untranslated regions of mitochondrial mRNAs (24).

Metabolic roles for the NADP$^+$-specific enzymes, homodimers with no allosteric regulatory properties, are less clear. Expression patterns in mammalian cells (25, 26) suggest possible contributions to production of reducing equivalents for sterol or fatty acid biosynthesis. Also, the existence of a single, NADP$^+$-specific isocitrate dehydrogenase in E. coli suggests possible contributions to TCA cycle functions. Thus, in addition to examining structural/functional relatedness, current research on the yeast isozymes seeks to clarify the metabolic roles of each enzyme and the importance of allosteric regulation and RNA binding to cellular respiration in vivo.

A. Genes and Mutant Phenotypes

NAD$^+$-specific isocitrate dehydrogenase purified from S. cerevisiae is an octomer containing four each of two subunits, designated IDH1 and IDH2, with respective molecular weights of 38,000 and 37,800 (Table I) (27). To date, there is no evidence for a third subunit, although the mammalian enzyme has an $\alpha_4\beta_2\gamma_2$ quaternary structure (28). Antiserum against the yeast holoenzyme and enzyme assays were used to identify independent complementation groups in the ACN collection lacking activity and with defects in each of the subunits (29). Significantly, these complementation groups and subsequent mutants with gene disruptions in *IDH1* and *IDH2* (30, 31) exhibit the energetic acetate$^-$ phenotype also observed for *MDH1* and *CIT1* mutants. This phenotype and patterns of mitochondrial respiration provided the first experimental proof that the NAD$^+$-specific isozyme is the TCA cycle enzyme in eukaryotic cells and that the NADP$^+$-specific mitochondrial isozyme does not significantly contribute to cycle function, at least in yeast cells that lack a mitochondrial transhydrogenase (32). The mutant phenotypes also indicate that both IDH1 and IDH2 subunits are essential for catalytic activity.

Because the X-ray structure of the homologous E. coli isocitrate dehydrogenase has been determined (33), it is possible to assign putative functions to conserved residues in IDH1 and IDH2. In particular, serine residues in both subunits correspond to Ser-113 of the bacterial enzyme. This residue is involved in isocitrate binding and is also the target residue for phosphorylation and inactivation (34), an important mechanism for control of carbon flux between the TCA cycle and glyoxylate pathway in bacteria. To test functional conservation, kinetic properties were compared for mutant yeast enzymes containing either an S92A replacement in IDH1 or an S98A replacement in IDH2 (35). The altered IDH1 enzyme exhibits a 6-fold decrease in V_{max} (versus a 60-fold decrease for the altered IDH2 enzyme) and a loss of cooperativity (versus no change for the IDH2 enzyme), but little change in

$S_{0.5S}$ for isocitrate (versus a 2-fold reduction for the IDH2 enzyme). In addition, the alteration in IDH1 results in loss of activation by AMP, whereas the change in IDH2 produces no effect on activation. These kinetic data led to formulation of a model for primary function of the IDH2 subunit in catalysis and of the IDH1 subunit in regulation, i.e., in cooperative interactions and AMP activation.

The *IDP1* gene encoding yeast mitochondrial $NADP^+$-specific isocitrate dehydrogenase was cloned using polymerase chain reaction (PCR) following purification and amino-terminal sequence analysis of the enzyme (36), and the *IDP2* gene for the cytosolic isozyme was similarly cloned following purification of the unstable enzyme from an *IDP1* disruption mutant (37). Disruption of either or of both *IDP1* and *IDP2* was found to produce no detectable growth phenotype under a wide variety of cultivation conditions. Elimination of both isozymes also eliminates measurable cellular activity (Fig. 6) despite the existence of a close homolog (designated *IDP3*) identified by genome sequence analysis.

Some metabolic functions have been clarified by analysis of a collection of mutants with various combinations of disruptions in *IDH2*, *IDP1*, and *IDP2* genes (Fig. 6; Table II) (38). The absence of all three enzymatic activities correlates with glutamate auxotrophy on fermentable or nonfermentable carbon sources, whereas the presence of any one isozyme confers glutamate

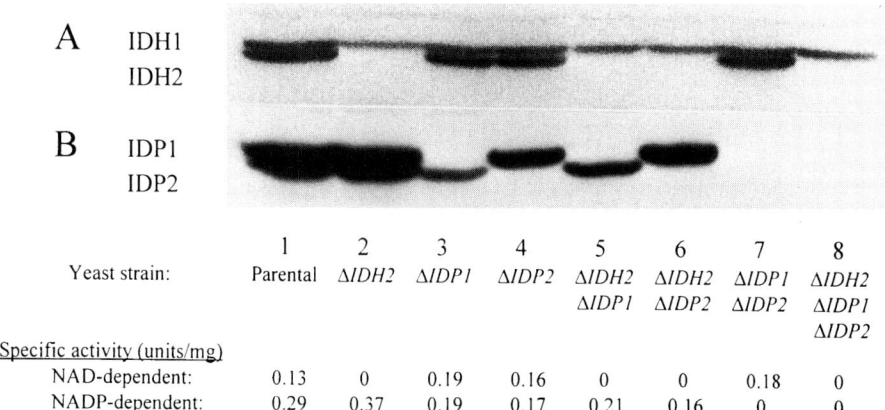

FIG. 6. Immunoblot analysis of yeast isocitrate dehydrogenase gene disruption mutants. Haploid strains containing different combinations of disruptions in *IDH2*, *IDP1*, and *IDP2* genes, as indicated for each lane, were grown in rich medium with glycerol and ethanol as carbon sources. Cellular protein extracts were used for determination of specific activities, as indicated, and for immunoblot analysis with antisera to detect the IDH subunits (A) or IDP1 and IDP2 (B). (Reprinted with permission from W.-N. Zhao and L. McAlister-Henn, *Biochemistry* **35**, 7873 (1996). Copyright 1996 American Chemical Society.)

TABLE II
Growth Phenotypes of Yeast Isocitrate Dehydrogenase Disruption Mutants

Yeast strain	Rich (YP) medium			Minimal (YNB) medium	
	Glucose[a]	Acetate[b]	Ethanol[c]	Glucose (with/without glutamate)[d]	Ethanol (with/without glutamate)[e]
Parental	+	+	+	+/+	+/+
ΔIDH2	+	−	+	+/+	+/+
ΔIDP1	+	+	+	+/+	+/+
ΔIDP2	+	+	+	+/+	+/+
ΔIDH2ΔIDP1	+	−	+	+/−	+/+
ΔIDH2ΔIDP2	+	−	+	+/s	+/s
ΔIDP1ΔIDP2	+	+	+	+/+	+/+
ΔIDH2ΔIDP1ΔIDP2	+	−	+	+/−	+/−

[a]Culture doubling times of 2–2.5 h are indicated by "+".
[b]Culture doubling times of 4–6 h are indicated by "+"; "−" indicates no doubling over a 24 h period.
[c]Culture doubling times of 4–5 h are indicated by "+".
[d]Culture doubling times of 4–5 h are indicated by "+" and of 7.5–8 h are indicated by "s"; "−" indicates no doubling over a 24 h period.
[e]Culture doubling times of 7–7.5 h are indicated by "+" and of 9–9.5 h are indicated by "s"; "−" indicates no doubling over a 24 h period.

prototrophy under conditions favorable for expression of that isozyme. For example, codisruption of genes encoding the mitochondrial isozymes (IDH2 and IDP1) produces glutamate auxotrophy on glucose, a condition associated with minimal expression of cytosolic IDP2. This establishes that all three enzymes can contribute to cellular pools of α-ketoglutarate whereas only the mitochondrial NAD^+-specific enzyme delivers reducing equivalents to the respiratory chain. Current experimental efforts are directed toward assessing contributions of IDP1 and IDP2 to cellular pools of NADPH, and to assessing expression, localization, and potential metabolic functions of IDP3.

B. Structures and Expression

The IDH1 and IDH2 polypeptides are very similar, having an overall 42% identity of aligned amino acid sequences. These subunits also show significant sequence homology with yeast β-isopropylmalate dehydrogenase, which is also an oxidative decarboxylase, and with an open reading frame (YNL009W) for a gene product of unknown function. Interestingly, despite differences in cofactor specificity and allosteric properties, IDH1 and IDH2 are much more closely related to the *E. coli* $NADP^+$-specific enzyme (approximately 32% residue identity) than are the yeast $NADP^+$-specific enzymes (less than 15% residue identity). This suggests some possible conservation of structure related to function in the TCA cycle. However, expression of the *E. coli* enzyme in mitochondrial or cytosolic compartments of yeast mutants has been found to restore glutamate prototrophy associated with IDP1 or IDP2 function but not growth with acetate associated with IDH1/IDH2 function (38), suggesting that cofactor specificity may be the critical attribute *in vivo*.

The IDP1 and IDP2 polypeptides (molecular weights of 46,400 and 46,600, respectively) share 71% residue identity and are similarly related to the *IDP3* open reading frame; all are closely related in primary sequence with corresponding mammalian enzymes (39, 40). As previously mentioned, IDP2 expression is apparently repressed by glucose to levels undetectable by enzyme assay or immunoblots. IDP1 levels, on the other hand, appear to be constitutive (36). This is a very unusual pattern of expression, in that glucose repression is commonly observed for yeast mitochondrial proteins, including the IDH1 and IDH2 subunits of the NAD^+-specific enzyme. IDP1 expression is also unchanged, whereas IDH1 and IDH2 polypeptide and mRNA levels are significantly reduced by the presence of glutamate (41).

C. Cellular Compartmentation

The IDH1 and IDH2 polypeptides have amino-terminal presequences of 11 and 15 residues, respectively, which are proteolytically removed on mitochondrial import (27). To examine structural requirements for import and assembly of the NAD-specific isocitrate dehydrogenase, mutagenesis and transformation were used to produce yeast strains expressing subunits lacking either or both presequences (42). The IDH2 presequence was found to be necessary for mitochondrial localization of that subunit, but localization of IDH1 was found to be influenced not only by its own presequence but by the compartmental localization of IDH2, suggesting that IDH2 may drive holoenzyme assembly. Removal of both presequences results in cytosolic localization of both subunits and in assembly of an active enzyme, indicating that mitochondrial factors are not essential for assembly or activity.

The cytosolic form of the NAD^+-specific enzyme was also found to be competent for restoration of growth on acetate of a mutant lacking mitochondrial activity. As previously described, similar results were obtained with a cytosolic form of MDH1. One possible explanation for these results is that reducing equivalents produced by the cytosolic forms of the dehydrogenases may be directly delivered to the electron transport chain. This possibility is attributed to the existence in yeast of two NADH dehydrogenases (43), one of which is apparently responsible for the utilization of exogenous NADH by isolated yeast mitochondria in oxygen consumption assays (44) and which may allow cytosolic delivery of reducing equivalents *in vivo*. In any event, mitochondrial localization is apparently not essential for, although it may enhance, cellular function of these two TCA cycle enzymes.

The IDP1 polypeptide is preceded by a 16-residue presequence that is absent on IDP2 and that is necessary for mitochondrial localization (45). Yeast IDP2 appears to be a soluble cytosolic enzyme. Interestingly the *IDP3* open reading frame contains a carboxy terminal Cys-Lys-Leu tripeptide, indicating the possibility of peroxisomal localization. The amino acid sequence of a mammalian counterpart also terminates with an Ala-Lys-Leu tripeptide (40). Tests to determine localization and function of these enzymes are in progress.

III. Citrate Synthases

The condensation reaction catalyzed by citrate synthase is the rate-limiting step for oxidation via the TCA cycle. The reaction also channels two-carbon units into biosynthesis of many cellular components, including amino acids, fatty acids, and sugars. The activity is highly regulated, allosterically by ATP, and, in many organisms, by alterations in cellular levels in response to environmental conditions. Recent research has focused on *in vivo* function and the dramatically regulated expression of compartmentalized isozymes of citrate synthase in yeast.

A. Genes and Mutant Phenotypes

The *CIT1* gene encoding mitochondrial citrate synthase was initially cloned following a screen for mRNAs enriched in polysomes bound to mitochondria (4). Disruption of the gene was found to result in increased levels of a nonmitochondrial activity (46) subsequently attributed to expression of the *CIT2* gene encoding a peroxisomal isozyme. Disruption of *CIT1* produces an energetic acetate$^-$ growth phenotype and a lag in attaining parental growth rates on nonfermentable carbon sources, whereas disruption of *CIT2* alone has no apparent effect on growth. Mutants with disruptions in both genes exhibit glutamate auxotrophy (5, 47) and a reduction in cellular activity to undetectable levels (48). Thus, as with the isocitrate dehydrogenase isozymes, the presence of either citrate synthase activity is sufficient for glutamate synthesis. Revelation of a third highly related locus (designated *CIT3*) by genome sequence analysis suggests possible expression of a third isozyme; however, the *CIT1/CIT2* disruption studies suggest that cellular levels of any residual activity are low and insufficient for metabolic needs of the cell.

The residual levels of citrate synthase activity in a *CIT1* disruption mutant were determined to be severalfold in excess of those necessary to support state 3 respiratory rates (48). Also, cellular levels and function of the citrate transporter from mitochondrial membranes, tested following reconstitution into proteoliposomes, are unaffected by the disruption (49). However, rates of citrate transport for mitochondria isolated from the *CIT1* disruption mutant are dramatically reduced relative to parental controls, suggesting that phenotypes associated with loss of CIT1 may be due to effects on transport rather than to the catalytic defect per se. This possibility was tested by expression of engineered catalytically inactive forms of CIT1 in the *CIT1* disruption mutant (50). The mutant enzymes were found to restore rates of mitochondrial respiration, citrate oxidation, and growth on acetate to near parental levels, suggesting that interactions among mitochondrial components, presumably in this case between the synthase and transporter, are critical for TCA cycle function.

B. Structures and Regulation of Expression

The mature CIT1 and CIT2 isozymes are homodimers with similar subunit molecular weights (Table I). The polypeptides share residue identities and similarities of 80% (51); translation of the *CIT3* open reading frame predicts a less related gene product (40% similarity). CIT1 has an amino-terminal presequence of approximately 20 residues that is removed on mitochondrial import (52); the exact amino terminus of the mature polypeptide has not been determined. CIT2 has a canonical carboxy-terminal Ser-Lys-Leu tripeptide sequence and peroxisomal localization is dependent on the presence of this sequence (53). The translated CIT3 sequence lacks the long amino-terminal extension of CIT1 and contains a carboxy-terminal Asn-Lys-Leu; thus, compartmental localization cannot be predicted from the sequence. However, a recent report suggests that CIT3 can be imported into mitochondria *in vitro* and may contribute to TCA cycle function (54).

Regulation of expression of the citrate synthase isozymes is complex. Levels of CIT1 are reduced approximately 6-fold by glucose and an additional 10-fold by glutamate. Distinct upstream activation sequences mediating glucose repression and derepression of *CIT1* expression have been identified (55). The HAP2,3,4 transcriptional activator, which also regulates expression of genes for electron transport proteins, was shown to be required for derepression but apparently does not influence glucose or glutamate repression (56). The molecular mechanism for glutamate regulation of *CIT1*, of *IDH1* and *IDH2*, and of the *ACO1* gene encoding aconitase (57) has not been defined.

Analysis of *CIT2* expression has provided evidence for a remarkable retrograde path of communication, i.e., for changes in nuclear gene expression elicited by dysfunctions in mitochondrial processes. Levels of *CIT2* mRNA are elevated 6- to 10-fold in yeast strains with disruptions in either *CIT1* or *MDH1* and a dramatic 30-fold in p^o strains lacking mitochondrial DNA (58). This response appears to be unique for *CIT2* among genes encoding peroxisomal enzymes (59), suggesting a key role for CIT2 in metabolic and regulatory interactions between the TCA and glyoxylate cycles. The increased retrograde expression of *CIT2* in p^o cells is mediated by a novel upstream activation site (UASr) in the 5′ flanking region of the gene and by the products of at least two genes, *RTG1* and *RTG2* (60). *RTG1* encodes a basic helix–loop–helix transcription factor that binds the *CIT2* UASr. *RTG2* encodes a polypeptide of unknown function. However, transcriptional activation by RTG1, measured as transactivation of a *lacZ* reporter gene by a GAL4/RTG1 fusion protein, is defective in an *RTG2* disruption mutant, suggesting functional interdependence of the proteins (61).

Yeast strains with disruptions in *RTG1* or *RTG2* genes demonstrate glutamate and aspartate auxotrophy and are unable to growth with acetate as a carbon source (60), defects associated with dysfunctions in both carbon assimilation and the TCA cycle. The metabolic basis for these phenotypes is unclear; associated 50% reductions in activities for CIT1, acetyl-CoA synthetase, NAD^+-specific isocitrate dehydrogenase, and pyruvate dehydrogenase suggest multiple minor, perhaps cumulative, effects on glutamate/aspartate production (62). The *RTG* mutants are also defective in growth with oleate as the carbon source (59), a phenotype associated with defects in peroxisome assembly (63). Oleate-induced expression of mRNAs for some peroxisomal proteins is blocked in an additive fashion in *RTG1* and *RTG2* mutants whereas oleate-induced expression of mRNAs for glyoxylate pathway enzymes is significantly reduced only in an *RTG1* mutant.

IV. Aconitase and Fumarase Isozymes

Nonmitochondrial activities have been reported for two other yeast TCA cycle enzymes, aconitase and fumarase. These enzymes share the common feature of differential cellular localization of isozymes postulated to be products of the same gene.

A. Aconitase

Yeast aconitase is a monomer with a molecular weight of approximately 80,000 and contains an evolutionarily conserved FeS center involved in the hydration/dehydration reaction. To date, only the *ACO1* gene has been demonstrated to encode a catalytically active product (57), although a close homolog of the gene has been identified by genome sequence analysis (64). Disruption of the *ACO1* gene produces a respiratory-deficient phenotype and glutamate auxotrophy on glucose; it also eliminates cellular aconitase activity. These results would indicate that if the *ACO1* homolog is expressed, cellular levels are below detection by assays and apparently unable to compensate for loss of ACO1. However, a complementation group with aconitase defects was identified in the McCammon acetate⁻ mutant collection; mutants in this group grow with glycerol as a carbon source and are not glutamate auxotrophs (7). Thus, some mutant alleles may produce phenotypes other than those observed for the gene disruption mutant or, in analogy with other TCA cycle enzyme defects, these phenotypes may indicate alternative isozyme function under various growth conditions.

Interest in the existence and origin of cytosolic aconitase or of a related protein in yeast is fueled by the discovery of an iron-regulated RNA binding protein with aconitase activity in higher eukaryotes (65, 66). The human protein interacts with the 5' untranslated region of ferritin mRNA and with the 3' untranslated region of transferrin receptor mRNA, and shares a residue identity of 33% with yeast ACO1. In addition to potential regulatory features, nonmitochondrial aconitase activity in yeast is presumably required for function in the glyoxylate pathway (Fig. 2). As is the case for some of the other enzymes in this pathway in yeast, aconitase does not appear to be associated with peroxisomes.

Mechanisms of mitochondrial import or of cytosolic partitioning have not been examined for yeast ACO1. The first 24 residues of the polypeptide sequence are rich in hydroxylated and basic residues characteristic of mitochondrial targeting sequences and the polypeptide size predicted from the sequence (molecular weight, 85,685) is significantly larger than that of the purified protein (67). However, the amino terminus of the mature polypeptide has not been reported.

Expression of *ACO1* is repressed by glucose and by glutamate in a synergistic manner (57). The 5' untranslated region of the gene contains an upstream activation sequence corresponding to a sequence shown to be the HAP multiprotein regulatory site of the *CYC1* gene (68). Thus, *CIT1* and *ACO1* may share similar mechanisms for glucose- and glutamate-mediated regulation.

B. Fumarase

The catalytically active form of yeast fumarase is a homotetramer. Activity is found in both cytosolic and mitochondrial cellular fractions, with cytosolic activity representing approximately 70% of the total. Both cellular activities are eliminated by disruption of the *FUM1* gene, which contains an open reading frame for a polypeptide of molecular weight 53,000 (69) and no close homologs of the gene have been revealed by genome sequence analysis. Mutants with a *FUM1* disruption accumulate fumarate (70) and exhibit an auxotrophic requirement for aspartate, asparagine, or serine for growth on glycerol as a carbon source (69).

It is clear that information for mitochondrial localization of fumarase is contained within the amino terminus of the FUM1 polypeptide: expression of a polypeptide lacking 17 amino-terminal residues reduces mitochondrial activity (69) and 92 residues from the amino terminus are sufficient for mitochondrial localization of a *lacZ* fusion protein (71). However, the mechanism for partitioning of the protein to the cytosol is more controversial, due at least in part to the absence of amino-terminal sequence analysis of purified isoforms and to discrepancies in reported sizes for the cytosolic isozyme (72, 73).

There is evidence for differential localization of FUM1 polypeptides due to transcription of two mRNA species, the longer containing codons for the mitochondrial targeting sequence and the shorter containing an alternative downstream site for translation initiation (69). Similar mechanisms have been described for differential localization of yeast invertase (74), α-isopropylmalate synthase (75), and histidyl-tRNA synthetase (76). Multiple transcripts of the *FUM1* gene have been detected by Northern blot, S1 nuclease, and RACE-PCR analyses (69, 77). The most abundant of the shorter transcripts have 5′ termini 57–68 nucleotides downstream of the initial ATG codon. Interestingly, the longer transcripts are three to four times more abundant despite the lower relative activity attributed to the mitochondrial isozyme. Also, altering the putative second initiation codon at residue position 24 (producing an M24I mutant enzyme) eliminates cytosolic but not mitochondrial activity (77). In accordance with this proposed mechanism for isozyme distribution, respective subunit molecular weights of 53,000 and of 48,000 have been reported for mitochondrial and cytosolic fumarases isolated from yeast (72).

There is also evidence for partitioning of fumarase isozymes from a single translation product (78). A single high-molecular-weight precursor polypeptide, equivalent in size with a mutant M24V polypeptide, is detected in cells treated with an inhibitor of mitochondrial import and processing. This observation and demonstration of rapid cotranslational processing led to proposal of a new model for subcellular distribution of isozymes. Large fumarase precursor molecules of the same size are proposed to be partially translocated across the mitochondrial membrane and processed in a cotranslational manner; a subset of the processed polypeptides is subsequently imported but a larger population is released and folded in the cytosol. Supporting this model is a report that yeast fumarase purifies as a single species with a subunit molecular weight of approximately 50,000 (73). Further tests of these two proposed mechanisms for cellular partitioning are awaited with interest.

V. Conclusions and Perspective

The three remaining TCA cycle enzymes (Fig. 1) represent unique mitochondrial activities, and genes encoding subunits of the enzymes have been cloned or identified by genome sequence analysis. Disruption of genes encoding subunits of α-ketoglutarate dehydrogenase (79–82) or of succinate dehydrogenase (83–88) generally produces a glycerol⁻ respiratory-deficient phenotype. Exceptions include the strain-dependent phenotypes obtained with disruption of the *KGD1* gene encoding the α-ketoglutarate dehydrogenase component of the complex (81). Also, disruption of *SDH4* encoding a membrane attachment protein results in inefficient membrane association and reduced activity but does not eliminate growth with glycerol (87). Probable genes encoding two subunits of succinyl coA synthetase have been identified but not examined at the level of function.

The expectation has thus been that codisruption of all genes for a TCA cycle isozyme family in yeast would produce respiratory deficiency. To date, this has not proved to be the case. For the isocitrate dehydrogenase and citrate synthase families, identification of previously unknown genes potentially expressed as additional isozymes may provide an explanation. Expression of these isozymes was not detected by biochemical methods in mutants with disruptions of the known genes, but it is entirely possible that conditions necessary for growth of multiple disruption mutants may be less than optimal for expression of residual genes. However, for the malate dehydrogenases, no unknown homologs have been revealed by genome sequence analysis, yet the triple gene disruption mutant retains some capacity for growth on nonfermentable carbon sources. Thus, there appear to be additional metabolic bypass reactions, definition of which will likely require both good guesses and genetic approaches, including screens for synthetic lethal mutants.

The basis for the acetate⁻ growth phenotype associated with disruption of *MDH1*, *CIT1*, and *IDH* genes is not entirely clear. This phenotype is described above as an energetic defect because it is displayed on rich as well as minimal medium. The phenotype is also specific for acetate and not ethanol, suggesting that the NADH reducing equivalents produced during two reactions in metabolic conversion of ethanol to acetate are sufficient compensation for the defect and that a C_2 carbon source can suffice for growth. Also suggesting that acetate utilization per se may not be the deficiency in the TCA cycle disruption mutants is the observation that acetate addition to otherwise permissive glycerol medium dramatically inhibits growth (7, 29). For example, growth of an *MDH1* or *IDH2* disruption mutant on 2% glycerol plus 2% acetate is reduced 5- to 10-fold below growth rates on glycerol alone, whereas parental strains show little difference in growth rates under both conditions. Acetate inhibition of growth is also noted for ACN mutants with defects in glyoxylate and gluconeogenic enzymes (7). Thus, the effect of acetate may be a direct inhibition of utilization of other carbon sources or, perhaps, an indication of some mechanism for preferential use of acetate as a carbon source. Resolution of these questions is of interest in understanding basic aspects of metabolic regulation.

Clearly there are many interesting avenues for future research in this area. Most immediate will be an understanding of the roles of isozymes not yet analyzed at a biochemical level. Questions about assembly of multisubunit complexes and about interactions between TCA cycle enzymes and other mitochondrial components that may contribute to metabolic regulation are now accessible. Also, patterns for coordinate and retrograde regulation of gene expression are emerging and are of significant interest. Perhaps most interesting is the prospect of simultaneous application of genetic methods and quantitative measurements to assessing metabolic flux and regulation *in vivo*.

Acknowledgments

Work from this laboratory was supported by Grants GM33218 and GM51265 from the National Institutes of Health.

References

1. W. Duntze, D. Neumann, J. M. Gancedo, W. Atzpodien, and H. Holzer, *Eur. J. Biochem.* **10**, 83 (1969).
2. P. S. Perlman and H. R. Mahler, *Arch. Biochem. Biophys.* **136**, 245 (1970).
3. P. A. Srere *in* "Energy Metabolism and the Regulation of Metabolic Processes in Mitochondria" (M. A. Mehlman and R. E. Hanson, eds.), p. 79. Academic Press, New York, 1972.
4. M. Suissa, K. Suda and G. Schatz, *EMBO J.* **3**, 1773 (1984).
5. K.-S. Kim, M. S. Rosenkrantz, and L. Guarente, *Mol. Cell. Biol.* **6**, 1936 (1986).
6. L. McAlister-Henn and L. M. Thompson, *J. Bacteriol.* **169**, 5157 (1987).
7. M. T. McCammon, *Genetics* **144**, 57 (1996).
8. K. I. Minard and L. McAlister-Henn, *Mol. Cell. Biol.* **11**, 370 (1991).
9. J. S. Steffan and L. McAlister-Henn, *J. Biol. Chem.* **267**, 24708 (1992).
10. M. T. McCammon, M. Veenhuis, S. B. Trapp, and J. M. Goodman, *J. Bacteriol.* **172**, 5816.
11. C. van Roermund, Y. Elgersma, N. Singh, R. J. A. Wanders, and H. F. Tabak, *EMBO J.* **14**, 380 (1995).
12. S. L. Roderick and L. J. Banaszak, *J. Biol. Chem.* **261**, 9461.
13. M. D. Hall, D. G. Levitt and L. J. Banaszak, *J. Mol. Biol.* **226**, 867 (1992).
14. L. M. Thompson, P. Sutherland, J. S. Steffan, and L. McAlister-Henn, *Biochemistry* **27**, 8393 (1988).
15. S. Subramani, *Annu. Rev. Cell Biol.* **9**, 445 (1993).
16. L. McAlister-Henn, J. S. Steffan, K. I. Minard, and S. L. Anderson, *J. Biol. Chem.* **270**, 21220 (1995).
17. H. Holzer, *Trends Biochem. Sci.* **1**, 178 (1976).
18. K. I. Minard and L. McAlister-Henn, *J. Biol. Chem.* **267**, 17458 (1992).
19. K. I. Minard and L. McAlister-Henn, *Arch. Biochem. Biophys.* **315**, 302 (1994).
20. L. M. Thompson and L. McAlister-Henn, *J. Biol. Chem.* **264**, 12091 (1989).
21. T. Joh, H. Takeshima, T. Tsuzuki, C. Setoyama, K. Shimada, S. Tanase, S. Kuramitsu, H. Hagamiyama and Y. Morino, *J. Biol. Chem.* **262**, 15127 (1987).
22. M. Ciriacy, *Mol. Gen. Genet.* **154**, 213 (1977).

23. J. A. Hathaway and D. E. Atkinson, *J. Biol. Chem.* **238,** 2875 (1963).
24. S. D. J. Elzinga, A. L. Bednarz, K. van Oosterum, P. J. T. Dekker, and L. A. Grivell, *Nucleic Acids Res.* **21,** 5328 (1993).
25. H. M. Farrell, Jr., J. T. Deeney, K. A. Tubbs, and R. A. Walsh, *J. Dairy Sci.* **70,** 781 (1987).
26. G. T. Jennings and P. M. Stevenson, *Eur. J. Biochem.* **198,** 621 (1991).
27. D. A. Keys and L. McAlister-Henn, *J. Bacteriol.* **172,** 4280 (1990).
28. N. Ramachandran and R. F. Colman, *J. Biol. Chem.* **255,** 8859 (1980).
29. D. A. Keys, Ph.D. Dissertation, University of California, at Irvine, California, 1990.
30. J. R. Cupp and L. McAlister-Henn, *J. Biol. Chem.* **267,** 16417 (1992).
31. J. R. Cupp and L. McAlister-Henn, *J. Biol. Chem.* **266,** 22199 (1991).
32. J. Rydstrom, J. B. Hock, and L. Ernster, *Enzymes* **13,** 51 (1976).
33. J. H. Hurley, A. M. Dean, J. L. Sohl, D. E. Koshland, Jr., and R. M. Stroud, *Science* **249,** 1012.
34. P. E. Thorsness and D. E. Koshland, Jr., *J. Biol. Chem.* **262,** 10422 (1987).
35. J. R. Cupp and L. McAlister-Henn, *Biochemistry* **32,** 9323 (1993).
36. R. J. Haselbeck and L. McAlister-Henn, *J. Biol. Chem.* **266,** 2339 (1991).
37. T. M. Loftus, L. V. Hall, S. L. Anderson, and L. McAlister-Henn, *Biochemistry* **33,** 9661 (1994).
38. W.-N. Zhao and L. McAlister-Henn, *Biochemistry* **35,** 7873 (1996).
39. R. J. Haselbeck, R. F. Colman, and L. McAlister-Henn, *Biochemistry* **31,** 6219 (1992).
40. G. T. Jennings, S. Sechi, P. M. Stevenson, R. C. Tuckey, D. Parmelee, and L. McAlister-Henn, *J. Biol. Chem.* **269,** 23128 (1994).
41. R. J. Haselbeck and L. McAlister-Henn, *J. Biol. Chem.* **268,** 12116 (1993).
42. W-N. Zhao and L. McAlister-Henn, *J. Biol. Chem.* **271,** 10347 (1996).
43. C. A. M. Marres, S. de Vries, and L. A. Grivell, *Eur. J. Biochem.* **195,** 857.
44. T. Ohnishi, K. Kawaguchi, and B. Hagihara, *J. Biol. Chem.* **241,** 1797 (1966).
45. R. J. Haselbeck, Ph.D. Thesis, University of California, Irvine, California, 1993.
46. T. M. Rickey and A. S. Lewin, *Mol. Cell. Biol.* **6,** 488 (1986).
47. G. Kispal, M. Rosenkrantz, L. Guarente, and P. A. Srere, *J. Biol. Chem.* **263,** 11145 (1988).
48. E. V. Grigorenko, W. C. Small, L.-O. Perrson, and P. A. Srere, *J. Mol. Recogn.* **3,** 215 (1990).
49. A. Sandor, J. H. Johnson, and P. A. Srere, *J. Biol. Chem.,* **269,** 29609.
50. G. Kispal, C. T. Evans, C. Malloy, and P. A. Srere, *J. Biol. Chem.* **264,** 11204 (1989).
51. M. Rosenkrantz, T. Alam, K-S. Kim, B. J. Clark, P. A. Srere, and L. P. Guarente, *Mol. Cell. Biol.* **6,** 4509 (1986).
52. T. Alam, D. Finkelstein, and P. A. Srere, *J. Biol. Chem.* **257,** 11181 (1982).
53. K. K. Singh, G. M. Small, and A. S. Lewin, *Mol. Cell. Biol.* **12,** 5593 (1992).
54. A.-M. Bécam, Y. Jia, P. P. Slonimski, and C. J. Herbert, *in* "Abstracts of the 17th International Conference on Yeast Genetics and Molecular Biology," p. 300. Lisboa, Portugal, 1995.
55. M. Rosenkrantz, C. S. Kell, E. A. Pennell, M. Webster, and L. J. Devenish, *Curr. Gen.* **25,** 185 (1994).
56. M. Rosenkrantz, C. S. Kell, E. A. Pennell, and L. J. Devenish, *Mol. Microbiol.* **13,** 119 (1994).
57. S. P. Gangloff, D. Marguet, and G. J.-M. Lauquin, *Mol. Cell. Biol.* **10,** 3551 (1990).
58. X. Liao, W. C. Small, P. A. Srere, and R. A. Butow, *Mol. Cell. Biol.* **11,** 38 (1991).
59. A. Chelstowska and R. A. Butow, *J. Biol. Chem.* **270,** 18141 (1995).
60. X. Liao and R. A. Butow, *Cell* **72,** 61 (1993).
61. B. A. Rothermel, A. W. Shyjan, J. L. Etheredge, and R. A. Butow, *J. Biol. Chem.* **270,** 29476 (1995).
62. W. C. Small, R. D. Brodeur, A. Sandor, N. Fedorova, G. Li, R. A. Butow, and P. A. Srere, *Biochemistry* **34,** 5569 (1995).
63. R. Erdmann, M. Veenhuis, D. Merkens, and W.-H. Kunau, *Proc. Natl. Acad. Sci. U.S.A.* **86,** 5419 (1989).

64. B. Purnelle, F. Coster, and A. Goffeau, *Yeast* **10**, 1235 (1994).
65. T. A. Rouault, C. D. Stout, S. Kaptain, J. B. Harford, and R. D. Klausner, *Cell* **64**, 881 (1991).
66. S. Kaptain, W. E. Downey, C. Tang, C. Philpott, D. Haile, D. G. Orloff, J. B. Harford, T. A. Rouault, and R. D. Klausner, *Proc. Natl. Acad. Sci. U.S.A.* **88**, 10109 (1991).
67. H. Sholze, *Biochim. Biophys. Acta* **746**, 133 (1983).
68. J. Olesen, S. Hahn, and L. Guarente, *Cell* **51**, 953 (1987).
69. M. Wu and A. Tzagoloff, *J. Biol. Chem.* **262**, 12275 (1987).
70. E. Kaclíková, T. M. Lachowicz, Y. Gbelská, and J. Šubík, *FEMS Microbiol. Lett.* **91**, 101 (1992).
71. Y. Peleg, J. S. Rokem, I. Goldberg, and O. Pines, *Appl. Environ. Microbiol.* **56**, 2777 (1990).
72. J. Mangan, N. M. O'Sullivan, and S. Doonan, *Biochem. Soc. Trans.* **19**, 185 (1991).
73. J. S. Keruchenko, I. D. Keruchenko, K. L. Gladilin, V. N. Zaitsev, and N. Y. Chirgadze, *Biochim. Biophys. Acta* **1122**, 85 (1992).
74. D. Perlman and H. O. Halvorson, *Cell* **25**, 525 (1981).
75. J. P. Beltzer, S. R. Morris, and G. B. Hohlhaw, *J. Biol. Chem.* **263**, 368 (1988).
76. G. Natsoulis, F. Hilger, and G. R. Fink, *Cell* **46**, 1163 (1992).
77. M. Wu, S.-M. Wong, H.-M. Tan and R. Ting, *Biochem. Biophys. Res. Commun.* **215**, 578 (1995).
78. I. Stein, Y. Peleg, S. Even-Ram, and O. Pines, *Mol. Cell. Biol.* **14**, 4770 (1994).
79. K. S. Browning, D. J. Uhlinger and L. J. Reed, *Proc. Natl. Acad. Sci. U.S.A.* **85**, 1831 (1988).
80. J. Ross, R. A. Graeme, and I. W. Dawes, *J. Gen. Microbiol.* **134**, 1131 (1988).
81. B. Repetto and A. Tzagoloff, *Mol. Cell. Biol.* **9**, 2695 (1989).
82. B. Repetto and A. Tzagoloff, *Mol. Cell. Biol.* **10**, 4221 (1990).
83. A. Lombardo, K. Carine, and I. E. Scheffler, *J. Biol. Chem.* **265**, 10419 (1990).
84. K. M. Robinson and B. D. Lemire, *J. Biol. Chem.* **267**, 10100 (1992).
85. K. B. Chapman, S. D. Solomon, and J. D. Boeke, *Gene* **118**, 131 (1992).
86. N. Schülke, G. Blobel, and D. Pain, *Proc. Natl. Acad. Sci. U.S.A.* **89**, 8011 (1992).
87. B. L. Bullis and B. D. Lemire, *J. Biol. Chem.* **269**, 6543 (1994).
88. B. Daignan-Fornier, M. Valens, B. D. Lemire, and M. Bolotin-Fukuhara, *J. Biol. Chem.* **269**, 15469 (1994).
89. E. Kopetzki, K. Entian, F. Lottspeich, and D. Mecke, *Biochim. Biophys. Acta* **912**, 398 (1987).

Index

A

ACO1, see Aconitase, yeast
Aconitase, yeast
 gene, 333
 mutant phenotypes, 333–334
 regulation of expression, 334
 signal sequence, 334
 structure, 333
Adenovirus, polyadenylation sites in late transcription unit messenger RNA, 59
Amyloid precursor protein, polyadenylation sites in messenger RNA, 58
Antisense oligonucleotide
 analogs, 97–102
 brain targeting, 112, 114–115
 cellular uptake, 105–110
 circular oligonucleotides, 102–103
 clinical applications, 117–122
 degradation, 114–116
 delivery strategies
  adenovirus release from vesicles, 110
  erythrocyte ghosts, 114
  hydrophobic modification, 111–112, 121
  liposomes, 110–113
  mannose 6-phosphate receptors, 112
  polylysine conjugates, 112, 121
  pore-forming agents, 110
  reconstituted viral envelopes, 113–114
  transferrin receptor, 112
 D-loop formation, 103–104
 effectors, 102
 length optimization, 101
 mechanism of action, 116–117
 proof of efficacy, 116
 ribonuclease activation in therapy, 100, 119–120
 RNA hairpin targeting, 102
 Systematic Evolution of Ligands by Exponential enrichment, sequence optimization, 101
 target folding and binding, 100
 tissue distribution, 114–115

 toxicity, 116
 triple-stranded complexes, 102–103, 120–121
 vesicle entrapment, 110
Apoptosis, E-cadherin role, 203–204
A protein, regulation of messenger RNA polyadenylation, 62–63
Autoantibody, oligonucleotides, 128–129
Azidothymidine (AZT), DNA hypermethylation and drug resistance, 233
AZT, see Azidothymidine

B

B cell, stimulation by CpG oligonucleotides, 130–131

C

Cadherin, see also E-cadherin
 cell-specific recognition, 199
 classification, 188
 functional domains, 193
 genes, 190–191, 193
 processing, 193
 sequence homology between species, 188, 190, 207
 structure, 188, 190
 tissue distribution, 198
 types, 188
Calcitonin gene-related peptide (CGRP), splicing of polyadenylation site in gene, 62
α-Catenin
 E-cadherin binding, 194, 196–197, 201–202
 gene, 196
 loss in cancer, 206
 modulation of adhesion, 198
CD4, inhibition by oligonucleotide binding, 123–124

Cell–cell junctions
  structural changes during tumorigenesis, 204–207
  types, 188, 198–199
CFI$_m$, see Cleavage factor I$_m$
CGRP, see Calcitonin gene-related peptide
Cholesterol
  antisense oligonucleotide delivery, 111–112, 121
  protein inhibition by modified nucleotides, 125
Chronic myelogenous leukemia (CML), antisense oligonucleotide therapy, 118
CIT, see Citrate synthase, yeast
Citrate synthase, yeast
  genes, 331
  mutant phenotypes, 331–333, 336
  regulation of expression, 332–333
  structures, 332
Cleavage and polyadenylation specificity factor (CPSF)
  cytoplasmic protein, 55–56
  RNA binding, 49–50, 53, 55
  structure, 49
Cleavage factor I$_m$ (CFI$_m$)
  RNA binding, 50, 53
  structure, 49
Cleavage stimulation factor (CstF)
  RNA binding, 49, 53, 62
  structure, 49
CML, see Chronic myelogenous leukemia
α-Complex proteins
  discovery, 271
  incorporation into α-complex, 273–274
  messenger RNA stabilization, 271, 273–275, 278, 283
  types, 271, 273
αCP-1, see α-Complex proteins
αCP-2, see α-Complex proteins
CPSF, see Cleavage and polyadenylation specificity factor
CstF, see Cleavage stimulation factor
Cytokine, see also specific cytokines
  general features, 74
  receptors, 77
  signal transduction, see Growth hormone receptor
m^{5}Cytosine-DNA glycosylase
  catalytic properties, 226
  expression in embryogenesis, 226
  models for formation of specific methylation patterns, 229, 231–232
  substrate specificity, 226–227

D

Dam
  DNA replication role, 174–175
  synchrony regulation, 176
DnaA
  ADP inhibition, 167–168
  aggregation, 156, 170
  ATP-dependence of initiation, 158
  binding sites in oriC, 157
  concentration autoregulation, 169
  DnaB interactions, 157, 172
  initiator titration, 164
  plasmid replication role, 172–174
  primase loading, 172
  strand opening, 158–159
  synchrony regulation, 176–177
DNA methylation, see also DNA methyltransferase
  binding proteins to methylated DNA, see MeCp1; MeCp2; Methylated DNA binding protein 2 H1
  de novo methylation, 220–222, 231–232
  demethylation, see also m^{5}Cytosine-DNA glycosylase
    active mechanism, 222, 224
    enzymes, 226–227
    genome-wide demethylation, 224–225
    nonhistone protein 1 role, 227
    passive mechanism, 222
    site-specific demethylation, 224
  determination factors, 228
  gene silencing mechanisms
    DNA structure alteration, 234
    repressor protein binding, 235–241, 243
    transcription factor binding inhibition, 234–235
  hypermethylation, 233–234
  methyl-directed maintenance methylation, 219–220
  p53 gene, 218
  structure-induced methylation, 222, 232–233
  tumorigenesis effects, 218–219

DNA methyltransferase
  de novo methylation, 220–222, 231–232
  lesion effects on methylation, 222
  models for formation of specific methylation patterns, 229, 231–232
  modulators, 232
  mutation effects in mice, 217–218, 243
  regulation of expression, 220
  replication foci association, 220
  sequence specificity, 219

E

E-cadherin
  apoptosis role, 203–204
  catenin binding, 194, 196–197, 201–202
  cDNA structure, 191, 193
  cell-specific recognition, 199
  cellular polarization effects, 201–202
  colonic disease, spatial changes *in vivo*, 204
  embryogenesis role, 199–201
  epithelial cell adhesion, 190, 203
  functional domains, 193
  gene
    promoter, 191
    structure, 190–191
  interaction with other adhesion molecules, 194
  metastasis role, 202–203, 207, 209
  phosphorylation in signal transduction, 196–197
  processing, 193
  regulation of expression, 191
  sequence homology between species, 188, 190
  signal transduction via cytoskeleton, 194, 196–198
  tissue differentiation role, 201
  tumorigenesis changes
    loss of heterozygosity, 204–205
    point mutations, 205
    posttranscriptional processing, 205
    posttranslational processing, 206
    transcriptional regulation, 205
Erythrocyte ghost, antisense oligonucleotide delivery, 114
Erythroid cell
  differentiation, 250, 253–255
  terminally differentiating cells
    destabilization of nonglobin messenger RNA, 256–257
    globin messenger RNA accumulation, 255–256

F

Fluorescence resonance energy transfer (FRET), oligonucleotide hybridization analysis, 108
FRET, *see* Fluorescence resonance energy transfer
FUM1, *see* Fumarase, yeast
Fumarase, yeast
  cellular compartmentation and signal sequence, 334–335
  gene, 334
  structure, 334

G

G·A, *see* Guanosine–adenosine base pair
Gene silencing, *see* DNA methylation; Methylated DNA binding protein 2 H1
Glg1p
  domains, 305–307, 309
  gene, homology between species, 302, 304–307, 309
  glucosylation substrates, 300
  primary structure, 294
  role, *in vivo*, 310
  self glucosylation, 290, 299–300
Glg2p
  domains, 305–307, 309
  gene, homology between species, 302, 304–307, 309
  glucosylation substrates, 300
  overexpressing mutants and glycogen synthesis, 311
  primary structure, 294
  role, *in vivo*, 310
  self glucosylation, 290, 299–300
α-Globin
  gene cluster, 257
  gene switching, 250
  messenger RNA
    Constant Spring mutation and thalassemia, 252, 262–263, 265–266

α-Globin (cont.)
  deadenylation, 274
  erythroid cell accumulation, 255–256
  half life, 256
  ribonucleoprotein complex formation and stability, 269, 274
  structural determinants of stability
    conservation between species, 276, 278
    experimental determination, 263, 265
    sequence of stability element, 267, 269, 276
    3' untranslated region, 259, 265–267, 269, 276, 282
  trans-acting stability factors
    α-complex proteins, 271, 273–275
    identification, 271
β-Globin
  gene cluster, 257
  gene switching, 250
  messenger RNA
    erythroid cell accumulation, 255–256
    half life, 256, 280
    stability elements in 3' untranslated region, 280, 282
γ-Globin, messenger RNA destabilization, 260–261
δ-Globin, messenger RNA stability, 260
ζ-Globin, messenger RNA stability, 259–260
Glycogen
  biosynthesis, see also Glg1p; Glg2p; Glycogenin
    elongation, 291–293, 295–298, 301
    initiation, 290–291, 311
    rate determining step, 311
  proglycogen, 291–293
  structure, 289
Glycogenin
  diabetes candidate gene, 313–314
  discovery, 291
  domains, 305–307, 309
  gene homology between species, 302, 304–307, 309
  glucose effects on activity, 312–313
  glucosylation substrates, 300
  glycogen synthase interactions
    effects on activity, 312
    elongation, 295–298, 301
  messenger RNA, tissue distribution, 302, 304
  phosphorylation, 311–312

plants, 304–305
primary structure, 293–294
regulation, 311–313
roles, in vivo, 309–310, 313
self glucosylation, 290–291, 298–299
subunit structure, 294–295
Growth hormone
  categorization of effects, 75–76
  receptor, see Growth hormone receptor
  regulation
    messenger RNA polyadenylation, 63
    secretion, 74–75
    transcription of other genes, 76
  structure, 74
Growth hormone receptor
  binding sites, 79
  crystal structure, 78–79
  dimerization, 79–80
  intracellular domain, 80
  signal transduction
    cellular phosphorylation pattern, 80–81
    Janus kinase 2, 81, 85–86
    mitogen-activated protein kinase, 86–87
    serine protease inhibitor 2.1, 85, 87–89
    signal transducer and activator of transcription, 81–83, 85–86, 89–90
  stoichiometry of binding, 79
G·U, see Guanosine–uridine base pair
Guanosine–adenosine base pair (G·A)
  DNA conformation, pH dependence, 15
  flanking bases and stability, 19
  hammerhead ribozyme, 17–18
  nuclear magnetic resonance, structural studies of RNA
    helical twist, 20
    imino signal, 19
    pairing mechanism elucidation, 19–20
    tetraloops, 20–21
  pairing arrangements, 15–16, 18–19
  sequence content and stability, 22–23
  stacking and stability, 17, 20
  X-ray structures
    DNA, 15–17
    RNA, 15, 17–18
Guanosine–uridine base pair (G·U)
  distribution in RNA, 2–3
  function in tRNA^{Ala}, 34
  glycosyl bond angles, 3
  nuclear magnetic resonance, structural studies of RNA

INDEX

hydrogen bond determination, 6
imino signal, 6–7, 11
manganese binding, 13–14
proton exchange and stability, 7–8, 11–13
stacking and chemical shifts, 7–10
orientation and stability, 12–13
stacking effects and sequence context, 3–5, 7–11, 13
water stabilization, 5, 8–9
X-ray structures of RNA duplexes, 4–5

H

Hemoglobin, *see* α-Globin; β-Globin
Herpes simplex virus, inhibition by oligonucleotides, 124
HIV, *see* Human immunodeficiency virus
Human immunodeficiency virus (HIV)
 inhibition by oligonucleotides, 123–124
 Rev response element structure, 21–22, 30

I

IDH, *see* Isocitrate dehydrogenase, yeast
IDP, *see* Isocitrate dehydrogenase, yeast
Immunoglobulin, splicing of polyadenylation sites in genes, 58–60
Interferon, induction by double-stranded RNA, 126–128
Interleukin-6, release stimulation by cellular binding of DNA, 129
Isocitrate dehydrogenase, yeast
 allosteric modulation, 326–327
 cellular compartmentation, 330–331
 genes, 327–329
 mutant phenotypes, 327–329, 336
 signal sequences, 330–331
 subunit sequence homology
  IDH, 327, 329
  IDP, 330
Iteron, *see* Plasmid P1

J

JAK 2, *see* Janus kinase 2
Janus kinase 2 (JAK 2), growth hormone signal transduction, 81, 85–86

K

Ku, *see* Nonhistone protein, 1

L

Liposome, antisense oligonucleotide delivery, 110–113

M

Malate dehydrogenase, yeast
 cellular localization and functions, 325–326
 isozymes, 319
 MDH genes, 319–320
 metabolic flux, 319
 mutant phenotypes, 319–320, 326, 336
 regulation of expression, 321
 sequence homology of subunits, 321
 signal sequences, 325
MAPK, *see* Mitogen-activated protein kinase
MDBP-2-H1, *see* Methylated DNA binding protein 2 H1
MDH, *see* Malate dehydrogenase, yeast
MeCp1, binding to methylated DNA, 235–236
MeCp2
 binding to methylated DNA, 236–237
 embryogenesis role, 237
Messenger RNA (mRNA)
 binding to methylated DNA, 235, 237–239
 3′-end processing
  coupling to transcription and splicing, 64–65
  endonucleases, 48
  nuclear processing, 48–54
  polyadenylation, *see* Polyadenylation, RNA
  principle, 41–42
  protein complexes, 48–50, 53
  sequences directing processing, 42–47
  in vitro systems, 48
  yeast, 56–57
 estradiol control, 237–240
 gene silencing, 241, 243
 stability, *see also* Globins
  half-lives, 251
  ribonucleoprotein complex stabilization, 253, 269, 274, 283

Messenger RNA (mRNA) (cont.)
   structural determinants, 250–251
   therapeutic applications, 283–284
   trans-acting factors, 253
Methylated DNA binding protein 2 H1 (MDBP-2-H1)
   phosphorylation/dephosphorylation and binding, 240–241, 243
   regulation of expression, 238
Methylation, see DNA methylation
Mismatch base pair, see also Guanosine–adenosine base pair; Guanosine–uridine base pair; Uridine–cytosine base pair; Uridine–uridine base pair
   adenosine–adenosine, 31
   adenosine–cytosine, 31–32
   biological roles, 33–35
   cytosine–cytosine, 32–33
   discovery, 1–2
   guanosine–guanosine, 30–31
Mitogen-activated protein kinase (MAPK)
   growth hormone signal transduction, 86–87
   substrates, 86
   types, 86
mRNA, see Messenger RNA
Myc, antisense oligonucleotide inhibition, 117–118, 120–121

N

NHP-1, see Nonhistone protein, 1
NMR, see Nuclear magnetic resonance
Nonhistone protein, 1 (NHP-1), DNA demethylation role, 227
Nuclear magnetic resonance (NMR), structural studies of RNA mismatched base pairs
   guanosine–adenosine, 19–21
   guanosine–uridine, 6–14
   uridine–cytosine, 27–29
   uridine–uridine, 27–29

O

2′,5′-Oligoadenylate synthetase, activation by double-stranded RNA, 126–127
Oligonucleotide, see also Antisense oligonucleotide

analogs
   binding affinity enhancement, 97–98
   nonionic, 97, 108
   α-oligonucleotides, 97
   peptide nucleic acids, 98, 103–104, 112, 121
   phosphorothioate, 97, 109, 115
   positively-charged for accelerated hybridization, 101–102
   reactive derivatives, 98–99
cellular receptors, 109
endocytosis, 105–109
immune response, 128–131
nuclear uptake, 107–108
protein binding in drug therapy, 96, 104–105, 122–126
ribozyme attachment, 99–100
OriC
   DnaA binding, 157, 174
   initiator titration, 164
   replication role
      Dam, 174–175
      DnaA, 157, 172, 176–177, 179–180
      SeqA, 174–175
   strand opening, 158–159

P

P1, see Plasmid P1
Par, plasmid replication role, 177–179
PKR, activation by double-stranded RNA, 127
Plasmid P1
   basic replicon, 149–150
   copy number limitation, 146–147, 159, 168
   initiator proteins, functions, 155–158
   iterons
      cooperativity of initiator binding, 171–172
      evolution, 150–153, 155
      initiator titration, 161–165
      inverted repeat sequences, 153, 155
      pairing, 157, 165, 171
      RepA binding, 150–152, 156
      sequence conservation, 151, 155, 180
      structure, 151–152
   Par locus, 177–179
   preparation, 149

replication
  cell cycle coordination, 168–179
  mode, 158–159
  negative control, 159, 161–165, 167–168
Polyadenylation, RNA
  cytoplasmic polyadenylation, 54–56
  deadenylation, 274
  functions, 42
  in vitro systems, 48
  poly(A) binding protein II functions, 52–53
  poly(A) polymerase
    activation, 50
    cytoplasmic enzyme, 54
    mechanism, 50–52
    RNA affinity, 51–52
    yeast enzyme, 56–57
  protein complexes, 53
  regulation
    multiple polyadenylation sites, 58–60
    quantitative regulation, 62–64
    splicing of polyadenylation sites, 60–62
  sequences directing processing
    AAUAAA, 42–45, 47
    animals, 43–44, 47
    effects of mutations, 43–45
    plants 46
    secondary structure, 44
    yeast, 44–47
  translational control in development, 54
  universal conservation, 42
  viruses, 42
Protein kinase C, antisense oligonucleotide inhibition, 117

R

Ras, antisense oligonucleotide inhibition of mutated proteins, 117–118
Reconstituted viral envelope (RVE), antisense oligonucleotide delivery, 113–114
RepA
  cooperativity of iteron binding, 171–172
  initiator titration, 163–165
  iteron binding, 150–152, 156
  promoter autoregulation, 169
Retrovirus
  antisense oligonucleotide therapy, 120, 122

polyadenylation sites in messenger RNA, 59–60
Rev response element (RRE)
  mismatched base pair functions, 35
  nuclear magnetic resonance of human immunodeficiency virus structure, 21–22, 30
Reverse transcriptase, inhibition by oligonucleotide binding, 122–123, 125
Ribonucloprotein complex (RNP), messenger RNA stabilization, 253, 269, 274, 283
Ribosomal RNA (rRNA)
  mismatched base pair functions, 34–35
  nuclear magnetic resonance, structural studies of mismatched base pairs, 7–8, 21–22, 28–29, 31
Ribozyme
  manganese binding, 14
  nuclear magnetic resonance of P1 helix, 10
  oligonucleotide therapy application, 99–100
  X-ray structure of hammerhead ribozyme, 17
RNA, *see* Messenger RNA; Ribosomal RNA; Transfer RNA
RNP, *see* Ribonucloprotein complex
RRE, *see* Rev response element
rRNA, *see* Ribosomal RNA
RVE, *see* Reconstituted viral envelope

S

SELEX, *see* Systematic Evolution of Ligands by Exponential enrichment
SeqA
  DNA replication role, 174–175
  synchrony regulation, 176
Serine protease inhibitor 2.1 (SPI 2.1), growth hormone and promoter responsive elements, 83, 85, 87
  signal transduction, 87–89
Signal transducer and activator of transcription (STAT)
  dimerization, 81–82
  growth hormone signal transduction by STAT, 5, 85–86, 89–90
  phosphorylation, 80–81
  types, 82
SPI 2.1, *see* Serine protease inhibitor 2.1

Streptolysin O, antisense oligonucleotide delivery, 110
Systematic Evolution of Ligands by Exponential enrichment (SELEX)
  antisense oligonucleotide sequence optimization, 101
  oligonucleotide ligand for protein, identification, 104–105, 125–126

T

TAR, see Transactivation response element
TCA cycle, see Tricarboxylic acid cycle
Thalassemia
  globin gene mutations, 262, 261–263
  therapy, 283–284
Transactivation response element (TAR), nuclear magnetic resonance of equine infectious anemia virus structure, 9–10
Transfer RNA (tRNA)
  mismatched base pair functions, 34
  nuclear magnetic resonance, structural studies, 7–9, 14
  X-ray structure, 5
Transferrin
  receptor
    messenger RNA stability, 252–253
    uptake of oligonucleotides, 112
  regulation of messenger RNA polyadenylation, 63
Tricarboxylic acid (TCA) cycle, see also specific enzymes
  compartmental localization of enzymes in yeast, 317–318
  metabolic flux, 317
tRNA, see Transfer RNA

Tumorigenesis, see DNA methylation; E-cadherin

U

U·C, see Uridine–cytosine base pair
Uridine–cytosine base pair (U·C)
  nuclear magnetic resonance, structural studies of RNA, 27–29
  pairing arrangements, 23–24, 29
  pH dependence of structure, 28–29
  water stabilization, 25, 27
Uridine–uridine base pair (U·U)
  glycosyl bond angles, 26
  nuclear magnetic resonance, structural studies of RNA, 27–29
  pairing arrangements, 23–26
  water stabilization, 25–26
U·U, see Uridine–uridine base pair

V

Vascular smooth muscle cell (VSMC), antisense oligonucleotide inhibition of proliferation, 118–119
Vasopressin, regulation of messenger RNA polyadenylation, 64
VSMC, see Vascular smooth muscle cell

X

X-inactive specific transcript (XIST)
  DNA methylation role, 228
  models for formation of specific patterns, 229, 231–233
XIST, see X-inactive specific transcript